Beyond Competition: The Future of Telecommunications

Beyond Competition: The Future of Telecommunications

Edited by:

Donald M. Lamberton
Urban Research Program
Research School of Social Sciences
Australian National University
Canberra, Australia

Emerald

United Kingdom – North America – Japan
India – Malaysia – China

Emerald Group Publishing Limited
Howard House, Wagon Lane, Bingley BD16 1WA, UK

Reprints and permission service
Contact: booksandseries@emeraldinsight.com

Printed and bound by CPI Group (UK) Ltd, Croydon, CR0 4YY

British Library Cataloguing in Publication Data
A catalogue record for this book is available from the British Library

ISBN: 978-0-44-482252-9

Preface

The papers included in this volume were prepared for the Tenth Biennial International Conference of the International Telecommunications Society (ITS), held in Sydney, Australia in July 1994.

The ITS is an association of professionals (academics and practitioners in operating companies, consultancies and government agencies) with an interest in the growing field of telecommunications planning, policy formulation and economic decision analysis. The aim of this worldwide network of professionals is to systematically encourage distribution of information, discussions and research concerning telecommunications issues and legislative and policy decisions.

Of the 138 papers listed in the Sydney program, some have appeared already in journals such as *Information Economics and Policy*, the official journal of the ITS, and *Telecommunications Policy*. This volume provides a substantial selection of 23 papers, grouped into five sections. These papers reflect the content of the program and the concerns of the speakers and participants who gathered together from many countries for another very successful ITS conference.

As Conference Chair, I wish to place on record my thanks to the many individuals and institutions, members of the Program Committee, the ACTS secretariat in Australia, plenary speakers, sponsors, presenters of papers, conference participants and enthusiastic helpers, who together contributed so much to the ITS'94 occasion. In particular, I thank the authors of papers selected for inclusion in this volume for their ready co-operation. Finally, thanks go to Siemens Limited for a special financial contribution towards the publication of this volume.

ITS has an important role through its biennial and regional conferences, *Information Economics and Policy* and its other publications. Developments in telecommunications and related information activities are playing a major part in world events. In these circumstances independent research that crosses national, regional and cultural boundaries and fuses the efforts from the variety of relevant disciplines has an important contribution to make to the policy process.

Donald M. Lamberton
Tenth ITS International Conference Chair

Contents

Preface v

Introduction

1. Technology, information and institutions
   D.M. Lamberton 3

Part I: Inventing the future

2. Beyond competition: where are we in the dialog about policy for telecommunications?
   D. Allen 13
3. A model for forecasting data communications towards the year 2003
   B. Svendsen and J.P. Saether 33
4. Mobile data: an emerging telecommunication giant
   S. Hultén, B.G. Mölleryd and A. Nyberg 47
5. Implications of infrastructure competition for private networks
   Y. Kurisaki 57
6. From data to wisdom: does IT contribute to the Gross National Happiness?
   M.J. Menou 71

Part II: Management and policy

7. The impact of structural changes in telephony prices
   D. Cracknell 97
8. Network management and the interconnection of networks. From the exogenous to the endogenous: the strategic dynamics of interconnection policy
   H. Williams and J. Taylor 111
9. Network evolution and prospects for switched video service in UK subscriber networks
   Y. Sharma 127
10. The sustainability of the Broadband Network
    D. Joram 147
11. Telecommunications demand: a new approach to media choice problems
    H. Ouwersloot 161

12. An activity-based model for communication-intensive companies as an approach to appraising information technology outsourcing
J.P. Briffaut and F. Spitz 187

Part III: Regulation

13. The AT&T divestiture: a 10-year retrospective
M.S. Snow 207
14. The economic consequence of the introduction and regulation of international resale of telecommunications services
M. Cave 227
15. The regulation of unnatural oligopoly: appropriate criteria for regulators where the goals of regulation are economic progress
J. Nightingale 249

Part IV: Growth and development

16. The changing telecommunications environment in CEE
N. Holcer 279
17. Telephone penetration: industrial countries vs. LDCs
D. Bowles 299
18. Growth dynamics under shifting telecommunications and trade regimes in Australia
N.D. Karunaratne 315

Part V: Some macroeconomic aspects

19. The macroeconomic effects of new information technology, with special emphasis on telecommunications
G. Eliasson 335
20. Managing international research and development activities: the role of communications and information technologies
P. Hagström 359
21. Technology and systems competition in mobile communications
S. Lindmark and O. Granstrand 377
22. Employment implications of information and telecommunication technologies in the Danish banking sector
K.E. Skouby and A. Henten 405

Index of authors 417

Keyword index 419

Introduction

Beyond Competition: The Future of Telecommunications.
Edited by D. Lamberton.

CHAPTER 1

Technology, information and institutions

Donald M. Lamberton
Urban Research Program, Research School of Social Sciences, Australian National University, Canberra, Australia

The purpose of my opening remarks as Conference Chair of the Tenth Biennial ITS Conference – the first to be held in Australia and only the second in the Asia-Pacific region – is 3-fold: some elaboration of the theme, *Beyond Competition*, a suggested viewpoint and a request aimed at furthering the conference and publication contribution to the policy process.

Planning of this conference started from the notion that the convergence of telecommunications, mass media and computer technologies had brought spectacular developments of ubiquitous intelligent interconnected systems. But it underlines the evolutionary nature of these changes, their path dependency, if we recall that in the 1960s television was hailed as the *deinotatos*, the most powerful force ever invented.

The recent debate about privatization and regulation or deregulation has pushed the technological, informational and social aspects off centre stage. In so doing it has made the discussion of telecommunications matters subservient to a broad pendulum swing in socioeconomic thought. Just as Jewkes' (1948) *Ordeal by Planning* captured the frustrations of the immediate post-World War II, so recent writings and speeches are once again turning attention towards greater reliance upon the role of the market.

Nevertheless, the underlying role of new information continues to generate "creative destruction" with "new" industrial structures with calls for new ways of organizing, new international relationships needing new methods of coordination, new ways of working and living that demand new training and education, and new research initiatives to aid management and policy making.

Major elements of competition have been brought to what was largely a noncompetitive sector but the dynamics of technological change and continuing economic internationalization compel the question: What lies beyond competition?

Over the years there have been many concepts of competition: perfect, pure, imperfect, monopolistic, workable, open, potential, sustainable, dynamic and acceptable to name some. Through all these intellectual efforts and rhetorical exercises, analysis and policy framing were motivated by a search for measures giving greater social gains than social losses.

Address for correspondence: Donald M. Lamberton, Urban Research Program, Research School of Social Sciences, Australian National University, Canberra, ACT 0200, Australia. Tel.: +61-6-249-3884. Fax: +61-6-249-0312.

The theme, *Beyond Competition*, implies that competition is not a stable, enduring state. Perhaps the most obvious reason for its somewhat precarious nature is that competition is interwoven with the dynamic forces shaping both the development of industries and their restructuring. Imprecise as this word restructuring is, I prefer it to "convergence" which is better applied to the changing technologies themselves rather than to the industries in which they are applied. When dealing with general purpose technologies like communication and computing, it does not help very much to say that everything is converging with everything else.

While there remains quite a great diversity between countries, we have witnessed much reorganization in the search for operating efficiency in telecommunications, with resort to policy-aided new entry, growth in selling costs, and, in rich variety, what *The Economist* called corporate couplings, i.e., acquisitions and mergers, R&D collaboration, strategic alliances, networking, etc. All this activity can be viewed as part of the process of "creative destruction", but we should not overlook the place Fellner's classic title, *Competition Among the Few*, deserves in current vocabulary. We should also consider the challenge Thurow (1977) issued to his colleagues:

> "Economics needs a theory of oligopolistic behavior and especially a theory of oligopolistic price determination. There are economists who deny the meaningful existence of oligopolies; there are economists who glorify their usefulness; there are economists who blame them for almost everything. But there are no economists who understand them".

Has his challenge been met? Have we over the last 15 years disposed of this problem? Or are we still trying to create the kind of markets we think we understand?

This brings me to my second purpose: a suggested viewpoint. Despite all the advances recorded in the *Handbooks in Economics*, we have not finished the struggle to break out of what Shackle (1967) called the perfectly competitive prison. It is my belief we have failed to pay sufficient attention to the role of information. It was not the assumption about the number of firms that mattered, but rather the assumption about rich and sure information and the capability of acquiring and using that information. This is the viewpoint that is central to information economics, which is proving to be possibly the major intellectual gold strike by the economics tribes of this century.

It is no accident that the journal *Information Economics and Policy* became the official journal of the ITS. The spectacular technological events, massive industrial changes, social concerns and policy initiatives and responses have positioned the telecommunications sector at the leading edge of change and have made it a primary work site for all those engaged in the development of information economics.

There are many entry points to a discussion of the transformation these efforts are bringing to economics:

- The "firm" is ceasing to be the mathematical point it has been in economic theory. It is a complex and often very large organization. Its information-handling capabilities and its embodied information are capital and both are essential parts of the "infrastructure". Even the neglected concept of entrepreneurship can be reinterpreted in terms of information handling; not just of R&D information, so beloved of econometricians because it is an available statistic, but of the entire range of business intelligence.

- An economics of internal organization, of the ways firms acquire and use information and the costs to them of doing so, is needed to complement the economics of markets, because information can be a commodity, although as Arrow (1984) warned us, only to a limited extent.
- Competition only exists in a world of incomplete information. When wants are known and either stable or controllable, uncertainty disappears and monopoly power becomes possible (Lamberton, 1971).
- Regulation has to cope with the costs of information and with the strategic advantage of an uneven distribution of information (information asymmetry). It needs also to give more attention to the potential gains from positive action to provide information as an additional weapon in the regulatory armoury.
- Much of the broad challenge to "mainstream" economics revolves around interrelated information and strategic behaviour problems. Why then was information marginalized in the literature and practice of economics? One suggestion is that "information plays the same role for neoclassical economics that rent did for classical economics. It is the problem case in which the internal tensions of the discipline come to the surface" (Boyle, 1992).

Whatever interpretation one places on this aspect of the history of economic thought, it seems clear that the role of information and its strategic importance has implications for major policy issues that arise for both economic policy generally and telecommunications policy in particular, e.g., issues relating to employment, equity and the design of institutions.

I do not suggest that this information perspective will solve all our problems but I believe it holds great promise. It should lead us to think about information rather than information technology; about markets and other forms of organization; and about information-handling capability which has several components. First, there are the skills of individuals created by educational processes, either formal or of the work experience type; second, there is organizational capital in business, governmental agencies, nonprofit activities and other institutions. Research efforts therefore need to give priority to the interdisciplinary study of information and its characteristics and to the design of organizations. This latter task seems now to be all-too-often subject to fads and fashions.

In addition, my request aimed at furthering the research process and its contribution to policy formation. The ideal is a continuing interaction between policy deliberations and research activity, although this must not be dominated by the seeming need for immediate answers. There must be dissemination of conference papers through our journals. However, the potential productivity of that research will not be realized unless there is recognition on both sides that data collection is vitally important, that research is often a hard, slow process, and that there are two significant kinds of barriers to be broken down.

The first barrier exists in the research community itself because that community is fragmented into disciplines, schools and tribes, each with its institutional trappings. We are then separated spatially, meeting occasionally through conferences, being dependent upon old-fashioned correspondence and now the NET. The second barrier can be found among the policy makers. They too are fragmented into political parties, government departments, industry peak councils and various networks – again with added spatial separation into nations, regions and cultures.

It should occasion no surprise in such complex circumstances that it is difficult to ensure that research findings and policy making are well matched and that they are blended together to produce wisdom. So I repeat that dissemination of conference papers through journal publication is only the first step in realizing the value of research embodied in papers.

The evolving nature of competition in telecommunications and information activities is interwoven with the technological trajectories, changes in the institutional arrangements, investment in information-handling capabilities and general infrastructure, shifts in demand for information goods and services and policy fashions.

I have mentioned the book *Ordeal by Planning* published in the 1940s but one might go back further in history to a more famous work. Did Adam Smith's classic not mark an earlier pendulum swing of far-reaching consequences? *The Economist* would have us believe that the election of President Clinton can be viewed through this lens, appearing as a celebration of the positive role that government can play in the economy: "After 5 years of being told by George Bush that the economy was fine – and where it was not, free enterprise would provide – all this was undoubtedly what a large number of voters wanted to hear" (The Economist, 1992).

This analysis may seem already to have missed the mark, but it was correct in sensing a shift in the rhetoric and reality of market economics. The *Beyond Competition* theme adopted for ITS'94 sought to pursue this line of thought. ITS'94 was to be an exercise in the tradition of J.M. Clark and Joseph Schumpeter in evolutionary economics.

I have taken the title for Part I, Inventing the Future, from a book by Gabor (1963), to suggest the positive role of people and government in shaping the technologies, the information-handling capabilities and the institutions. For David Allen, competition is the watchword in telecommunications policy, although the evidence to date is of mixed results. Turning back to intellectual foundations, he noted the cultural contrasts between the US, the Asia-Pacific and Europe. From these he drew out the information processing perspective, taking as his focus the network and its technology which will co-evolve with the communities and communications of its users. Competition will remain one of the mainsprings but the network of the future must cater for the "telecommunications dispossessed", alternating between competition and consensus. Analytically, this requires that the economics of information should not only address asymmetry as it has been doing but also that it should widen its scope to include the processes of communication.

Berit Svendsen and Jan-Petter Saether sought to forecast one element of communication activity: data communication, which, of course, proved to be very complex. It takes in fax and electronic mail, extends to information retrieval and electronic market places, and has a future with virtual reality and remote education. Staffan Hulten, Bengt Molleryd and Anna Nyberg challenged expectations that the mobile data industry, which builds in a major way through strategic alliances, is destined to become a telecommunications giant. They considered the industry to be still in its system-building phase and noted that only four of the 87 alliances they studied are market oriented.

Yoshiko Kurisaki examined the implications of infrastructure competition for private networks, especially those based on leased circuits. She found that the introduction of competition in the supply of LCs and usage liberalization had to be undertaken simultaneously. Importance should be placed on the vanguard nature of LC users as forerunners of advanced telecommunications use that may be made available for a wider public. The mismatch between economic and political logic, with the former extending beyond national boundaries while the latter is confined within them, is more important for LCs and corporate activities than for PSTN.

Michel Menou took us back directly to Allen's concerns by posing fundamental questions about the significance of computers and telecommunications as indicators of the economic and social conditions of societies. The changes that are taking place may be more a far-reaching transformation of knowledge and cognitive processes than a jump in the material capacity for handling and circulating information across time and space. What differences do these changes make? He reported on an important research initiative by the International Development Research Centre in Canada.

Part II brings together papers dealing with a variety of management and policy issues. David Cracknell examined the impact of structural changes in telephony prices, drawing on BT experience. Howard Williams and John Taylor explored the implications of the diffusion of network management technologies for both the structure of the industry and regulatory approaches. They argued that the major telecommunications operators are seeking to reshape their businesses not around the ownership of network elements but around the manipulation and control of network management technologies, and in particular of the new information flows these technologies support. This implied a need for a change in the focus of regulatory policy.

Yogesh Sharma discussed some of the cost implications of switched video services in the UK and investigated the cost structure of fibre-based local networks that could provide a host of broadband services. Competing technologies present both economic and regulatory challenges. Denis Joram asked if the broadband network can be sustainable. He listed the main phenomena that can put a brake on the emergence of broadband and compared the development of videotex and ISDN in France. Critical mass contestability was seen as a difficulty. Hans Ouwersloot tackled the media choice problem faced by both the supplier and the consumer of information, concentrating his attention on the motives and reasons which influence supplier decisions. This is an exercise in the dynamics of information exchange, and links with Allen's plea for an economics of communication. J.P. Briffaut and F. Spitz rejected the traditional microeconomics assumption of competence, even in communication-intensive companies. Outsourcing was seen as a solution to these inadequacies of IT management. They presented an activity-based business model for rational outsourcing.

Regulatory issues have intruded already in the earlier parts but Part III endeavours to group some papers more directly concerned with the discussion of regulation. Marcellus Snow provided a 10-year retrospective on the AT&T divestiture. He recalled Ithiel de Sola Pool's prediction that the 1982 Consent Decree would by 1992 be seen as short-sighted and ill-considered in light of technological developments. He concluded that the very competitive forces unleashed by the divestiture ensured a built-in obsolescence rivalling the half-life of the latest generation of computer game software. Martin Cave analyzed the impact which international resale, especially international simple resale, may have upon the tariffs charged to subscribers for international telecommunications traffic and on commercial relationships between operators in the originating and terminating countries. He discussed also how these effects are influenced by the regulatory approach to international simple resale adopted by the authorities in the two countries involved, or adopted by countries acting on a multilateral basis.

John Nightingale argued that an evolutionary perspective on firm and market behaviour suggested a framework of analysis in which price-cap regulation has the natural outcome of stimulating the rate of productivity change and innovation. He contended that a market-based system of oligopolistic rivalry is likely, on the evidence of other industries, to best promote rapid innovation and its diffusion. During the transition to such a situation, the simulation of effective market rivalry by a price-cap system is a means of stimulating adoption of innovation and rapid productivity change.

In Part IV the papers address the vexed issue of the contribution of telecommunications to economic development, so returning to issues raised by Menou in Part I. Nikola Holcer examined the changing telecommunications environment in CEE countries, with a focus on the manufacturing industry in Hungary, Poland, Czech Republic and Slovakia: the challenges, solutions and some of the already achieved results. Three clear trends emerged: changes in the legal and regulatory environment are much more complicated, difficult and time consuming than originally anticipated; network expansion and service provision are progressing more or less satisfactorily; and the telecommunications manufacturing industry is struggling to restructure by downsizing and finding foreign partners to modernize and capture a slice of the domestic market.

David Bowles turned his attention to a comparison of the telephone penetration rates of industrial countries and LDCs using data from 75 countries. He found that a large part of the variance across countries can still be attributed to differences in incomes. However, the results also supported the hypothesis that differences in the benefits of the telephone, which, in turn, are tied to the structures of economies and abilities of users, explain telephone penetration. The varying influence of nonincome variables, e.g., schooling, household size, etc., implies that the level of development matters, with the increasing sophistication of the telephone network and its uses affecting demand. The supply equation told a similar story, with the quality of the labour force being an important part of the explanation of supply across countries at dissimilar stages of development. There is, therefore, a danger of under- and overbuilt networks if suppliers do not take into account the wealth, abilities and economic structure and diversity of individual countries.

Neil Karunaratne analyzed the dynamic interactions between growth, trade and telecommunications in the Australian information economy during the period 1971–1992. This period had witnessed a major reversal in the country's protectionist trading policies and a liberalization of the trading regime. There had been a veritable revolution in information technology and dramatic deregulation in the world's leading information economies. Diffusion of changes to Australia, he argued, may have been impeded by slow adoption of procompetitive policies and deregulation of the state owned telecommunications monopoly.

His econometric results suggest that it is in the long run that the impact of the mutual interactions amongst growth trade and telecommunications become more fully apparent.

The papers in the concluding Part V discuss some macroeconomic aspects. Gunnar Eliasson started from a basic recognition that information use and the diffusion of knowledge are dominant economic activities subject to technical change. He examined the impact of new information technology and increased receiver competence. He argued that learning about economic structures through search and experiments has to be explicitly modelled. He reported on the attempt to look at the use of such analytical information and communication methods in the Swedish micro-to-macro model to make the financial market arbitrage process more efficient.

Peter Hagström explored the challenges of managing international R&D activities in the light of existing research, and framed the question of R&D coordination with regard to usage of corporate communications networks. He reported empirical findings, from interviews and questionnaires from 18 Swedish MNCs, that international information systems have an important and growing role; problems and shortcoming have surfaced and many benefits evolve often in unanticipated ways. In terms of implications for management of international R&D, he reported improved knowledge sharing in general, improved working of joint projects and greater local acceptance of new ideas in the MNC.

Ove Granstrand and Sven Lindmark link the technologies of mobile communications to the diverse and evolving forms of cooperation discussed in the Introduction and Part I. They focussed on competing, and sometimes complementary, technologies and systems competition at three levels: between wired and wireless telephony at the user end; between analog and digital wireless technology; and among various cellular systems. For them, systems competition refers to "hard" systems as well as to a set of different competitive processes operating interdependently at different levels in an encompassing system of entities (typically individuals, companies, nations, products and technologies). They applied their model to the substitution of analog for digital cellular systems and offered hypotheses for further research.

In the concluding paper, Knud Erik Skouby and Anders Henten discussed some of the effects of the new information and telecommunications technologies on employment, demand for qualifications and organizational structures in the Danish banking sector. The main trends reported were:

1. Presently the ITT technologies have a limiting effect on the number of jobs. The decrease is primarily felt among the least qualified groups, especially women working part-time.
2. The structure of organization in branches is becoming less hierarchical because the individual staff members can perform more functions.
3. Banks are aiming at more self-service in simple transactions and information services both for private and business customers, but are emphasizing counselling functions, implying a higher level and range of qualifications among the staff.
4. There will be increased demand for on-line information systems, e.g., to broaden and refine the use of EDI applications. Counselling functions will use decision-support systems. General office functions will be increasingly automated.

References

Arrow, K.J., 1984. *Collected papers of Kenneth J. Arrow, Vol. 4. The Economics of Information*. Oxford: Blackwell.

Boyle, J., 1992. A Theory of Law and Information: Copyright, Spleens, Blackmail, and Insider Trading. *California Law Review*, 80:1413–1540.

Economist, The, 1992. *After the Market*. 14 November:15–16.

Gabor, D., 1963. *Inventing the Future*. London: Secker & Warburg.

Jewkes, J., 1948. *Ordeal by Planning*. London: Macmillan.

Lamberton, D.M. (ed), 1971. *Economics of Information and Knowledge*. Harmondsworth, UK: Penguin Books.

Shackle, G.L.S., 1967. *The Years of High Theory Invention & Tradition in Economic Thought 1926–1939*. Cambridge: Cambridge University Press.

Thurow, L., 1977. Economics 1977. *Daedalus*, 106:79–94.

Part I: Inventing the future

Beyond Competition: The Future of Telecommunications.
Edited by D. Lamberton.

CHAPTER 2

Beyond competition: where are we in the dialog about policy for telecommunications?

David Allen
Concord, Massachusetts, USA

*"The efficient economic agent of traditional theory is essentially a 'superior statistician,' as (Kenneth) Arrow put it, whereas I argued that she also has considerable gifts of...*communication...*that will enable her to affect economic processes."* (p. 158)

"It is now becoming clear why, in spite of our lip service to the dialectic, we find it so hard to acknowledge that contradictory processes might actually be at work in society...But after so many failed prophecies, is it not in the interest of social science to embrace (such contradictions), *be it at some sacrifice to predictive power?"* (p. 139)

Albert Hirschman
Rival Views of Market Society
1992, my (reverse-)italics

1. Competition

Where is the current dialog about *competition* in telecommunications?

Competition has become *the* watchword in telecommunications policy, around the world. With AT&T's divestiture a dozen years ago, and the formal advent of competition policy in the US, the notion of competition as the right choice became perhaps the principal US telecommunications export globally. One result has been a set of trials – experiments with competition in telecommunications – across whole societies. Where have these trials so far come out?

1.1. Competition in telecommunications? National experiments

Japan

With several active and successful competitors for various parts of its basic network services, and with more than a thousand competitive entrants overall, NTT seems to be in the very thick of competition. Long distance is the pivotal business, and NTT has lost dramatic market share on its most heavily trafficked corridors. In fact

Address for correspondence: David Allen, 316 Heaths Bridge Road, Concord, MA 01742, USA. Tel.: +1-508-287-0433. Fax: +1-508-287-0434. e-mail: davida@ksgbbs.harvard.edu.

however, the Ministry itself has carefully maintained a differential between the long distance rates NTT may offer and those that the New Common Carriers/NCC's offer in Japan. The NCC's have been assured a price advantage. That surely is not competition as conceived neoclassically.

UK

BT, despite a decade of competitor entry into UK telecommunications, continues to command a greater than 90% share there. That also is not neoclassical competition.

USA

In the aftermath to divestiture, two steadfast and healthy competitors came to share the US long distance market with AT&T, ultimately to compete with it for global partners, even global markets. Long distance prices declined in the US, which was the desideratum that perhaps drove breakup in the first place. But recent research has converged from several different quarters, each analysis demonstrating how US long distance is really a price-umbrella oligopoly, led by AT&T (Huber et al. (1992); Noll (1993); Taylor and Taylor (1993)). Price declines, in this analysis, have been a function of regulators' fiat, as access charges are shifted from carrier to user, or also are the product of technology cost declines continuing from predivestiture days. Recently prices have even risen. This also is not neoclassical competition.

Those are the early-established trials. What of the experiments that have got underway only now?

Australia

The results from Australia are too fresh to be in any way conclusive, but the structure put in place does reflect careful hedging between full competitive freedom for the actors versus explicit and restrictive safeguards.

New Zealand

New Zealand is also very new, but already substantial doubts have been raised about the sustainability of competition for telecommunications in New Zealand.

Some newly industrializing economies are instituting competition for parts of their telecommunications, such as in Chile. These by and large are also too new to yield usable conclusions. But anecdotal reports suggest these cases have encountered difficulties implementing competition, difficulties which directly parallel the experience of the earlier experiments.

Is there no where that competition has succeeded in telecommunications?

Finland

Finland is a case like almost no other in telecommunications, where relative equals face each other. Initial competition has lowered some prices; it has also raised local prices, and residence prices more than for businesses. Here competition appears to operate closer to theory.

According to theory, the main experiments with competition in telecommunications — the largest already a decade into their trials and implanted, by now organically, into some of the world's most powerful and developed economies — are a failure, a *miserable* failure is really the only accurate summary.

Yet if you asked the AT&T executive responsible for long distance services, that person would surely declaim about the heat and rigors of contending with MCI's marketing prowess, or Sprint's thrust into internetworking. Similarly, NTT executives will surely tell you that DDI, and the others in Japan, have pinched NTT's cash flow and put them very much on their toes — look at the steady declines in total number of NTT personnel over these years, they might say, if you should need confirmation. (Or in Germany where competition has hardly arrived, officially anyway: DBP Telekom officials will recount the number of world-class competitors who have already set up shop on German soil.)

Or: The "word" in Washington telecommunications circles, today, is that there has swelled up a very broad consensus among US policy protagonists — enormously significant since, to date, any rewrite of the outdated 1934 Communications Act has been stymied by literally decades of gridlock and failure to reach agreement. What is this "broad consensus", reportedly to be enshrined in the first substantial legislation to issue in 60 years? Namely: full and open competition is the way forward for US telecommunications.

Clearly, our use of the notion *competition* is, at the best, subject to ambiguities. Analysis tells us competition does not exist, but practitioners say they can tell it when they have to live with it. The policy debate, in the protean US, embraces a future of competition when the record says a decade, and more, of trying has not got there. What, indeed, do we *mean* by competition? Can we add substance to our interpretation of the notion and thereby do fuller justice to what now seem to be sharply conflicting uses? (We can ask the parallel question about our use of *market*[1] — if competition is process, market is its structure, the flow and the stock which together give us conceptual grasp.)

How is the notion of *competition* framed in different cultural settings? The perceptual purchase we may get by viewing from alternative perspectives can help to triangulate on confusions. Here I will attempt to encapsulate the view from three, separate cultural vantage points.

1.2. Competition — the view from a given culture

The first view, the natural starting point is seen often as the Western, neoclassical position — more accurately though, it represents a strain of economics that has been perfected largely in the US. The tenets of methodological individualism call up the crux of the position perhaps most parsimoniously, I will suggest here. According to methodological individualism the unit of analysis should be carefully limited to the individual; certainly no miasma, no "ghost of a collectivity" qualifies as a realistic subject for analysis. So also when we turn to competition — from this perspective — competition is the free interaction of individuals each in pursuit of individual goals, none of whom has power to control another. As soon as power and influence intrude, so that some entity may dominate individuals' actions, competition begins to be squelched and extinguished.

Of course even in the US economics establishment, there is significant range and nuance in views. More to the point, the problem being tackled here is precisely a disagreement and confusion over the basic definition. So necessarily, I am highlighting and characterizing only one position among many – but if that position distills an essence which is useful for comparison and contrast with an even *more* different set of views, then our purpose is advancing.

When I turn to contrasting positions – in the East – I naturally face the same difficulty, but on a much larger scale. The number of views of the matter, there, is large indeed. For the purposes here I will limit the question to views of competition in Japan, since Japan is one well-developed case in question (but then note that the next step along this track is to enrich the picture with other key cases across Asia and the Pacific – some similar but some vitally different relative to Japan).

Since I am not Japanese, I face the dilemma of trying fairly to convey fundamentals from a culture where I personally lack inculcation in its mainsprings – but again, that only recapitulates the underlying problem that is in hand. My best sense issues from an amalgam of queries to/talks with friends from industry, the government bureaucracies, and universities in Japan. Appropriate English phrasing, such as managed competition, surely does not do justice to the experience in Japan,[2] inevitably borrowing as it does the Western presumptions which the words carry. Whatever the translation, apparently the experience is a mix of what the essential US economist would see as intense competition together with, at the selfsame time, a sensibility for the welfare of all competitors taken as a group.

Can the experience in Europe, and the culture(s) there, contribute a useful third perspective on our question of *competition*? Europe, too, is a welter of different cultures and approaches to competition. But Europe is most remarkable, in this context anyway, for its steps toward bridging across those very differences. For me the essence is captured by the tensions between DG IV, the competition directorate in Brussels, and practices in individual countries. As the EU moves to establish common practices for competition across Europe, subsidiarity also works to maintain some differences continuing, from ingrained past practice in member states.

These three hip-pocket characterizations of different regional engagements with competition – in the US, Japan, and Europe – would be no more than caricatures, unless they capture essential features which prove useful to compare and contrast. And, a sample of three is surely a bare minimum. Is it enough, does this set a stage where we can begin sorting out the ambiguities which beset our grasp of *competition*? (We should also note: we left the realm bounded just by telecommunications, when we asked about general notions of competition – we may return to telecommunications, after groundwork has been laid at the level of competition itself.)

To try effectively to utilize this stage setting, of contrasts and complements, I am going to reach back to what I see as some fundamentals. For inspiration that is even prior to the usual concerns with scale versus dominance versus access, I will take my cue from the information processing perspective. Specifically, I will here posit two fundamental challenges that I see facing any human, cognitively. Economics is at base, after all, a human science and the operative assumptions about cognitive functioning are really all determining. The aim, in looking to these fundamentals, is to find a purchase for commensurating among the three thumbnails of competition that have just been sketched from the corners of the triad.

I invite your own assessment of the two basic building blocks which follow.

2. Cognitive antecedents — two fundamental challenges

The first cognitive challenge might be called "getting beyond oneself, or beyond one's own horizons". Ironically though, while the emphasis seems to be on the "getting beyond", the principal concomitant appears to be solid establishment of a "sense of self, separated and independent".

2.1. Conceptual solipsism and self definition

The lurid *counter*-cases come all too easily to hand, where the prototype is a dictatorial leader intent on impressing his/her view of the world onto others. History is replete, of course, with political demagogues who take any ruthless measures to stamp out dissenting opinion, who mobilize nationally to proselytize for their viewpoint among the neighboring duchies, frequently through war. Or, religious figures whose entire lives are spent trying to corral a widening horde into accession to a delineated world view. While more than a world view may be at stake in the political leader's war, or in the holy crusade fomented by a religious zealot, the cognitive facets are also compelling, and instructive for our purpose.

In our own personal lives – at least in mine, and I believe certainly in others – the tendency to treat others simply as instrumental to our own goals seems remarkably near the surface most the time. Whether in pursuit of survival basics such as sustenance, shelter, reproduction, or of loftier pleasures, the capacity to see another person as means to the end is almost a natural refrain in the repertoire. Again, more than the cognitive usually will also be at stake each time this occurs. If the job aspirant excitedly chats up a member of the company hierarchy, the "real" job is part of what is at stake. But if the aspirant has mistaken and the other party is actually a rival who can undercut, we see that cognitive clarity is also at stake. *Viewing* the job landscape *with oneself as the frame* may help with target acquisition, but *the cloud generated from hopes and fears for self likely distorts that vision.* The conceptual solipsism of framing everything from one's own view has the most far-reaching impact.[3]

How else to frame? We only have ourselves, right? Surely one of the mysteries is the human ability to contemplate itself – consciousness – so that indeed we *do* have some, (though, in the end, limited) choice of frame. The first cognitive challenge I identify is to escape conceptual solipsism – seeing the job scene, for instance, not just from one's own view but also from other, "detached" perspectives. An unavoidable prerequisite seems to be a well-firmed sense of self, and the reason is fairly evident: an amorphous self, without reasonably defined boundaries, cannot be pinned down in any attempt to establish other frames for perceptual reference. So the first challenge, "to escape the frame of one's own horizons", depends on establishing "a solid sense of self, separated and definite". Separated and definite relative to what and whom? That takes us to the second challenge.

The stamp of self definition, then separation of view from self – hard won – is a first step. But, overall, the cognitive task is a clear view of one's world. The second part of this task, the second challenge might be called "assembling, and reassembling, the picture, together with others".

2.2. Communication for information

Even with clarity relative to oneself, the cognitive job just begins. For the ever-evolving picture that we carry around in our head, we are overwhelmingly dependent on our interactions with others. Without our social interactions, we would not have even language. The experience of the feral child – brought up among animals, without other humans around – speaks profoundly to our dependence on others for the organizing machinery of language. That we take for granted, of course.

Much "information" we gather and use – certainly in the modern world of numerous organized media, overflowing in plenty – is in fact the product of another's efforts. Some information is primitive, input directly from our senses – the sound of the alarm clock in the morning. Beyond that, though: When we search an Internet database, read a newspaper, watch the evening news on television or ask our spouse about the Little League schedule, we may experience the event as "gathering information". In fact, in each case, someone else was there before and stored what we then came along and plucked in our own time, effecting finally the communication between us and them. In a quite basic sense, communication (albeit, sometimes on a "store and forward" basis) is prior to information.

(To return momentarily to telecommunications: We see then that policy planners – first in the US, now around the world – might more constructively be thinking in terms of a National (now Global) *Communications* Infrastructure, rather than of an NII. The *policy* frame of reference will mightily shape the outcome, and communications is the force that underpins access to information.)

The second cognitive challenge I identify is to "exercise the web of communications essential to build and maintain the accurate view". Our job aspirant, after adopting a frame which recognizes the uncertainties of the interview, must rely on those who are familiar with the company, to discover which executives can in fact go to the heart of the job decision. Perhaps the aspirant finds a better fit in the organization because of consultations all around.

We each have an experience of the world not strictly available to anyone else. As such, we have *our* contribution to a larger picture, and we depend on others to allow us *their* contributions for enlargement of *our* view.[4] The result is unavoidably a shared view – our picture, and that of others, is necessarily the product of joint effort. Each person's view will differ from others in some way (necessarily, because of individual experience), but each view will also share with others. Finally, the process is iterative. The picture(s) evolve through repeated flows along the network of communications. These dynamics, though they make the story complex, lie at the heart of understanding this cognitive fundamental, I believe.

Together, the two challenges which everyone faces are, first, clarity for one's own vision through distinct definition of self and, then, building and evolving a both separate and joint picture through repeated exchanges with others. Now – if you have limited taste for psychological construction, you may be chafing, asking what purpose this has served. Well, to keep on track let us recall that competition, and its various interpretations, are the present question.

3. Communities of competitors

The position on competition identified as neoclassical above – the strict focus on individual action – surely is responsive to the first of the two cognitive challenges. Choice by individuals, to serve each person separately, parallels the purpose of a defined self seeking a clear view. But that position, attached here to the West and specifically to the US, has got us only "half way to paradise". The position identified as Japanese, however, brings together the seeming contradictions necessary to meet *both* challenges. There, competition encourages self-definition and clarity of individual view; at the same time, concern for a group underpins the exchange necessary to maintain an evolving separate/joint picture.

Whether that is compelling for you will depend on the force of the "challenges", for you – but it will also depend on how comprehensible you find the idea of combining apparent opposites. How might such a seemingly contradictory arrangement work (not necessarily how does the Japanese culture work, but at least how *could* one such system work)?

3.1. Competition to consensus — evolving ideas

A group – a "community" – actively stay in touch with each other, conveying different impressions of some question in hand. The differences in view *compete* with each other for the picture that is shaped. Through some mechanism (marked, at this point in the narrative, only as a "black box") a *consensus* eventually produces a composite where certain differences are resolved, and shared elements of cognition are confirmed – the shared part of the picture gels. With time, new experience encourages a re-examination, so there is reiteration of the process...and reiteration, indefinitely.[5]

What of the differences not resolved? There is a "nested" structure to the community, so that any individual holds simultaneous membership in concatenated (if not concentric) groups, some more global in character, some more local. The more local – the more "private" – groups permit the maintenance of differences as local variety (with the uniqueness of individual cognition as the ultimate "privacy"). At the same time, a question goes no higher than the nested level for which it has some relevance – an issue is only as "public" as breadth requires. The ability of any individual to move viewpoint fluidly back and forth, among levels in the "nest", is central to the dynamics of the iterative process; these dynamics are also integral to any sensible description of the structure.

What is "community" (and what is it at each of its nested levels)? Community is only a piece of the picture in each, and every, member's head. As such, it is *inevitably perceived differently from individual to individual* – but that *does not materially diminish the force of its effect on individual behavior*, and ultimately on outcomes for the group. As fundamentally, the definition of community boundary – who is in and who is out – is also subject to the evolutionary process, and may shift as the process works.

How are differences resolved, to reach the periodic consensus – what is the "black box"? Do subjects acquiesce to perceived authority, does a violent strongman prevail, is there a "Quaker meeting" (an admittedly Western image, but invoked to convey more complex consensus processes)?[6] What – and especially *how* – differences resolve to a common view, and which of them are maintained as ongoing distinctions, strike me as one of the determining hallmarks of culture. Though this description based on community has been cast in the cloak of an Eastern culture, characteristic balances between competition and consensus are I believe evident in virtually any human endeavor – even, maybe especially, in a supposedly "Wild West" culture such as the US. Reinforcement, or change, in these norms points to evolution of the culture itself.

What are the metrics – the indicia of payoff? *Speed* of iteration through a cycle tells how quickly new conceptions will be available (limited, of course, by capacity to assimilate new experience into usable new consensus). *Size* of the participating community/"nested" subcommunity shows who may partake.[7]

This is exclusively the world of thought – what of the world of action, where economic results actually play out? I have chosen from the start to build from the cognitive, in part because that is where the human interplay is so pointedly intricate. But parallels can be drawn for economic events, the data of the discipline, I believe.

3.2. In the real world

First: do arguments beyond competition – arguments for group-based, consensual endeavor – also apply in the world of affairs? (Though the arguments are surely fundamental, I will only summarize here, since space has already been devoted to the comparable question "on the Platonic plane".)

The question seems hardly to need asking. Cooperative behavior, the Janus face to competition, must have been central to human outcomes from the time the possibility first dawned. (Evidence from the animal world certainly seems unambiguous, despite the popular notion that evolution means only competition.) Even in cultures ideologically committed to competition, the economic behavior inside firms – the larger part of economic behavior, in other words, by practically any measure

— is actually a mix of competition and cooperation. Certainly, the finest technical achievements of man — and the living benefits they may confer — could not have diffused without some of the most intimate cooperation, whether that be in a lab, a factory, a distributor or in the "backyard gossip" forum at the retail and service level.

More interesting perhaps, the increasing complexity[8] that is the legacy of technical evolution positively demands an increase in cooperative behavior, if there is hope to enjoy benefits — at all steps: to conduct the research, to build the machinery (soft or hard), to learn it, service it and evolve it. Initially the industrial revolution may have, for a time, freed individuals somewhat from their direct interdependence on each other, even for survival. But the proliferation of increasingly complicated tools — of every sort — has injected a reversing trend and made closer cooperation also a new necessity. The question is not *whether* cooperation or not, but *how*.

So, now: how does this apply? Again I will not argue specifically, but I believe machinery for the real world of economic events can be drawn parallel that which is just above for the "world of ideas". In this "community" tableau, the focus is on evolution of "technical" ideas which may enhance life (including ideas about the inner life of the human) — in other words, the dynamics are the starting point, not a derivative. Relations are an inextricable mix of competition and cooperation, with an iteration where competition among alternative new ideas is periodically supplanted by some consensus to gain benefit from those ideas — but variety also is maintained across subgroups.

Rather than a world view with homogenous commodities that enable competition, the focus is on balancing commonalities against differences. Related to this balance, the maintenance of commonalities from the past — "traditions" — provides the security of stability, which serves then as a foundation necessary to enable flexibility and more rapid uptake of new differences in other areas — more "innovations". When competition *is* employed, that is in the world of the different, rather than of the same[9] — principally to test one new idea against another. Again, rather than either competition or cooperation, the mix of the two is inextricable — it is a mistake to try and think in terms of just one or the other, despite their seeming antagonism. And again, the specific cultural mechanisms by which differences characteristically resolve into cooperative consensus — especially, how "spoils are shared" — mark, and distinguish, a society. The ultimate issue is the human capacity for change.

3.3. Performance and fragility — underpinnings

The likely first response from a Western perspective asks: but how can the bad behavior which typically is to be expected from ganging-up, and dominance, be prevented, if groups are not constrained by competition? Will not individuals allow their perquisites from the group to substitute for old-fashioned quality of performance?

The answer appears to be as compelling as it is simple: In Japan for example, performance is paired absolutely with group membership — a failure to perform is no more thinkable than is dissolution of group affiliation (but if one is threatened, so may be the other). In an historical antecedent, the formation of literally hundreds of thousands of local cooperatives throughout Japan, as the Tokugawa period came to a close and Meiji arrived, were founded on twin equally immovable pillars: trust, and so group membership; and exactitude of commitment, and so performance (Najita, 1994). A culture of personal commitment to excellence can be compelling, too – then the self is pushing for results, not just the heat of another's success.[10] (Personal honor for each individual also powerfully reinforces self-generated performance.) And the complex of cross-cutting obligations, which membership in "nested" groups produces, can work to secure that excellence for a wide constituency.

The cross-cutting obligations are vital. A virulent form of the Western response warns: decisions made by a group will always favor just the group (and so, freeze out others). Well, yes – actually that is the whole idea, to work out balances which *are* broadly acceptable to the group. But the "nested" structure means that smaller groups, ultimately individuals, also have their say, and their place. The "cross-cutting" balance, really among constituencies at different levels, of commonalities and differences, is the structurally integral check-and-balance against abuse. Both part *and* whole have their place, and get their due – no mean trick.[11]

Analogous to the Western neoclassical concern that dominance spoils the competitive ideal, there also seems to be a "spoiler" theorem in the community conception. Since community process is balanced fairly delicately on a shared acceptance for protocols to compete and then agree, a group member can deadlock the proceeding by holding aloof, or otherwise unraveling the tacit path. Just as antitrust or regulation is conceived to deal with "market failure", so will "shunning" (for the Amish) or "sending to Coventry" (in a British factory) seek to isolate the disruptive element – but none of these remedies, neoclassical or otherwise, proves so satisfactory. Whether outsiders may also act as "spoilers", for the community,[12] will below lead to a pivotal issue.

Eastern ways of thinking seem to underpin – to facilitate – this community conception. Western thinking is schooled to see opposites as contradictory, exclusive of each other; but in Eastern thought, opposites seem to be juxtaposed easily and productively. So cooperation intermixed with competition feels natural. In a related thought model, Buddhism holds as one of its bases a cyclicality – first generation, then recession. That would be encouraging to a naturally dynamic approach with, first, a flowering and competition of ideas; then, consolidation around some amalgam.

This is a most brief outline that meets both cognitive challenges, as well as works in the real world too – one that situates competition and the individual, however contradictorily, in the natural setting of consensus and community process. So far, Western proclivities have been typified around the neoclassical ideal of competition. What does a more sensitive, nuanced view of the West reveal?

4. The Western economic frame

The Western concern with methodological individualism surely proves correct – to mistake different individual's conceptions of the community for something transcendent only clouds the understanding. But, now we see how complex is the real task of understanding – somewhat differing views of community have somehow all to be accounted. But we may also sense the power that could be available to a complete analysis. If humans will naturally – unavoidably – conduct their economic lives from the perch of some group membership,[13] power goes to the analysis that so accounts.

The position staked out in the neoclassical view – defense of the individual, so that individual choice should not be unduly swayed – is hard won, clearly. Overbearing pressure from would be dominant demagogues[14] was the sober predicate underlying the first cognitive challenge. There *is* good reason to defend individual choice – the reason is imbedded in the human makeup. But having stood the individual upright, the next – I believe, necessary – step is to situate that person in the social sea where everyone does swim.

Where specifically is Western/US economic thinking, inspected in this strong light?[15]

4.1. A reprise

Externalities are the principal outcropping of community,[16] in the neoclassical system of thought. Externalities might, on occasion, be described as the paradigm's Achilles heel – they could also be an invitation to new thinking. The effects of scale – though less commonly conceived in demand, especially those in demand – are at a similar place. To cope with the pervasiveness of differences set against commonalities, there arose monopolistic competition (and its progeny) and then economies of scope. Concerns with agency have given us information asymmetries and moral hazards. From the Eastern perspective, we see a "second order" *symmetry* of information asymmetries – *everyone* has a privileged view, relative to others, on *some* matter. In that approach, it is a *strength*, rather than a flaw, that we can pool our separate contributions to share a jointly more robust picture. Moral hazard *could* refer to the "spoiler" who would disrupt the productive community process. But surely it is the notion of choice articulated by a group that will give Western economists the greatest pause, with a long preliminary struggle already underway.[17]

Though orthodox economists have lately begun to focus specifically on evolutionary processes, their attention has generally got as far as growing complexity. Another set, of heterodox economists, long fixed on economic evolution, are more concerned with network styles of economic organization. The network is surely the Western/US analogue to, and likely the step before full-blown consideration of, community.

Joseph Schumpeter was perhaps the original evolutionary economist. Though he was to change (or at least shift) his mind on some specifics, he established the idea of a dynamic cycle. Among today's heterodox, one senior statesman is Albert Hirschman (already quoted of course). Using notions of "exit" and "voice", he consciously intended to suggest coexisting opposites, with exit representing the side of competition. In fact, I believe the two ideas, together, also nicely set up the dynamics around transiting the boundary between "inside" and "outside" a group.

4.2. Whither creativity?

Where do the new ideas, the spurs to competition, come from? Though not addressed previously, both these styles of thought, from East and from West, depend on some flow of new experience or new thinking to fuel prospects for a better life. The West may openly celebrate novelty, but the new is no less in evidence, say, in Japan. The community conception is built, in fact, from competition among new ideas.

Creativity, it is commonly observed, is a special gift of individualistic cultures, such as in the US or the UK. Do individualistic cultures have a superior capacity to generate innovation, when compared with the community conception? – the prospect does not seem too farfetched. Some would even say that, after innovations are created, the more community-oriented economies are better at adapting and commercializing them.

Until this point, the thrust underlying, here, has sought synergy for the Western framework when brought together with the community conception. Implicitly though, that entails an expansion of community, a fairly dramatic expansion. Large societies with rather different sets of rules for resolving, and maintaining, differences – protocols for competition, on the one hand, and communities of competitors, on the other – would "open their (cultural) borders" to find new, now common cultural ground with each other. As put previously, the "individual would find its place in the social sea". What – then – is the fate of creativity? Is it possible, in a "joint culture", to spur creativity *and* to effect common purposes? – or, is the most sparkling creativity a "spoiler" that necessarily abjures community? The question is pivotal to the conclusions that can be drawn here.

And, as we try to look both East and West, expanding community becomes the pivot between the two views (before we can then turn back to telecommunications).

5. When community expands

Europe, as observed earlier, is today's living laboratory for the expansion of community – the widening of the sense for what is inside as against what is out. The European experience suggests the process is exactly the tug-and-pull that is now familiar, between reaching new common ground and maintaining differences. The almost innumerable pan-European agreements that have been effected, along with local variations still continued, are testimony. Compared with "normal" (nonexpanding) operation though, there is a key difference: what is now at stake are the protocols of the process itself.

Those same rules may evolve day-to-day also, but in expansion they themselves become the grist for any competition/consensus machinery, since different whole sets are now colliding. With the rules themselves at stake, there *are no* accepted guides for what will be kept and what will be melded. The going is gingerly, at best.

Why people continue to "push out", expanding the boundary, is a question not asked so often, but basic. The would be hegemonic demagogue does account for important eras in history. But the balance of the record leads us back to the question asked earlier – what underlies the "urge to merge"? Many nation states have traversed the same amalgamation, joining together what then became the component elements of the new state. Europe is just the most recent and visible case. Perhaps the ties are less strong, however, the wider the loop that is thrown. Devolutions – recently the USSR, or Yugoslavia – make clear that forces also move in both directions.

Is the community conception viable, in a world where different groups' different sets of rules truly collide – where they cannot "agree to disagree"? That seems to be the question, and in an increasing number of fora around the world. A tentative answer seems to be:

5.1. Collision or expansion?

The social nature of the human means that virtually all people operate by reference to a set of decision norms iteratively shared with others – "community" is unavoidable (the exception, the vanishingly small number of true hermits serving to highlight the universality of the principal case). Technology – communication and transportation, especially – have nudged and pushed together communities which might otherwise stand more alone. But also there is something else in the human (if not well-explored intellectually) which reaches out, and without which the technical nudge would largely be ineffective (perhaps triggered by the technology, but the human element nonetheless is determining).

In a world of *communities* interacting, the idea of externalities, rather than acting mainly as a "patch" for the conceptual system, has instead a useful role. Whether a rule is internal or only external to a given community becomes vital. The stability of a community's evolutionary trajectory, though not a simple equilibrium, becomes central to any analysis of how two communities may interact together, particularly the prospect for a mutual expansion. The Western/US case of "competition only"[18] becomes a special case of the general set.[19] To agree on "competition only", apparently the community has tacitly (or otherwise) also agreed to forego some of the more intricate satisfactions which would require cooperation, perhaps in part out of desires to guard zealously the sanctity of the individual, perhaps to spur even greater creativity as discussed already.

The prospects? To reiterate the obvious: Failures to reach new common ground for two clashing sets of rules may beget the classic wars to silence murderously "the other side", whatever the weapons and the battlefield. *Now* we can ask about viability for the community conception: to take the horrors of genocide as the extreme case, are whole communities lost, or, does the "style" of the community often live on, to re-assert itself another day? The alternative prospect – the alternative to violence as the resort – the prospect for mutual expansion is more hopeful, but it does also seem to come in many shades of gray.

With this question as a tentative answer, policy has surely become a more complicated subject. And we have moved from trying to sort out confusions about competition alone, to policy as a broader matter. Have we been positive/descriptive, or have we been normative in this quest? To ask about competition and policy is, in the first place, normative. But along the way we had to be as accurately descriptive as possible – in this case, to try and ferret out the individual in its social milieu. Most importantly, the normative conclusions are *self-defining* – norms that are the evolutionary product of community function. So, this paper has necessarily been both positive and normative, when taken in this special sense.

With the focus broadened from competition to policy, as we turn back to telecommunications – what are the implications? What might be the future framework for our dialog about telecommunications *policy*?

6. Telecommunications: framework(s) for policy

Telecommunications, it turns out,[20] demonstrates the cycle between competition and consensus quite precisely. Competing technical innovations will, by definition, not interconnect exactly; only a consensus which eventually melds together the various innovative proposals can create the network, seamless to the flow. (And the cycle reinitiates when the *next* innovations also will not interconnect with the standard which eventually establishes, from the first round and its consensus.)

More profoundly, the tools of communications – the network and its many associated media – become integral to the human web of communication, which sits at the base of the community's reiterations and its unfolding forward. The protocols by which the community debates and then parcels between agreement and diversity – the rules of the "black box" – ride upon the web of communications. Community process itself, in other words, is at stake when we make choices about policy for the network.

It is people who do the communicating, not the tools – but the network and its users will co-evolve together.[21] The modes and patterns by which people connect to each other will shape and mold – sometimes subtly, sometimes not so – as their tools for conversing take new form (intertwined also, of course, with new possibilities in transportation). And typically the network progressively shapes, to account for the community's experience with the system. So our policy choices for the evolution of the telecommunications system ramify to the capacity of the community, itself, to operate – ultimately the network and our understanding of the community itself will co-evolve.

To look specifically to framework for future dialog: What is the policy view of telecommunications which might come from a "community of competitors", as outlined above?

6.1. Telecommunications in a "community of competitors" — a policy framework

6.1.1. Competition and consensus — innovation and standardization

In a "community of competitors", network evolution is a repeated cycle between a competition among innovations and consensus around a new standard.[22] The capacity to reach consensus has policy billing that parallels the ferocity of competition; and consensus and competition inextricably intermix, particularly since different technologies will be at different points in their development. With technologies also interrelated, also expect cycles within cycles, and correlatively local variety within standards.

The market for the network, like the market for any new technology, can only emerge as the technology and its users mold – co-evolve – to fit each other. The market does not exist but emerges, evolves. Because a network serves some whole community, adoption of *network* technology depends on transiting a threshold, a critical mass[23] (what in the West would be ascribed to network externalities).

In a process of *social* innovation, users negotiate a cycle that parallels the alternated regimes on the supply side. Because of the threshold, users cannot even conduct *their* experiment with a network until the supply side has installed a critical mass of sufficient size for some community.[24] (The providers in supply also cross a "softer" form of the threshold.) The canonical form of user variety is the demand for private networks.[25] The "nested"/private communities are the basic building blocks.[26]

6.1.2. Dynamics of industry structure

Industry structure is integral to the conception. The appropriate structure is "nested" – a global network likely requires the complexity of at least three major levels, with internal subdivision. But "nesting" has meaning only when taken dynamically. When consensus is required, the layered components operate more monolithically, up to the layer affected; when competition is required to experiment with new possibilities, the pieces fragment. A musical accordion is not an inappropriate image – tracked across time, the pieces come together, then move away from each other; come together, then... An individual component in the "nest" must both define itself distinctly, for separation, *and* be able to link effectively with the other components, for consensual effort – concerns for *both* part and whole are explicitly articulated.[27]

Universality is then a simple idea. Universality is the connection of a given community (the membership at one of the concatenated levels) via some standard capability, that is, as codified at a specific cycle in technical development. Universality has meaning only for a given group and a particular technical moment. If we become serious about a global system for a global community – if community actually enlarges that dramatically – then universality at the global level becomes a distant goal, with many millions still a day's walk from any node on the electronic network.

(Does this complete the story? By no means. For instance mechanics such as pricing have not been touched here; the cases that *do* appear to be commodities (so that there is true excludability), such as reseller offerings, have not been covered. Nonetheless, the case overall is unmistakable.)

In the Western case – this time particularly the US – what are the continuities, or the exceptions?

6.2. Western/USA parallels/contrasts in telecommunications?

Individualism has been famously on display in US telecommunications, but in recent decades less so the capacity for consensus. For instance in ISDN, when Bellcore finally announced a national ISDN, two RBOC's found themselves unable to join the standard, at least for some time – and this when ISDN had been nationally available in countries from both Asia and Europe for several years. Or in a different way: The recent collapse of large mergers – originally mooted to converge telephone and cable and so break ground for a new "superhighway" – raises sober questions about the US ability to deliver the consensual interconnection that its grand vision would promise. These are just two more conspicuous cases of the US hesitation with consensus, despite its vaunted capacity to produce exciting new ideas.

At the same time, there appears to be a "heterodoxy" also at work in US telecommunications. Internet developers have nurtured a "gift" culture which rapidly assembles consensus for an evolving set of standards. The HDTV Grand Alliance was itself an historic consensus among what had been, to that moment, fierce competitors. Vice President Gore's call that the US telecommunications private sector should assure connection for education and medicine, though a naked return to cross-subsidy on a massive scale, has been adopted explicitly to serve community goals – and the industry has openly embraced the call.

Examples from computing typify the US dilemma, particularly since compared to telecommunications, computing does not have strictly to interconnect. The Open Doc movement, for "document-centered computing", is such a case. A group of companies have moved to make software applications modular, in the sense that a module from one developer would seamlessly interface with another. So a user of a word processing program could call up the graphics package that s/he thought best, though it may not have been written by the developer of the word processing program. This is precisely the "variety enabled within a standard" which sits at the heart of the conception above. But success depends upon convincing dominating Microsoft to integrate its parallel approach, Object Linking and Embedding, into the effort – standards are "open" only to the extent that there is true consensus.

Europe, perhaps partly as the product of its historical circumstance, has practically enshrined competition and consensus as two thrusts for its telecommunications policy. Though the two may coexist uneasily, competition is intended for instance to bring down cross-border rates, while consensus is necessary to move beyond the bewildering quilt of separate national systems. Sweden, along with other Nordic/Scandinavian countries, has demonstrated how other means besides competition may be effective to deal with the problems of monolithic, dominant organizations in telecommunications. (And globally, structure has begun to move toward three-level "nesting", but with no indication yet that the necessary dynamics are appreciated.)

Such, I believe, are rough outlines for a future dialog about policy in telecommunications. In the prospective framework, regimes alternate between competition and consensus, for the evolution of network and community between innovation and standards, with a dynamic "nested" industry structure. Though key ideas have arisen from experience in the East, both thought and practice in the US have opened to the possibilities and Europe seems actively to pursue a working model, if still largely unaware that it does so.

The practitioners in the business have now found a place for their observation that, by their own lights, they live daily with competition in telecommunications – their experience of competition is central to this dialog (when, anyway, the competitors are each in pursuit of some new, better method). This dialog is indeed beyond competition, all the while it situates competition as one of its mainsprings.

7. Economics of communications

Much of this paper has flowed from an information processing perspective on people. Does that simply confirm the information processing view that now is traditional to an economics of information? I think, instead, it propels us beyond. Yes, individual human processing of information is basic to the scenario – but I believe that the scene described here is different in kind from the now-standard picture. Outcomes depend upon (tacitly) shared decision rules which are applied and evolved in iterative communications, communications which in a fundamental sense are prior to the availability of usable information.[28] I believe that encourages us to an economics of communications.

Surely an economics of communications incorporates information (and surely it will be even more complicated). The intent is to move to a more accurate picture and a better world. Whether that would actually be so – both whether more accurate and whether an improvement – will now have to be discussed, debated. Is not that the process – iterative exchange about some new idea – we commit ourselves to?

Notes

1. In his Preface to the Paperback Edition, p. vi et seq., Hirschman (1992) (quoted to open this paper) also marks how "there always has been a *large variety* of 'market societies'". (my italics)
2. And certainly would not make happy Japan's trade negotiators, who now characterize the *US* position in just these terms.
3. The same story can be told countless ways of course, the more compelling the drive, the more telling the effect. The young swain, coming back from a first date, convinces his partner to a goodnight kiss in what may initiate the mating cycle, a proximate "real" cause. But he likely *also* does so because his *view* sees this woman as a potential mate, when *her* view *may remain completely unaccounted*. (Or, if the genders reverse in this exchange: she succeeds with her heartthrob.) Those later convicted of date rape may have some sense how sorely their vision grew from internal imperative, and not from a realistic assessment of the world around. Or: those sadly re-assembling the pieces after a divorce, now with small children to raise, are in a more complex situation, but may also have reason to re-assess their earlier perceptions.
4. If I do not, someone else will here refer to the blind men and the elephant.
5. Such a repeated sequence is in the same family of shifts that Thomas Kuhn first made evident – I clearly am indebted to him.
6. Do the systems avowedly committed to "hearing all views" in fact work? Consider Japanese politics

during the long reign of the LDP after the war. Could the Socialists, for instance, say that their position was heard in the policy outcomes? Or did the votes which perennially returned a Socialist delegation to the Diet in fact weigh in the final balances?

7. I resist the temptation to construct a concrete case as parallel illustration until we return to telecommunications. But then telecommunications itself comes naturally to hand as a "pure" and whole working example.
8. In the quote I used to open the paper, but modified, Hirschman actually said, "...is it not in the interest of social science to embrace *complexity*...?" (my italics)
9. To jump the gun on a look back to the West: Michael Porter's seminal work on corporate competition has as a principal thesis that strategy should differentiate the firm, and thereby *escape* the only competition which neoclassical thinking, strictly interpreted, sees possible (namely, competition among the same, homogenous products). Some then see irony in the title of the book, *Strategy of Competition*.
10. In *this* world, another's success, rather than a threat to be destroyed and eliminated, is instead a beacon, a call to even greater success by the self.
11. What of those outside the community's *external* border? That *is* a basic issue: see below the discussion about "expanding the community".
12. In the 19th century in the USA, when settlers who headed into the West were forced to "circle wagons" against a threatening foe, was it necessary to suspend the more deliberate exchanges in community life?
13. Some nonetheless see membership as tenuous – they might ask: Is there actual pleasure in group affiliation? Or do we flee into inevitably temporary alliances only under pressure of common external threat? The wagon train, from the previous note, is depicted as well organized prior to the threat – was the threat of, say, poverty necessary as an earlier impetus, for those pioneers to band together in the first place?
14. Though Max Weber may not have intended it precisely this way, his charismatic leader whose vision is routinized into a bureaucracy depicts exactly the transition envisioned in the first challenge. (The transition from good new idea to consensual standard must also be recognized as another parallel!)
15. There is of course a tradition, extending from Marx, that excoriates capitalism for its destruction of community ties. The question here, though, is not discontinuities in thought, but the prospect for continuities.
16. Any serious discussion of community, from the West, must acknowledge antecedents such as Durkheim, or Tœnnies and his gemeinschaft and gesellschaft.
17. Marked perhaps at the one side by Kenneth Arrow's *Social Choice and Individual Values* (1951), and at the other by Amartya Sen's "Rational Fools" (1977 – Sen is of course Indian, working in the West).
18. Western views of trade policy already find a new context, in the community conception, and its "nested" economic structure. Now, in the expansion which joins groups together, the "trade bloc" takes on new meaning. The bloc is expression of the new group *when it operates as a whole* (but only then). The implications overall for trade are an area that will also likely feel inimical to the Western sensibility.
19. Analysis of potential merger between two *companies in* the "competition only" world is, of course, a case of the *general* set, with two cultures embarked on a prospective expansion.
20. Telecommunications is, after all, a construction in the real world of means to connect some community which exists essentially in thought – so that the usually nontangible is then given some evidence in the corporeal.
21. See Mazlish (1993) for a general treatment of co-evolution between people and their tools.
22. Notice that this is not the same as the currently popular dictum, "standardize to enable competition" – in fact it is, in logical terms, the *direct opposite*. The competition here is not among homogenous commodities but rather it is aimed to surface exactly the opposite, that is different ideas and different implementations. Rather than anticipating homogenous substitutes, this expects continuing variety to be enabled within any standard.

23. As described in the critical mass literature. My own work is an example; there are several other authors also.
24. France Télécom's Minitel shrewdly built geographical region by geographical region, but nonetheless it faced billions of francs in investment for what could start only as an *experiment* at the national level.
25. There is some tendency to confound moves toward privacy with the fragmentation necessary during experimentation (the discussion of a "network of networks" tends not to make the distinction, for instance). There is a relationship between privacy and fragmentation for experimentation, but the distinction is also fundamental.
26. For a more complete discussion of the basic cycle see Allen (1991).
27. For a detail discussion of this dynamic structure see Allen (1994).
28. I am indebted to Marcellus Snow, Professor of Economics, the University of Hawaii at Manoa, for sprightly discussion of this question, and indeed for discussion of some of the broader topics in this paper. Meheroo Jussawalla, Senior Fellow, the East-West Center, Honolulu, has steadfastly maintained an atmosphere that encourages committing the ideas to paper in the first place. I, however, am responsible for the conclusions here.

References

Allen, D., 1991. *Between innovation and standardization: the evolving network.* Manuscript.

Allen, D., 1994. Competing visions for world telecommunications: the global evolution of industry structure. In: Granstrand, O., Bohlin, E., (eds) *The Race to European Eminence.* Amsterdam: Elsevier.

Arrow, K.J., 1951. *Social Choice and Individual Values.* New York: Wiley.

Hirschman, A.O., 1992. *Rival Views of Market Society and other Recent Essays.* Cambridge, MA: Harvard.

Huber, P., Kellogg, M., Thorne, J., 1992. *The Geodesic Network II: 1993 Report on Competition in the Telephone Industry.* Washington, D.C.: The Geodesic Company.

Mazlish, B., 1993. *The Fourth Discontinuity: The Co-Evolution of Humans and Machines.* New Haven: Yale.

Najita, T., 1994. *Lectures on commoner economic thought and practice in modern Japan.* Manoa: University of Hawaii at Manoa, April.

Noll, A.M., 1993. *A Study of Long Distance Rates: Divestiture Revisited.* Annenberg School for Communication, University of Southern California, Los Angeles (CITI conference on localism, New York City, December 10).

Sen, A., 1977. Rational fools: a critique of the behavioral foundations of economic theory. *Philosophy and Public Affairs,* 6:317.

Taylor, W.E., Taylor, L.D., 1993. Postdivestiture long-distance competition in the United States. *American Economic Review,* 83(2):185 (May).

Beyond Competition: The Future of Telecommunications.
Edited by D. Lamberton.

CHAPTER 3

A model for forecasting data communications towards the year 2003

Berit Svendsen and Jan P. Saether
Telenor Research, Kjeller, Norway

Introduction

The dominant service in today's telecommunication network is voice, both in terms of traffic and income for the network operator. The network infrastructure is essentially developed for supporting voice traffic. Therefore there is no doubt that all telecommunications network operators need to develop the network infrastructure to support future data communication services. The question is rather when and how to start developing the infrastructure to support both broadband services (much traffic from single users) and a growth in total bandwidth usage (low to medium traffic from many users — aggregated traffic). In other words, what is a reasonable level of total supply of network capacity? Which applications and which requirements are necessary for future network capacity provision? The problem is to match the capacity in the network with the market needs. This paper focuses on forecasting methods for finding user demand for data communication services and capacity.

In the past forecasting methodologies for the number of subscribers to new and existing telecommunication services has largely been based on an anticipation of a saturation level (sites, white collars, etc.), fixed or variable, and in some cases with observations for the first years after market introduction for the telecommunication service. This gives the input for estimating growth curves; and the state of the art is to produce logistic curves. The estimates of market penetration have traditionally been given for every single telecommunication service without taking cross-impact effects in the demand between services implicitly into consideration, if they are considered at all. For new services without any market observations as support at all, we have realised that some comparisons have been made to existing services, but not necessarily with the objective of making a trade-off.

All the Nordic forecasts for mobile communication have been considerably higher than elsewhere in Europe, seen year by year. The background for these high forecasts is, among other things, a shared vision on future demand, cooperation in planning and

Address for correspondence: Berit Svendsen, Telenor Research, Instituttveien 23, P.O. Box 83, N-2007-Kjeller, Norway. Tel.: +47-63-80-91-00. Fax: +47-63-81-98-10.
E-mail: berit.svendsen@tf.telenor.no or jan-petter.sather@s.hk.telenor.no

developing NMT (Nordic Mobile Telephone) systems and input from a long history of public, manual mobile telephone systems, resulting in building almost enough capacity for the whole history of the automatic mobile telephone. The Nordic forecasts were 10 times higher than the forecasts made in comparable European countries in economic terms in the middle of the 1980s. Largely for these reasons the result is a much higher mobile telephone penetration in the Nordic countries. In the past there were few highly independent services and it is of less importance to consider the possible relationship among them.

However, in recent years an evolving tendency has been to consider all telecommunication services in relation to each other with interrelated forecasting models, but with usage of secondary indicator variables such as the number of terminals, number of connections, number of white collars, number of sites (workplace establishments), etc., in models more or less explained by the structure of the life of production (Chidambaram, 1978). In the Eurodata Foundation efforts have been made to construct a global model for application development and the subsequent need for services and capacity.

Our goal is to establish consistent global models for business communication with relationships representing feedback and where the substantial entities are the total capacities for communication and total realised usage of the given aggregate network capacities. Our next goal is to disaggregate total demand into demand for certain services or applications. Presently, a practical limitation is that we are only considering non-voice-traffic, and furthermore that our analysis is based on a study in Norway, however, we believe that these methods can be used in other developed economies as well.

Future broadband users will be recruited from two sectors: business and home. In the years to come it is probable that broadband communication will be driven mainly by business users (Hagenhaus and Von der Straten, 1993). This is the reason why we have limited our model to the business sector. But a future total model should aim at covering both business and private sectors. Extended causality, as the number of sites and their location, should also be attended.

Methodology

System dynamics is a methodology for understanding certain kinds of complex problems (Richardson and Push, 1981). The problems addressed from the perspective of system dynamics have at least two features in common. Firstly, they are dynamic: they involve quantities which change over time. They can be expressed by graphing variables over time. Secondly, the problems involve the notion of feedback. Another major advantage using system dynamic models is the possibility of running iterations for calibration in order to get values for unknown parameters instead of trying to estimate them. The third advantage is that system dynamics is now realised through simple graphic software, lowering the threshold nearly as far down as spreadsheet computing. However, with too many unknown parameters this is less applicable, and dynamic models will turn into a more qualitative tool for considering effects and not so much a quantitative tool for forecasting. It is therefore advantageous to try to estimate as many as possible of the single relations statistically. Involving both estimation and calibration will give the best opportunities to make proper forecasts or projections. Figure 1 shows a coarse-grained causal loop diagram/model for realised demand of data communication (e.g., economic ability to meet the demand for data communication outside a company site) for the business sector in public networks.

Before we describe the model blocks in more detail, we need to define what we mean by data communication. We define the total market for external data communication to be all kinds of telecommunication except voice telephony, picture phone and video conferencing. Data communication includes telefax communication, data communication in public data and telephone networks (e.g., modem traffic), and natural evolution of these services (including multimedia communication). These limitations are practical and tentative given the fact that the purpose of telephony and video conferencing is also the transmission of information.

We have focused on a time span of 20 years, the past 10 years for estimation and calibration of the model, and 10 years for forecasting or projection.

In the following sections we will explain the different building blocks in the model, we will show some of the graphs with historical data from the early 80s to the present situation, and we will explain how we will use the model to forecast the need for data communication in the business sector of the future.

Needs for telecommunications

Telecommunication service vendors have traditionally had limited proximity to the market. A consequence of this is that signals from the customers, if they exist, are not sufficiently picked up. Few customers are able to define their needs within the boundaries of possible technologies; and if they are, they are so competent that they

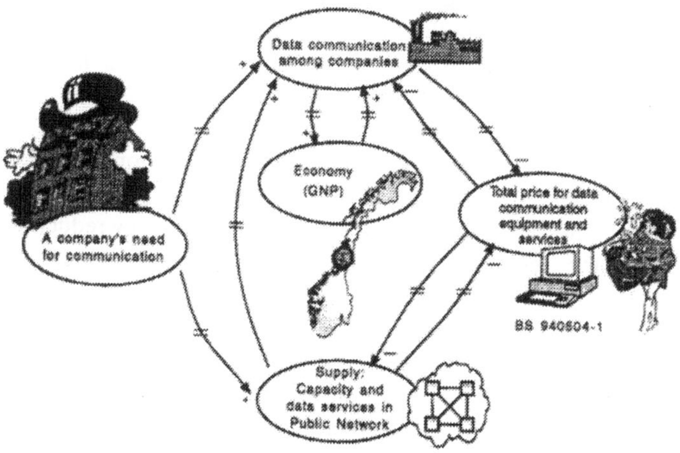

Fig. 1. Coarse-grained causal diagram for data communications.

tend to establish their own communication services, typically in large companies with specialised competence within the field. However, the picture will probably change in the future.

The time from introducing new services until the demand "takes off" is often long. This seems to be confirmed in the figures we have compiled. The year of introduction is also correlated with the market penetration.

There are many reasons why the time spanning from the initial market evaluations and technical considerations to an implemented service offered in the market is a long one. The process through the first ideas for a new service to specification, construction/programming, budgeting/financing, ordering, testing, production and finally to supplying the service normally requires several years. ISDN could serve as an example. Reluctance to introduce new services may arise from these facts, however, so far this has not been the case. Competition may increase the risk that the efforts to bring new services to the market will be avoided. In fact, the contrary has happened; many new telecommunications services have been launched, not always with success. However, falling back to reluctance could prevent the supply of potentially useful services in the market.

On this scene the vendors of telecommunication services may introduce services that the market may or may not accept. Market acceptance has different causes which should be classified in necessary and sufficient conditions. One such necessary condition is that there is a latent need at all, or that some variables over time contribute to establishing such needs. Sufficient conditions could, if specified, be such variables as user friendliness, user competence, etc. New services will generate new traffic, but also attract traffic from other services and networks. New services will also induce traffic, i.e., give positive cross-impacts between services/networks. Facsimile has had an observable attraction of traffic from postal services. On the other hand, e-mail might very well be a competitor to facsimile and post in the future, and to telephone calls for instrumental nondialogue usage (but here we should also expect traffic induction). Fax seems to increase the number of voice calls and vice versa, i.e., cross-impact effects. There is a considerable data communication traffic in the telephone network despite the fact that advanced dedicated data networks are established. Hence, the decisions regarding new networks must be considered thoroughly, taking into account the fact that decisions will have implications for the long term, and that the consequences for other services are considered. The effects for other services caused by the introduction of ISDN represent an example. Needs for broadband services have also to be considered within this framework. But even evolving technologies, as for instance redundancy reduction, have to be considered for the secondary need for transmission velocities and capacities.

Establishing new services without considering the more or less common total market for the different existing and new services is a pitfall into which the vendors have been falling many times since the great variety of new services started to appear in the 80s. With few services, as in the late 70s, there were fewer problems introducing new services and keeping up with demand, at least after some time of supply adjustments. Today, there rather seem to be overinvestments in new different services and networks in a somewhat limited total market.

Business and public authorities today exhibit a strong need for communication. A company's needs for communication depend on a lot of different factors, among other things the size of a company, where the company is located (with geographically dispersed locations), industry and so on. In statistical figures there is a distinction between enterprises and establishments, and in the following we will use these terms.

An enterprise may be defined as an "institutional unit comprising all economic activities engaged in by one and the same owner".

An establishment, on the other hand, is in principle defined as a functional unit which, at a single physical location, is engaged predominantly in activities within a specific activity group.

Several studies (Svendsen, 1993; Backhaus and Spath, 1993) show that most data communication in public networks today is between geographically dispersed workplace establishments (inter-establishment data communication) belonging to the same enterprise.

We have focused on the total number of workplace establishments in Norway in the model and are trying to find the total needs for data communication, both among establishments belonging to the same enterprise and among single and independent establishments.

According to the Central Bureau of Statistics in Norway, around 9,095 Norwegian enterprises (numbers from 1992) with an average of 5.45 establishments per enterprise represent the initial and primary market for inter-establishment data communication. Expressed as needs for logical communication links among the establishments (this must not be regarded as communication links in a telecommunication network), a potential of around 40,000 to 100,000 could be possible, depending on whether these logical communication links are of a star network topology ($(n-1)$ connections) or a meshed network topology ($(n(n-1))/2$ different connections). A star topology is logical communication links from all the branch offices (establishments) to one main office (centralised enterprise), and a meshed topology is logical communication links among all the establishments in an enterprise (decentralised enterprise). In terms of demand for subscriber lines for data communication services, each establishment may of course have more than one line or physical connection (or several logical connections on one physical connection). However, the average undoubtedly will not deviate much from one. In the Norwegian market for data communication today the installed subscriber base for data communication services is already at the level of around 40,000. Svendsen (1993) indicates that star networking data communication is still by far the most dominant topology. These facts may tell us that the market will soon be exhausted for the inter-establishment connections among the number of presumably cooperating geographical units of enterprise. But this can very well change if the companies are moving from a centralised to a more decentralised organisation by division into still smaller units.

We may expect that the market will increase, i.e., by new ways of cooperating, due to the new possibilities of geographical fragmentation of establishments supported by data communication, and of course due to still more bandwidth-consuming applications. The increasing number of employees with information-related jobs may also increase the need for data communication. While there are tendencies today to establish physical lines according to the number of different applications, the future will probably give us relatively fewer lines, but with more traffic on each line.

The 210,000 single and independent Norwegian companies will probably not represent a large market for data communication in the near future. However, new alliances and new ways of cooperating made possible by data communication may arise. "Virtual companies" is a phenomenon in focus today. The term is linked to smaller, cooperating companies that can deal with large projects by virtue of their common size and wide competence.

Economy

Much as for transport, the driving force for telecommunication demands lies in the social, organisational and economic sphere as a whole, both as a production input and in relation to consumption in the private sector. Increased incomes at individual, enterprise and aggregate levels are partly based on telecommunications for cost reductions and for market expansion, and partly on specialisation, division of labour and geographical dispersion, and on the interrelation/interaction between these effects. Telecommunication investments can explain changes in income and vice versa, even if the directions of causal effects are still under discussion.

Cronin et al. (1991) claim to have statistical evidence for stating a two-way, lagged causal relationship between GNP and investments in telecommunications infrastructure. They have tested and confirmed the following two hypotheses:

Hypothesis 1: The level of US economic activity at any point in time is a reliable predictor ("cause") of the amount of US telecommunications investment at a later point in time.

Hypothesis 2: The amount of US telecommunications investment at any point in time is a reliable predictor ("cause") of the level of US economic activity at a later point in time.

We are not going to discuss the basis and accuracy of these results, but the results intuitively seem to be realistic. However, we want to give a more precise interpretation of the relationships, and the conceptual basis for our own efforts in trying to make global models for forecasting telecommunications needs.

In a relatively balanced national economy without specially constricting bottlenecks, and with full utilisation of the productive resources, investments (including investments in telecommunications in public and private networks) may give increased productive power. Alternatively, GNP is used for private and public consumption, of which an amount represents education. Education is investment in human capital, increasing productive power of the workforce in the long run. Otherwise, consumption represents the reproduction of labour workforce. Important elements in our concept are the interactions between technology, organisation and competence. This is in accordance with modern theories of economic growth where technical progress is endogenous. Our conceptual model looks roughly as in Fig. 2, where the arrows indicate the directions of influence on the left-hand side, and usage of GNP on the right-hand side (consumption, investment, net export). (This indicates that we could also include the macroeconomic relationships in the model, but in the initial phase we want to go into a more limited approach, where GNP is exogenous in the initial steps.)

Early national estimates of production functions from the 50s and the 60s indicate that productive capital measured in market values and labour workforce measured in numbers of persons do not fully explain the growth in GNP. However, a residual factor denominated as "factor of technical/organisational conditions" does explain growth. For Norway estimates based on Cobb-Douglas models have been made indicating that the growth in GNP from a 1% growth in labour and a 1% growth in capital contributed by 0.76 and 0.2%, respectively, while the residual factor of technical-organisational progress gave considerably more effect (GNP increased by 1.8% per year attributable to this residual factor). Later theories of endogenous growth deal with the same problem, but in these theories there are attempts at specifying growth components for techno-organisational relationships. Learning by doing, research and development and externalities are brought into growth models, e.g., Young (1991) and Romer (1993).

Cronin et al. (1993) indicate that as much as 25% of the increase in productivity in the US economy is attributable to advances in telecommunications. Full benefits from investments are achieved after about 15 years.

As an illustration, the next figure (Fig. 3) could indicate that there is a relationship between income and telecommunication investments. Naturally, high incomes give the opportunity to buy more telecommunications equipment. On the other hand, if such investments also contribute to the productive power, optimal investments should be focused thoroughly.

Prices for computing, information storage and retrieval, data production, and data communication are crucial factors, together with the expanding wage costs in the

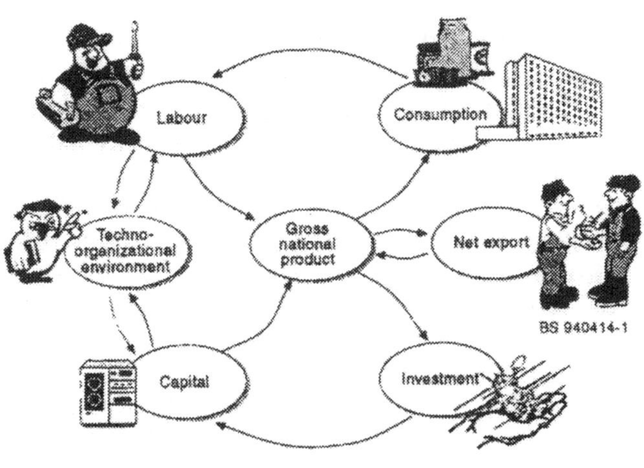

Fig. 2. GNP provision and economic circulation.

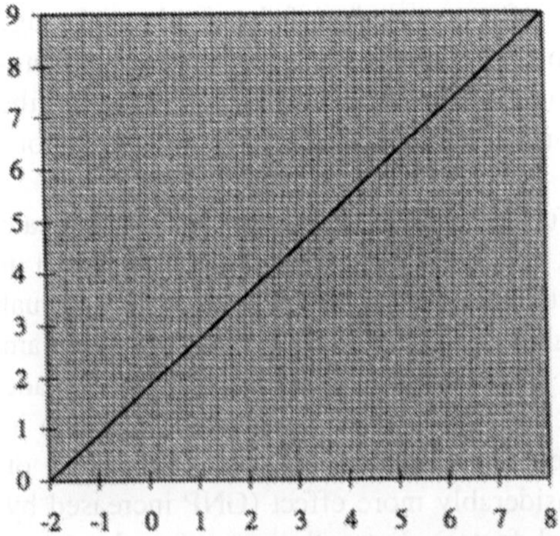

Fig. 3. Change in the number of business telephone mainlines per year related to the change in GNP in Norway.

information sector of the community and for the individual business establishment. As a basis, there is a tendency that fewer people produce more goods based on the facilities given by communication in general, and especially for our sector, telecommunications.

In the model that we are going to establish these variables will be introduced successively with few details in the initial phases. We should call this a model of aggregate, macroeconomic relationships. The aim is to break down the total forecasts into partial forecasts for different applications and services.

Total capacity and traffic in the public network

The growth of the capacity in the core networks in industrialised countries has been very high, and Fig. 4 shows this growth for the Norwegian core network. Core network is used in this paper as the part of the network which consists of a total amount of 64 kbit/s full duplex circuits longer than 20 km. A very rapid price fall for the single mode fibres has made it possible to introduce more capacity in the network in a very cost-effective way. The reason why we had a jump in capacity in 1988 and 1993 is the introduction of 565 Mbit/s and 2.5 Gbit/s fibreoptic systems. From 1986 to 1993, we had an average growth of 75% per year. As mentioned earlier, telephone traffic is still the most dominant service in the telecommunication network, and the telephone traffic is growing by 6–8% a year, but many people expect future growth to be for services other than traditional telephony. Kato (1992) showed that most of the traffic in today's networks is telephony. It is also estimated that the network load caused by advanced communication services (non-voice-services) will reach almost the same amount as the present network load (telephony) by the year 2000.

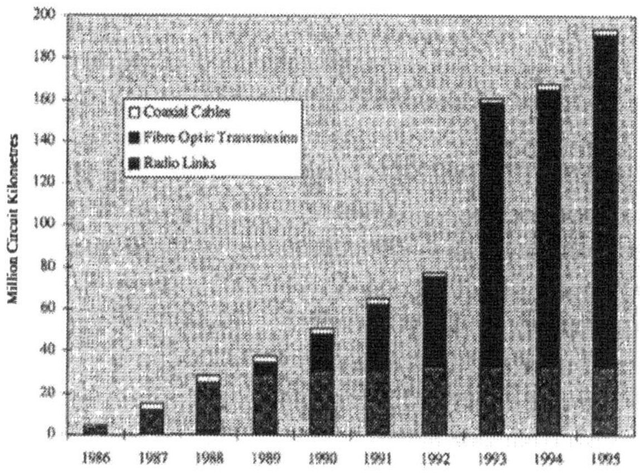

Fig. 4. Growth in total capacity in the Norwegian core network.

A very important issue is to find a status today for data communication in public networks. Without finding the status today, our prognoses for the future will not be reliable. We have different services for data communication in Norway. There is a line-switched data network (based on recommendation X.21), we have a packet switched data network (based on X.25) and we have a leased line (digital and analogue) service. ISDN is a commercial service from 1994. The traffic is measured in different ways for the packet switched and circuit switched data network. The reason for this is, among other things, that traffic measurements serve different tariff systems. For circuit-switched data networks the holding time of a call is measured, whereas for a packet-switched call the actual amount of bit transferred are measured. For leased lines there are no measures for traffic at all. Considering all sorts of data communication as parts of a total market is thus difficult.

We have used the total capacity as a tentative measure. The total capacity for the data services is defined as the following:

Total capacity = $\Sigma_j \Sigma_i v_i q_i$

where j refers to different services, i is the number of different transmission speeds for a data service, v is the transmission speed that depends on the data service and is in the range 300 bit/s, 600 bit/s, 2,400 bit/s, 4,800 bit/s, 9,600 bit/s, 19,200 bit/s, 64,000 bit/s and 2 Mbit/s (every data service does not support all these different speeds). q is the number of subscriptions for each transmission speed. Data traffic relates to total capacity as:

(Data traffic) = c(Total capacity)

where c is a factor varying between 0.003 and 0.01 for different data communication services.

One way to find the growth in data communication over the last 10 years is to look at how the capacity has grown for the different data services (Fig. 5).

One important amount of data communication that we do not include in this way is data traffic in the telephone network. It is a fact that not all the traffic in the telephone network is voice telephony. Considerable modem and fax traffic is transmitted in the telephone network. We have observed traffic in several telephone exchanges, and found that approximately 14% of all telephone calls were fax calls, and 1.6% of all calls were modem-calls. Registration of this traffic is the basis for estimates of the modem and fax traffic.

Another way to look at the growth for the data services is to measure the amount of data transferred through the different networks. The packet transmission services are paid by the amount of data transferred. For networks where we cannot measure the amount of data traffic directly, we have to estimate the amount of data transferred through the networks (e.g., in-line switched networks, where calls and conversation time, not data transferred, is registered). We should also include the data traffic in the telephone network. It is also important to find the growth for the data traffic over the last 10 years in this way.

Total price

There has been a dramatic fall in the prices of computers and storage media. Other products with a potential to support communication such as video cameras, scanners and digitizers are also maturing technologies with increasing performance and decreasing prices.

The prices are also decreasing for data communication in public networks due to the introduction of new technologies, but also due to increased competition. In addition, redundancy reduction based on different compression techniques for text, image and digitised video changes price/capacity relations in favour of more data

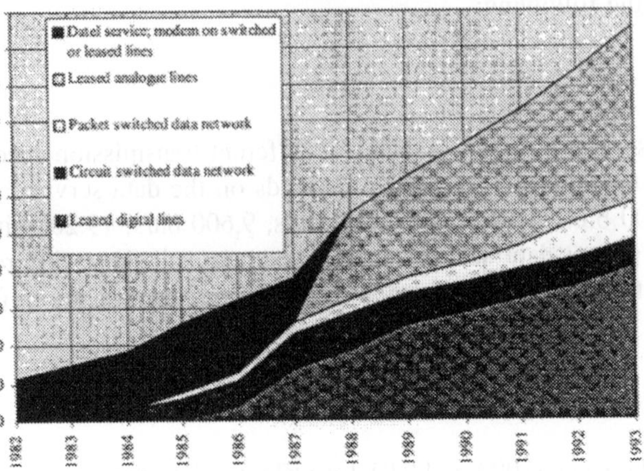

Fig. 5. Capacity development for data services in Norway (1981 = 100).

transmission.

On the other hand, the total price of information processing may rise due to the fact that the proportion of labour workforce in such an activity is increasing, and in many countries the wages have also increased in real terms over time.

The prices for different applications for the customers where data communications is involved are much more than the prices for data services supplied by public network operators. However, the cost of equipment used for interfacing public networks is relatively small compared to the prices of data services.

Further work on prices, i.e., total prices for certain applications, have to be investigated more thoroughly.

Applications

Initially, the model is established as an aggregate traffic model based on the apparently steady growth of total bit rates both in data communication capacity as well as in the usage of this total capacity. Applications are not building blocks in the model but, as mentioned, one of our goals in decomposing total demand into demand for certain services and applications by trying to find the growth rates for different services. The decomposing phase of the model development will turn it more into a simultaneous causal model. The different main applications of data communication can be classified into the following:

- Exchange of information person-to-person (or among groups). This area consists of, among other things, fax and electronic mail.
- Information retrieval, booking and payment, electronic marketplaces, bank transaction systems and travelling booking systems.
- Terminal networks (e.g., IBM networks).
- Backup and file-transfer.
- Technical applications, e.g., telemetry.
- Program-to-program services, e.g., client-server applications, etc.
- New applications, e.g., virtual reality, remote working, remote education, etc.

We are trying to estimate the size of the market and growth rate of all these applications. A very important aspect is to find the relationship between the main applications of data communication and the amount of data and data calls in our public network. This work is in its initial phase and will be included as saturation levels in the models.

Demand forecast for data communication towards the year 2003

After having gathered data for the different building blocks in the model over the last 10 years and expressed the building blocks in terms of graphs of variables over time, our objective is to find the relationship between the building blocks. We then have to calibrate the model and successively we then will have to change the model. We will at this time try to use the new model to forecast the demand for data communication in the future.

In this paper we have indicated what we want to be incorporated in the model, or rather the family of models. The job of gathering information and data for model input has been much more time consuming than expected, so the efforts in modelling so far have been limited to a few hours. However, we are going to outline the first steps, then say some words about different traditions in modelling in economics and in system dynamics.

In the first step we have not used feedback. We have defined GNP and data communication prices as exogenous variables, i.e., defined outside the model. Assuming that the supply has been sufficient, the model does not include discrepancies between demand and supply. Using elasticities (e.g., quasi-elasticities) as central terms, because the terms are well understood both by economists and beyond the circle of economists, and the differentiated demand function for "total data communication", the following expression in system dynamic language forms the core of the first model:

$$\Delta DATACOM = DATACOM\ [e(\Delta PRICE/PRICE) + E(\Delta GNP/GNP)]$$

which expresses that the change in data communication capacity, as defined previously, equals data communication capacity times the direct "price elasticity" (e) times the relative price change on data communication plus data communication times "income elasticity" (E) times relative change in GNP.

Total data communication capacity is, as mentioned earlier, correlated with total capacity for all data communication services including data communication in leased lines.

Assuming tentatively that there is a high degree of interdependency between data communication capacity and other input factors in production processes and constant elasticities over time, we observed the following and simulated figures for data communication capacity over time (Fig. 6). GNP and prices are also included in Fig. 6. All data are represented as indexes.

As we can observe from Fig. 6, data communication capacity will double from 1993 to 2003, presupposed a price reduction of 2.5% per year. If we assume a price reduction of 10% per year (that is mainly because of the liberalisation process), data communication capacity in 2003 will be a little more than 4 times the level of 1993.

The "elasticities" e and E have values of –2.9 and 3.3, respectively, and these values give the closest tracing to the observed development from 1980 to 1993. (In fact we have to work more with the price elasticities of data communication, the preliminary figures are some indications.) Some people may object to both the high absolute values of e and E, and some people may also question why the values are constant in the period of forecasting.

The high values could be argued for from the fact that data communication in our terms is a new service, and still a bit of a "luxury". Furthermore, not all the cost components for end-user equipment, personnel costs, etc. are included in the

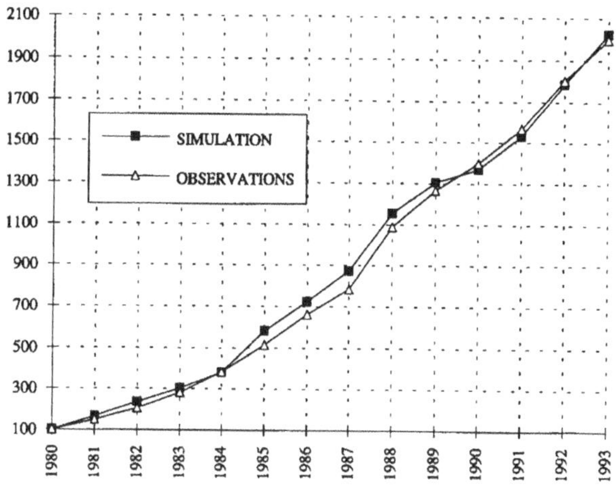

Fig. 6. Observed and simulated data communication (indexes).

calculations.

Constant elasticities could be applied when the "commodity" in question is not a constant one over time, but is a continuously changing one. Data communication in 1980 was a cumbersome affair. One page of text in 1980 was a far less number of bits than one page today, formatting procedures and document layout has changed very much for the better. Many other factors have changed in the same manner.

So far, we have used system dynamic tools in what we could call a traditional economist approach. We could extend the model by specifying partial models for the different data communication services, their inter-relations, total data communication, effects of relative prices, etc. We also could introduce feedback. However, such models will become very complex, with problems regarding precalculations of estimates for all the price elasticities, direct and cross-elasticities, and expenditure elasticities. This direction of model development could be fruitful purely for forecasting purposes, i.e., for giving the technical planning department some quantitative input. However, for the economists this way of attacking the problem is a traditional one in the sense that the model will turn into a correlational model more than a causal model. Our next highest priority will therefore be to construct models where causalities are given high attention to get qualitative analyses. So far, we have used the model tool as a tool for forecasting with exogenous GNP and prices for data communication.

The first steps are already presented for the advisory board of the project. The discussions after the presentation were fruitful, and the simplicity of the initial model contributed very much to this.

References

Backhaus, K., Spath, G.-M., 1993. *Results of the Study - Company-Wide High-Speed Networking, No. 16.* Munster: Institute for Major Systems Marketing (IAS).

Chidambaram, T.S., 1978. A forecasting model for evaluating the potential of a new communication service. *Telecommunications Journal*, 45(X):541–546.

Cronin, F.J., Parker, E.B., Colleran, E.K., Gold, M., 1991. Telecommunications infrastructure and economic growth. *Telecommunications Policy*, 15(6):529–534.

Cronin, F.J., Colleran, E.K., Herbert, P.L., Lewitzky, S., 1993. Telecommunications and growth. *Telecommunications Policy*, 17(9):677–690.

Hagenhaus, L., Von der Straten, G., 1993. Broadband Networks: Will there be users? *Telephone Engineer & Management*, 97(8):84–90.

Kato, T., 1992 Broadband network evolution for the next century – systems and technlogies. *Proceedings of the Technical Symposium*, Americas Telecom '92.

Richardson, G.P., Push, A.L., 1981. Introduction to system dynamics modeling with dynamo. *MIT Press*, ISBN 0-262-18102-9.

Romer, P.M., 1993. Increasing returns and economic growth. *Journal of Political Economy*, 94:1002–1037.

Svendsen, B., 1993. High speed data services for LAN interconnection in Norway. *Proceedings of the Integrated Broadband Communication Network and Services*, Copenhagen.

Young, A., 1991. Learning by Doing and the Dynamic Effects of International Trade. *Quarterly Journal of Economics*, 106:368–405.

Beyond Competition: The Future of Telecommunications.
Edited by D. Lamberton.

CHAPTER 4

Mobile data: an emerging telecommunication giant

Staffan Hultén, Bengt G. Mölleryd and Anna Nyberg
Department of Marketing Distribution and Industry Dynamics, Stockholm School of Economics, Stockholm, Sweden

Introduction

The mobile data industry and the mobile data networks are presently emerging as industries in their own right. In these two industries we see the merger of computer and software firms on the one hand and telecommunication equipment suppliers and telecommunication operators on the other hand.

In articles and reports mobile data has been identified as one of the principal future growth sectors of the telecommunication industry. The first mobile data networks were opened in 1983 in the USA (Ardis) and in 1986 in Sweden (Mobitex). In 1994 mobile data networks in the USA and in Europe had approximately 1 million subscribers. According to the most optimistic forecasts mobile data networks will have 10–20 million subscribers by the year 2000. Such growth predictions are extremely bullish compared with recorded growth figures for mobile telephony. If the mobile data networks should reach 20 million subscribers by the year 2000 the growth rate would be nearly twice as high as the growth rate of cellular telephony in the 1980s.

Presently the mobile data industry is propelled by large R&D projects financed by firms believing in the industry's growth potential. A substantial part of these projects are carried out in strategic alliances. These alliances cover nearly all aspects of the mobile data business. The aim of these strategic alliances will be presented and discussed in detail later in the paper.

The aims of this paper are: 1) to describe and analyze the goals of the strategic alliances in mobile data; and 2) to discuss possible future developments of this envisioned telecommunications giant.

Mobile data: an overview

A new industry emerges in mobile data based on systems, software and terminals invented and produced in its parent computer and telecommunication industries.

Address for correspondence: Staffan Hultén, Department of Marketing Distribution and Industry Dynamics, Stockholm School of Economics, Stockholm S-113 83, Sweden. Fax: +46-8-334-322.

The first mobile data systems were opened in the mid-1980s and constructed to meet the needs of the service organizations of Telia[1] (Mobitex) and IBM (Ardis). However, the two systems proved to meet the needs of other organizations and of individuals and they have come to dominate the market for mobile data (Table 1). A possible strong competitor to the purpose-built mobile data networks are the digital cellular networks that allow data transmission. The mobile data transmission in the cellular networks can either be services added to the system as in the case of the data applications to the GSM system, or be a specific packet data service like Cellular Digital Packet Data which is currently being adapted for the mobile cellular networks in USA. The drawback with the simple solutions is that they consume more air time and that the risks for transmission failures are higher. The drawback with the specific packet data services is the high costs related to R&D and the need to transform the cellular system.

Three types of terminals exist for the mobile data networks: 1) those which are purpose built for the networks like Mobitex terminals; 2) microcomputers like personal digital assistants (PDAs) or personal communicators adapted for mobile transmission; and 3) telecommunication terminals, e.g., fax machines, which can be connected to the mobile data networks. Consequently in the terminal market firms compete from the computer industry and the telecommunication industry. Table 2 gives examples of terminals developed for mobile data.

The mobile data industry does not thrive solely on new hardware. The establishment of standards and software development also play important roles. The standardization work is to a large degree carried out in committees at ETSI. They work on standards for a new mobile digital trunked radio system (TETRA) and a specification for a general packet radio service (GPRS). At the same time individual firms and alliances of firms in the market place claim that their systems or products will become the standards in the future. The most important alliances in the beginning of 1994 were EO, General Magic, Ardis, RAM Mobile Data, Viking Express, CDPD and AMSC. In addition, software is made to solve the problems of interaction between terminals and mobile data systems.

Table 1. A Selection of Mobile Data System in Service 1993.

Country	Operator	System	Speed (kbyte/s)	Launch
Canada	Cantel Inc.	Mobitex	1.2/8	1989
Finland	Telecom Finland	Mobitex	1.2	1990
France	France Telecom	Mobitex	1.2	1993
France	TDR	Mobitex	1.2	1993
Germany	Deutsche Telecom	Motorola	2.0	1991
Netherlands	Ram Mobile Data	Mobitex	1.2	1993
Norway	Norwegian Telecom	Mobitex	1.2	1989
Sweden	Telia	Mobitex	1.2	1986
UK	Hutchison	Motorola	2.0	1991
UK	Ram Mobile Data	Mobitex	1.2	1991
USA	Ardis	Ardis	2.0/19.2	1983
USA	Ram Mobile Data	Mobitex	8	1991

Table 2. Terminals for the Mobile Data Networks — some examples.

Producer	Terminal	System	Usage
Ericsson GE	Mobidem	Mobitex	Modem
Apple	Newton	Bi Bop, etc.	PDA
IBM	PC Radio	Ardis	PC with radio and modem
Eo	Personal communicator	Not specified	Personal communicator

Note: PDA = Personal Digital Assistant.

When a new technology is new

When a new technology is invented few persons can foresee its future applications and its future impact on society and on individual behavior. This was true for the three most important technologies underpinning the mobile data business. Computers were for many years mainly used for mathematical and statistical problems. One early important market for telephones were shop owners who wanted to connect a shop with the home.[2] Mobile telephones existed for more than 20 years before they reached a mass market.

The first users of mobile data were the producers of the first networks. The focus was directed towards the needs of the service organizations of a telecommunication operator and a computer firm. Similar customer groups continue to be important but the fastest growing market segment seems to be transportation firms. Still, there is a long way to go before 20 million persons will use mobile data services and many competing scenarios exist for the future mobile data market.

The lack of entrepreneurial imagination regarding the uses of a product leads to uncertainty both among the users and the producers. Furthermore, entrepreneurs and producing firms face technical uncertainties as well. These uncertainties are increasingly being handled in larger organizational entities. Consequently entrepreneurial activities in many modern technologies are organized differently from how they were carried out in established technologies when these were new. The monopolies that replaced the small scale entrepreneurial firms in bringing about innovations are now under way to be replaced by strategic alliances or networking firms. In the networks of firms a mixture of firms of diverging sizes and with diverging goals cooperate.[3] Despite the changes in the organization of the entrepreneurial activities their directions remain more or less unchanged. In the Schumpeterian conception of entrepreneurship innovative activities aim at resolving five types of entrepreneurial problems.[4]

1. The introduction of a new product.
2. The introduction of a new production process.
3. The use of a new input material.
4. The opening up of a new market.
5. The creation of a new organization of industry.

The emergence of networking as a generator of innovative activities and the entrepreneurial activities as envisaged by Schumpeter give the point of departure for our analysis of strategic alliances in mobile data.

Analyzing the strategic alliances in the mobile data business

The empirical material of strategic alliances in the mobile data industry and in the mobile data networks includes 76 alliances, involving a total of 113 firms. Information on these alliances has been obtained through articles in business and general news press and in telecommunication, computing and information technology journals.[5] The last items in the data base were entered in May 1994.

The strategic alliances have been classified in two ways. First we looked at what the firms stated that they wanted to achieve with the alliances and then we classified the alliances in accordance with Schumpeter's five types of entrepreneurial activities. All our classifications could of course be disputed. In some cases the motives presented by the participants were fairly general and in other cases it was difficult to understand the envisaged result of the alliance.

Before we present the results of the network analysis we describe some of the network measures that we have used and some other measures that we have not used.

The centrality concept[6]

The concept of centrality may be viewed in a number of different ways, taking on a variety of different meanings: 1) we may consider the centralization of a network as a whole; or 2) we may consider the centrality of a certain actor's position in the network. These two concepts are termed *graph centrality* and *point centrality*. Let us first turn to the issue of graph centrality.

Graph centrality

In its simplest form, graph centrality is defined as the ratio of actual relations between actors in a network, to the maximum possible number of relationships in the same network. Other more sophisticated measures exist, but they all share the problem that they are sensitive to the number of actors included in the network. Without some mechanism for normalizing the centrality measures, networks with different numbers of actors cannot be compared. Suggestions for comparisons have been made, although problems remain unsolved.[7] To our knowledge, no such measure of comparison is at present included in statistical packages available for quantitative network analysis, and we have therefore not included graph centrality measures in this paper.

Point centrality

Following Freeman (1978), three types of point centrality may be distinguished:

1. One actor in a network is more central than another actor if it has more direct relationships with other actors. The actor with more direct relationships has a more central position because it has a greater potential for communication and higher access to information. This aspect of centrality is referred to as *degree* of point.[8]
2. An actor that is situated on the shortest path between two other actors can block or tap into information that passes between these two actors. This aspect of centrality is called *betweenness*. An actor that is situated on many such shortest paths also holds a central position in the network. A way of measuring this type of centrality has been proposed by Anthionesse (1971). This measure captures the betweenness of points. The particular procedure used, is termed *rush.*[9]
3. An actor that has a relatively short average *distance* to the other actors in the network holds a central position. The more links that are direct, and the shorter the indirect links, make it easier for an actor to acquire information at low cost.

Centrality could also relate to *influence* or *status structures*. Although potentially of interest, these measures depend on information of relationship content and actor characteristics outside the scope of our empirical material.

The measures of point centrality used in the study

In this study two measures of the point centrality will be presented: the degree of point and rush. The first measure is used to give a description of the relative position of the strategic alliance actors in the whole network and in the 10 subnetworks. The second measure is used to find out if the actors have similar overall positions in the total network.

The results of the network analysis

We counted a total of 530 relationships in the 76 strategic alliances. Development of terminals and systems are the two most frequent reasons mentioned by the firms entering the strategic alliances. Software and services are also mentioned often. Sales activities is a relatively unusual motive (Table 3).

From a Schumpeterian perspective the dominant entrepreneurial activity performed in the strategic alliances in mobile data is the introduction of a new product (Table 4). The entering of a new market is the second most important entrepreneurial activity and in third place comes a new organization of industry. Few alliances seek to utilize a new input or to change a production process.

Table 5 shows the distribution of the strategic alliances in mobile data if we consider both the stated goals and the entrepreneurial activities. Also, Table 5 reveals some alliance patterns. Alliances seeking to introduce new products concentrate on systems and terminals. Alliances that aim at opening up new markets are focused on services and those that reorganize industry often have as a goal to build a system.

The strategic alliances also differ with respect to the average number of business firms active in the alliances. This is measured as the number of relationships in each

Table 3. The goals of strategic alliances in mobile data.

Goal	Number of alliances
Terminal development	22
System development	24
Service development	15
Software development	10
Sales activities	5

alliance type. Table 6 gives these figures. Three values in Table 6 (19.5, 25.5, and 27.3) are much higher than the average number of relationships (seven). In each of these cases, the high figure coincides with the fact that less than three alliances are found. Three of the strategic alliance types have only one relationship on average.

The leading firms in the mobile data strategic alliances

The strategic alliances in mobile data are dominated by a small number of very large firms. The most prominent firms are AT&T, Apple, IBM, Motorola and Matsushita. When we measured the degree of point of the firms we found that AT&T dominates the network of strategic alliances as a whole and is also dominant in half of the individual alliance types we identified. AT&T has the highest degree of point measures for the network of all strategic alliances and for five out of 10 of our strategic alliance types. In two of the alliance types AT&T is not active at all, namely in sales cooperation and in the opening up of a new market for an existing product.

Motorola and Apple are two other firms that play important roles in the strategic alliances. The two firms have the highest measure of degree of point of all firms in

Table 4. The innovative activities performed in strategic alliances in mobile data.

Innovation type	Number of alliances
The introduction of a new product	34
The entering of a new market	21
The creation of a new organization of industry	12
The use of a new input material	7
The introduction of a new production process	2

Table 5. Goals and innovative activities of strategic alliances in mobile data.

	System	Terminal	Service	Software	Sales
New product	12	15	3	4	0
New market	3	3	8	2	5
New organization	6	2	2	2	0
New input	3	2	2	0	0
New process	0	0	0	2	0

Table 6. The average number of relationships in the strategic alliance types in mobile data.

	System	Terminal	Service	Software	Sales
New product	3.1	12.3	4	6	—
New market	4	1.7	2.8	1	1
New organization	2.3	3.5	19.5	12.5	—
New input	27.3	1	3	—	—
New process	—	—	—	25.5	—

two out of 10 strategic alliance types. IBM and and Lotus Development Corp. are ranked highest in each one of two strategic alliance types. In one alliance type the two firms shared the top position. The Japanese firm Marubeni has a high overall degree of point but is not active in eight of our 10 alliance types (Table 7).

The indices of betweenness confirm to a large extent the degree of point pattern. The most striking difference is that a number of firms with relatively low degree of point indices have higher rush indices than firms with higher or much higher degree of point indices. Firms that move upwards are Ericsson, Rank Mobile Data, Ardis and Bell South. Firms that move downwards are Matsushita, Northern Telecom, NEC, Marubeni and Ericsson GE Mobile. The four top firms: AT&T, Apple, IBM and Motorola are ranked highest with both measures.

Table 7. Degree of point of the leading firms in the mobile data strategic alliances.

	Overall degree of point	Strongest alliance-type ranking	Weakest alliance-type ranking
AT&T	54	System, new product, new input, terminal, new organization (1)	New market, sales (-)
Apple	37	Software, new organization (1)	System, new input, sales(-)
IBM	37	Sales (1), new market, new input (2)	New process (-)
Motorola	32	Service, new market (1)	Sales, process (-)
Matsushita	32	New input, new market (2)	New process, sales (-)
Northern Telecom	24	Service, new organization (4)	Software, new process, new market, sales (-)
NEC	24	New process (2)	System, new inputs, new organization, market, sales, service (-)
Microsoft	23	New process (3)	New market, new input (-)
Marubeni	20	New product (4), terminal (5)	All other types (-)
Ardis	17	New process (2), software (3)	Terminal, new market, service, new product, new organization, sales (-)
Ram Mobile Data	16	New market, sales (2)	New process, new product, new organization, terminal (-)
Ericsson GE Mobile	14	New process (2), software (5)	All other types (-)

Note: (-) = no value of degree of point; the actor is not involved in this type of alliance.

Two observations can be made regarding this result:

1. The strategic alliances have as we pointed out different goals. Firms that are very active in one type of alliance may therefore get a high overall of degree of point without being connected to firms involved in other alliance types.
2. The American firms have high values for both indices while the Japanese firms have much lower values for the betweenness indice. We can recapitulate from the technical discussion of indices that "an actor that is situated on the shortest path between two other actors can block or tap into information that passes between these two actors. This aspect of centrality is called betweenness." Hence the Japanese firms, while being very active in the mobile data field, seem to have a relatively weaker possibility than the American firms to control the development of the totality of the mobile data industry (Table 8).

Conclusion

Mobile data has since its conception been an industry of networking firms. The first systems were the result of cooperation between system producers, terminal producers and system operators. The producers were also to a high extent the first users.

The mobile data business is still in its system-building phase. Nobody is certain about the future of the industry. To a large extent the problems of mobile data will be formulated and solved in other industries. The core problem is the user uncertainty that delimits the growth of the industry. The transition from the company internal use of the Ardis and the Mobitex networks to the diffusion to users outside the innovating firms was facilitated by the fact that the new user groups were similar to the first users. In the next transition stage mobile data will approach new types of customers that will demand changes in the products and in the services. Probably transmission speeds need to be higher. A mobile data network can transmit 1.2–19.2 kbyte/s which

Table 8. Ranking: degree of point compared with rush (betweenness).

	Degree of point	Rush
AT&T	1	1
Apple	2	4
IBM	2	3
Motorola	4	2
Matsushita	4	9
Northern Telecom	6	19
NEC	6	15
Microsoft	8	10
Marubeni	9	22
Ardis	10	6
Ram Mobile Data	11	7
Bell South	11	5
Ericsson GE Mobile	15	21
Ericsson	43	8

is a fraction of the working capacity of a computer and much less than the transmission capacity of a fixed network. Probably future computers will be significantly lighter and have more powerful batteries that offer longer autonomy.

These questions are not being addressed in the mobile data strategic alliances. Instead they focus on the imagined and easily satisfied needs of the next generation of mobile data users. There is a strong bias towards incremental development of existing terminals and systems. One way out of this dilemma could be to increase the number of market-oriented alliances. In our material only five of 76 alliances were directly focused on sale.

Postscript: will there be a fallacy of speculative investment?

In mid-1994 some of the key strategic alliances within mobile data started to dissolve. The operator alliance "Ardis" became a fully owned subsidiary of Motorola when IBM sold its stake in the network. EO was terminated when the key actors found that the market demand was weak and that EO may have used ideas that were patented by other firms. General Magic seems to have lost some of its magic to attract new venture capital and is running out of funding.

Three factors seem to be conducive when the strategic alliances are being questioned:

1. The firms participating in the strategic alliances are changing their position in the totality of the information technology industry. This happened for instance when IBM divested its share in Ardis and Motorola increased its stake. IBM is opting for the fixed network while Motorola is becoming increasingly involved in mobile networks – see for example Motorola's partnership in the satellite-based system Iridium.
2. The majority of the strategic alliances aim at developing new technology. If a firm like AT&T gets involved in too many alliances the overall costs of the investments in strategic alliances may be equal to or higher than in-house development.
3. Key segments of the mobile data market are growing slowly and many products sell poorly. Both the first version of Newton and EO's personal communicator have sold far less than expected.

Acknowledgements

We are grateful to Telia, The Swedish Transport Research Board and Ruben Rausings Fund for Research into Entrepreurship for financial support. We would like to thank Ulf Essler for help with organizing the material on strategic alliances and Hans Kjellberg for expertise on the network programs. Both work at EFI at the Stockholm School of Economics.

Notes

1. Throughout the paper we refer to the former Swedish state-owned telecommunication monopoly with its present name Telia.
2. For a history of the early uses of telephones see Marvin (1988).
3. The importance of interfirm collaboration in the telecommunication industry has been studied by, e.g., Antonelli (1992). Ioannidis (1994) reports the importance of strategic alliances in the telecommunication sector. Networking as a key feature of modern business is discussed in David (1992) and Mattsson and Hultén (1994).
4. Schumpeter (1934, 1983).
5. We have for example made a systematic manual search of Business Week, The Economist, Datamation, Byte and the Communications of the ACM. In addition we have asked an expert on mobile data at Ericsson to look through the list of alliances and suggest alterations.
6. This section draws on Sprenger and Stockman (1989), chapters 14–15.
7. See Freeman (1978) and Hoivik and Gleditsch (1970).
8. In social networks, if nonmutual links are also considered, in degrees and out degrees are distinguished. This distinction is obviously not relevant in the case of alliances.
9. Anthoniesse (1971).

References

Antonelli, C., 1992. The economic theory of information networks. In: Antonelli, C., (ed) *The Economics of Information Networks.* Amsterdam: North-Holland.

Anthoniesse, J.M., 1971. *The Rush in a Directed Graph.* Amsterdam: Mathematical Centre.

David, P.A., 1992. Information network economics: Externalities, innovations and evolution. In: Antonelli, C., (ed) *The Economics of Information Networks*, (pp. 103–105). Amsterdam: North Holland.

Freeman, L.C., 1978. Centrality in social networks – a conceptual clarification. *Social Networks*, 1:215–239.

Hoivik, T., Gleditsch, N.P., 1970. Structural parameters in graphs: a theoretical investigation. *Quality and Quantity*, 4:193–209.

Ioannidis, D., 1994. The internationalization of telephone operators. In: Bohlin, E., Granstrand, O. (eds) *The Race to European Eminence.* Amsterdam: North Holland.

Marvin, C., 1988. *When Old Technologies Were New.* Oxford: Oxford University Press.

Mattsson, L.-G., Hultén, S., 1994. *Företag och Marknader i Förändring - Dynamik i Nätverk.* Gothenburg: Nerenius & Santérus Förlag.

Schumpeter, J., 1934, 1983. *The Theory of Economic Development.* New Brunswick, NJ: Transaction Books.

Sprenger, C.J.A., Stockman F.N., 1989. *Gradap: Graph Definition and Analysis Package, Version 2.0*, User's Manual, ProGAMMA.

Beyond Competition: The Future of Telecommunications.
Edited by D. Lamberton.

CHAPTER 5

Implications of infrastructure competition for private networks*

Yoshiko Kurisaki
Société Internationale des Télécommunications Aéronautiques (SITA), Geneva, Switzerland

Introduction

This report is intended to examine the implications of infrastructure competition for private networks, especially those based on leased circuits (LCs). The following questions are investigated in the context of telecommunications policy for LCs in OECD Member countries; "whether and how does competition in infrastructure enhance users' interests?" and "what are the implications of competition in infrastructure provision and introduction of new technologies for the roles of private networks in relation to the PSTN?".

Users in this report include users of private corporate networks and service providers based on a private network, i.e., network composed of transmission lines leased from telecommunications operators (TOs) and switches provided by a party that uses this network (so-called value added services or VAS, such as electronic mailing services). These two types of users are not always mutually exclusive in practice. There are some cases in which a corporate network user sells excess capacity of the network to the third party. The resale of excess capacity to third parties by Reuters in the UK falls into this category. Operation of telecommunications services by resale of excess capacity of private networks based on LCs is allowed in Japan if a reselling party is registered as a Type II carrier.[1]

It should be noted that this report focuses on fixed-link infrastructure. This is because these networks are the least opened for competitive supply (as compared with wireless networks, such as those for mobile and satellite communications services) in the current telecommunications regulations in OECD Member countries.

Roles of LCs in the telecommunications market

LCs have played a substantial role in the liberalisation process of telecommunications.

Address for correspondence: Yoshiko Kurisaki, International Relations, SITA, 26, Chemin de Joinville, 1216 Cointrin, Switzerland.

*This paper has drawn upon a project report prepared by the author while she was in the OECD. The views expressed in this paper are those of the author and do not necessarily reflect those of the OECD, its member countries, SITA or its members.

Service competition via LCs started earlier than infrastructure competition

Liberalisation of LC usage was a forerunner of infrastructure competition. Service competition using LCs preceded infrastructure competition. The experience in Japan provides an example. In 1985, competition in the provision of infrastructure and services became possible by the enactment of the next Telecommunications Business Law. The level of competition in services based on different infrastructures developed, however, only gradually. This was mainly because of the time needed for new entrants to build and interconnect their network infrastructure with the existing PSTN for local services throughout the country. In the meantime, fierce competition started soon after both voice and data communications services via LCs were liberalised. The allowance of simple resale of LC capacity enabled swift growth of service competition.

Another important impact of usage liberalisation was that it stimulated exploitation of telecommunication resources. The number of computer-based services that provide a variety of information processing services using LCs interconnected with the PSTN has increased. Remote computing services, electronic-mail systems and on-line data base services are examples of these.

Infrastructure competition first began in the LC markets

In many cases where the competitive provision of infrastructure is allowed, competition started from LCs, then developed into the PSTN (which serves large and small users). In Canada, network interconnection between local LCs provided by Bell Canada (the PTO in most of the Provinces of Ontario and Québec) and long-distance services provided by CNCP telecommunications (a competitor to the existing monopoly carrier) was allowed in 1985[2] and may be viewed as the advent of infrastructure competition. The evolutionary process of competition in Japan was similar. When competition in infrastructure was introduced in 1985, two of NTT's competitors (the ex-monopoly PTO in the country) started service provision with long-distance LCs, followed by network interconnection between NTT's local PSTN and their toll switches. As of March 1993, amongst 18 Type I carriers[3] (including two former PTOs, NTT and KDD, but excluding providers of mobile and paging services), only eight of them provide both PSTN and LC services, while all provide LCs.

Entering the telecommunications markets with LCs first is a natural strategy for new TOs. LCs requires less facilities than those needed for publicly switched networks. Facilities connecting a limited number of customers are needed. LC customers are corporate users which are heavy users of telecommunications, hence large sources of income to the TO. The same strategy may apply to those countries that are introducing competition in infrastructure supply in the future.

Large (corporate) users' current demand leads that of small users' demand

The growth of public demand in the near future may be assessed by examining the viability of infrastructure competition from LC users' point of view. The liberalisation of telecommunications, that began with LCs as described above, has been primarily in response to strong and long-lasting requests by corporate users. Most of these corporate users require communications systems well integrated in their work. Industrial associations, such as Canadian Bankers Association in Canada and Data Communications Users Association in Japan are examples of corporate user groups which contributed to the liberalisation process in each country. These groups lobbied and used various opportunities to put pressure on telecommunications policy makers to request the relaxation of usage limitations in LCs and, further, the introduction of full competition in PSTN. As corporate users seek to fully exploit communications systems, they often viewed existing regulations as barriers hindering their corporate activities. It has been a natural consequence that corporate users, especially large ones, were the first to advocate the benefits of telecommunications liberalisation.

LCs are largely used by enterprises

The economic interests of users in telecommunications systems are much stronger in LCs than in the PSTN. While both the PSTN and LCs are instrumental to corporate activities and have economic implications, the former serve many other social values as well, for example for universal service obligations. Users of LCs, in contrast, view LCs as a means to undertake economic activities. Establishment of information systems that serve as a company's nerve systems is the major concern of large users of corporate network.[4]

LC is a fast-growing market

The growth of the LC market is higher than that of the PSTN. Table 1 shows the ratio of revenue from LCs in relation to revenue from telecommunications service as a whole and its growth rate in eight selected Member countries. While data were not available for all the Member countries, the existing data allow to assess that revenues from LCs are growing and are expected to be 10% higher than TO's revenue. In fact, in the USA, 30% growth in revenue is expected for LCs from 1992 to 1993, while the PSTN's revenue is expected to increase by 3%.[5]

LCs in the current telecommunications policy

LC service is as old as the PSTN, and has been inherently treated as a special service for exceptional demand that is not being met by the PSTN, typically to those users who could afford the service and for whom telecommunications service (the first telegraph service soon followed by telephone services) had a critical importance. This nature of the LC service has essentially remained unchanged to date. In Japan, it started in 1906 between the two major banks, 16 years after the commencement of the PSTN.[6] Similarly, the commencement of LC service in the Netherlands was in

Table 1. Revenue from Leased Circuits in selected countries.

	LC/Total revenue (%)				Revenue growth (%) (Upper: LC, Lower: PSTN)		
	1989	1990	1991	1992	1989–1990	1990–1991	1991–1992
Belgium		14.3					
France			6.9			 6.8	26(F)[a]
Germany[b]					12.9(F)[c] 4.1	 4	
Ireland	2.7	2.8	3.1		17.6 11.9	12 8.9	
Japan[d]		NTT:6.5 KDD:7.1	NTT:6.5 KDD:7.0			NTT:3 KDD:-0.4 Domestic:0.1 Int'l:0.8	
Luxembourg		12.1			38 12.3		
The Netherlands			7.9	8			11.3 6.8
Spain	3.5					46.8(F) 5.3	28.4(F) 4
Sweden (Telia)		7			3(F)[a]	68(F)[a]	
UK (BT)[e]	6.3	6.3	6.8		9.8 10.2	16.4 8.9	

Source: Compiled by the author from various sources.
(F): The number of facilities. [a]Digital leased circuits only. [b]Former West Germany only. [c]Leased circuits for data communications only. [d]NTT: Domestic only, KDD: International only. [e]Domestic only.

1892, 11 years after the PSTN. In the US, the first record of the leased line (it was for simultaneous telegraph service) appeared in 1887, 10 years after the first PSTN services in the country.[7] The first LC customer in the US was a broker who used it for the communication between New York City and Philadelphia. The first fixed local telephone lines were installed even earlier than the PSTN in Sweden, these lines were owned by users, not the PTO.

Now that LCs have strategies which are important in corporate activities, such a traditional idea for LCs has to be changed. LCs are a commodity for users, not a minor supplement to the PSTN.

A review of rules applied to LCs in the OECD Member countries indicates that their modernisation is in progress. Cases for monopoly provision with restricted usage regulations are being replaced by competitive provision with liberalised use. Interconnection with PSTN, third party traffic, shared use and capacity resale are gradually being allowed. There are, however, many conditions which still remain to be kept in many of OECD Member countries, such as exclusion of voice transmission, requests for permission by the telecommunications authority to build private networks and the existence of "a close business relationship" among the users concerned.

There is a close interrelationship between the general telecommunication policy framework and the treatment of LCs across the OECD Member countries. The level of liberalisation in LC usage corresponds to the ownership of LC facilities. The level of liberalisation in LC usage corresponds to the level of market liberalisation in the PSTN. Where competition in the PSTN takes place, LC usage and its interconnection with the PSTN tend to be liberalised, as is the case in Canada, Japan, New Zealand, Sweden, the UK and the USA. The level of liberalisation in LC usage and ownership is largely subject to the policy for voice telephony services.

"LC policy" had been *ad hoc* despite the role it has played in the introduction of competition in some countries. The traditional idea of many of the governments in Member countries seems to have been that all the telecommunications demand must be satisfied primarily by the PSTN. LC use is strictly regulated to avoid competition with the PSTN. For telecommunications regulators, LCs were a means to serve only exceptional demand, such as a large volume of traffic that takes place between a limited number of parties, such as intracompany communication and intercompany interconnection (i.e., shared and/or third party use) of LC-based private networks traditionally allowed in the financial and air-transport sectors. Although self-provision of circuits has often been allowed for public utilities (e.g., railways and electricity), interconnection of such networks with the PSTN has been prohibited.

A comprehensive OECD-wide policy for LCs is yet to be developed. Many of them are similar in nature but slightly different from each other. The usage regulation of LCs, unlike that for the PSTN, often leaves a large scope for case-by-case consideration by telecommunications authorities.

A substantial degree of ambiguity in LC regulation is, in itself, a barrier to service providers and corporate users, as the merger of computer and communications technologies proceeds. For example, how may potential VAS providers intending to enter the market in Spain know the criteria to obtain permission by the Spanish Administration, given the ambiguous regulation described above? How may a company with many co-working companies plan its private network systems to improve production systems if there is the following rule as regards the third-party traffic in Belgium: "no third party traffic is allowed via LCs unless a significant value is added either through interconnection between LCs or with a public network or offered directly to the user". By what criteria may "a significant value" be judged? Who is the appropriate body to judge this? Is it the user, the telecommunications authority or another party? Rules that are too general do not help users, who would be unable to know what the practice would be.

A large variation in usage regulation is reflected by ambiguity in the definition of terms that are used to refer to services transmitted via LCs. The meaning of, and consequently the regulatory boundary between "VAS", "data-only", "nonvoice only", are not unique among Member countries. In theory, the simple transmission of data and data communications that involves information processing via LCs belongs to a different usage category. The former is the same as the simple transmission of messages (such as voice and fax) in the sense that signals generated by a terminal on one end are transmitted to the other end unchanged. The latter is "value-added" since the signals originated on one end are processed by a computer interconnected with the LC and delivered in different forms to the other end. Both, however, sometimes seem to be categorised simply as "nonvoice" transmission without a distinction between the two. Such separation is only confusing to users, or is it meaningful in light of the practice in data communications?. Data processing may or may not be undertaken via private networks. Both "value-added" and simple transmission of data happen in the same private network. Whether or not to add "value" (i.e., information processing by computers) should be determined by users of the network, depending on the manners in which data are used in their work organization. Too much instruction by regulations may result in underutilisation of technological capacities.

The interpretation of terms referring to different modes of usage varies among Member countries. Examples include "shared use", "third party traffic", "capacity resale" and "traffic resale". A term "private network" may or may not include those networks interconnected with the PSTN. Countries seem to leave the interpretation of regulatory rules on a case-by-case decision by the telecommunications authorities, rather than defining them. Case-by-case judgement may allow a certain level of flexibility in the regulations.

Flexibility and full liberalisation are, however, very different in nature. Many social costs are incurred, as long as a large ambiguity in the regulation on LC usage exists in the decision-making process. These costs are borne both by the applicant (i.e., a user) and the government (i.e., the telecommunications authority). These may include the financial costs and time for the preparation of documents by the applicant, the negotiation between the two parties and for the review process by the authority.

A series of spontaneous decisions based on flexible regulation may undermine the existing LC regulations. The number of users and the usage variation of LCs are increasing. Many of such uses fall beyond the rule-setter's policy scope. If flexibility in regulation is the best means to meet with an ever-increasing demand in LC use, the regulators may well lift the existing rules. Rather than trying to define those usages which are and are not allowed, full liberalisation of LC usage with a list of well-defined prohibited usages may deserve consideration as an alternative policy instrument.

Asymmetry in LC regulation between countries creates additional difficulties for users in the case of international LCs. A situation in which there are differences between what is possible in Country A and in Country B does not allow the optimal use of LC-based private networks across countries.

LCs in a competitive market with new technologies

This section examines the usage and roles of LCs in the future in light of major factors that will shape LC service; i.e., market liberalisation, user needs, technologies and some regional movements aimed at unrestricted flow of information.

Further liberalisation of LC regulations

After service liberalisation, further liberalisation in LC usage and competitive supply of LC facilities are the steps towards further competition in telecommunications markets. This will stimulate the growth of a variety of businesses through the "smart use" of LCs, such as the creation of more options in capacity resale and VSA. The liberalisation will also be a factor that will enhance cost performance of corporate networks, and thus providing economic benefits.

LC users have more control over the price of LCs if there are more than two facility suppliers. There is empirical evidence from the countries where competition in the supply of LCs has been introduced that it leads to lower prices. According to the study undertaken by a major user of LCs worldwide, the average price of intra-European digital LCs is 10 times higher than the price of the equivalent LCs in the USA domestic market.[8] In the USA, LCs are supplied on a competitive basis, while in Europe they are on a monopoly basis, except in Sweden and the UK.

Competition in the supply of LCs brings many benefits other than price reductions. The improvement in quality, availability and reliability in LCs, service innovation (such as quick introduction of new technologies in markets) and the suppliers' quick response to users are examples of nonmonetary benefits. They form a crucial part of the improvement of cost performance in the total network systems.

Excluding voice telephony service from competition is a bottleneck for users who seek fully to exploit LC services. The strict limitation or prohibition of interconnection of an LC-based network with the PSTN has adverse effects on the cost performance of a corporate network system. Ambiguous or case-by-case regulation of LC usage may create insecurity among users in relation to the telecommunications authority. LCs may even be overcharged to subsidise local telephony services. The competitive supply of LCs may be seen, in such a case, against an underlying political purpose in LC regulations.

Introducing competition in the infrastructure for LCs will eventually accelerate the disappearance of the regulatory boundary between the VAS market, characterised by unregulated firms, and data communications services provided by the regulated PTO. As supply competition in infrastructure will naturally involve liberalisation of the network use (both private and public networks), existing PTOs are in the same position as their competitors. Regulatory distinction between the owner of the infrastructure and the service provider will no longer be needed where the users of the infrastructure are not "captured consumers" (i.e., users who do not have an option in the selection of the same type of infrastructure provided by different parties) any more.

The introduction of competition in LC supply will have some impact on rate rebalancing in the PSTN, where infrastructure competition is introduced and where simple resale of voice via LCs is allowed. Where noncost based distortions in local and long-distance telephone rates exist, this makes it economically attractive to use LCs connected to the local exchange switched network to provide alternative switched long-distance services. Rate balancing may reduce the attractiveness of the simple resale of voice.

The rate rebalancing in telecommunications services will require infrastructure suppliers to accentuate the roles of LCs in telecommunications service options. The comparative advantage of LCs in relation to the PSTN will have to be enhanced not in regard to price but in the quality and their capability to provide end-to-end service.

Users needs

Users needs should not have to be constrained by the means of communication, such as LCs, public data switches and the PSTN. The selection of technology is a means to an end, not an end in itself. If, for example, public data switches meet users' demand much more than LCs, users will simply select the former. In reality, however, users are likely to continue to need, even with various other means of communication available, for various reasons. Existing corporate networks are mostly composed of LCs. It is unrealistic to expect that a quick change from an LC-based network to other means will occur. The costs of scrap-and-build of the corporate network may be higher than the potential benefits.

Users needs are changing as corporate activities stretch over countries. They strongly need a corporate network to be a single-managed entity. The need for end-to-end network systems across many countries is rapidly increasing more than ever. This will require the liberalisation of LC usage in those countries where many restrictions are still imposed.

New technologies

A variety of technologies and applications are being introduced that may serve the PSTN and corporate private networks. Such technology development is likely to create further pressure towards the alleviation of the asymmetrical regulations between voice and data placed on LC.

Virtual private network (VPN)

One of the strategic objectives of VPN for PTOs is to compete with LC-based networks, including those supplied by VAS providers. VPN is based on the PSTN and allows a user to organise its own private network as if it were composed by LCs. VPN may have an advantage in relation to such an LC-based private network in which a large number of user sites are scattered over a wide geographical area. Currently VPN is mainly used for voice only (or fax through voice channels). It is estimated that VPN may efficiently transmit data traffic of moderate load, but that dedicated networks are more appropriate when high performance is requested for the handling of high volume traffic. If VPN based on the broadband integrated services digital network (B-ISDN) with high quality will be come widely available, LC-based private networks will have less of a competitive advantage in nonvoice (data, image, etc.) transmission. Competition between VPN and LCs will become more intense than it is now.

Asymmetry in the current regulation may emerge between LC usage and VPN. "Interconnection" between a VPN and the PSTN for voice transmission is one of the VPN's service options. It is a natural result of the fact that VPN service is one use of the PSTN. In other words, the PSTN is the physical facility used for VPN. Such free access from the PSTN to a private network, however, is not always the case for LC-based private network. As discussed, the interconnection is allowed only for data transmission in many Member countries. VPN is, however, an imaginary *private* network viewed from its users. Why should regulations for these two private networks be different? The introduction of VPN based on B-ISDN will nullify an existing restriction on LC in its interconnection with PSTN for voice.

The development of VANs, especially large-scale ones, will also undermine the regulatory difference between VPN and LC-based networks. In terms of a network usage from users perspective, a large-scale international VAN, such as IN (b IBM) and Infonet (led by MCI and joined by other PTOs), is very similar to that of VPN. Although VANs use for links, large ones serve as a common-use network for a number of users. This means that a large-scale VAN may be seen as a mini public network, or a network shared by "members of a club". If the number of members are infinitely large, this network will simply grow to be the public network. With current digital technologies, regulatory distinction between voice and data is only artificial. In reality, most of the major VAN services include voice telephony services where allowed (in closed user group basis, etc.).[9] Where two networks have similar utility to users, what is the legitimacy to maintain regulatory distinction between them?

The regulatory difference between VPN and LC-based private networks will have to cease. Technology development will demand further liberalisation in interconnection between LC and the PSTN. Two different technologies must compete merely on the basis of their technological advantages, not regulatory ones.

Frame relay and switched multimegabit data service (SMDS)

Both frame relay[10] and SMDS[11] technologies will bring maximum benefits where dynamic information flow between different networks is ensured. The adoption of the frame relay technology will facilitate interconnection between data communications networks provided by different operators, such as between VAN providers and the PTO, between private network users and the PTO, etc. Once SMDS is introduced in the markets, it is expected that frame relay systems may continue to serve the lower end of the interconnection between private networks, while SMDS will serve demand for very high speed networks.

Improvement of computing systems

The development of computing systems will create a demand for high capacity digital LCs. A configuration of networked computing systems is developing the distributed systems, where a number of high performance and intelligent terminals, such as workstations and personal computers (PCs), are connected with communications paths. The design philosophy behind the frame relay is to allocate intelligence to terminals. Thus the frame relay allows for distributed computing systems.

An increase in the number of the distributed computing systems stimulates demand for high-capacity LCs, those with more than 6 Mbit/s (the current maximum digital LC capacity among the standardised services available in many Member countries), even 600 Mbit/s. High performance computing systems need to be connected with communications links with high quality and availability for the full use of computing capacity. Only where such a network environment is achieved, does the entire network system perform efficiently, taking advantage of the full capacity of each computer. A typical example of such a system may be local area networks (LAN). A high speed LC will also be needed for the interconnection between LANs.

Policy issues

Enhancement of the economic roles of telecommunications

Government's role should be to ensure the best policy mix that allows maximum efficiency in the use of social resources, such as in this report's context, the combination of public and leased facilities. In order to allow this, users are placed in a position where they have a wide range of options in the selection and combination of communications systems. Through rational activities of individual users, maximum efficiency in resource use is achieved.

While telecommunications services, especially the voice telephony service, will continue to have the nature of a public utility, this does not prohibit current policy from being intended to enhance the economic roles of telecommunications and to institutionalise such an intention. Policy instruments that help users obtain the most efficient communications systems for their economic activities should not have to be automatically considered to have adverse effects on the public interest. A set of policy instruments that serve both economic and civil interests needs to be continuously reviewed in response to new economic and technological circumstances.

The roles of LCs for economic activities and technology development should not be overlooked. LCs are indispensable for computer communications systems, which have substantial economic implications for corporate activities in many economic sectors. When new communications technologies are first introduced they are usually first tested on LCs. Frame relay and asynchronous transfer mode (ATM) being the current examples. The LCs' flat rate system provides incentive to users to develop efficient use of its capacity, including VAS and resale. The liberalisation of LC usage should take into account the wide range of roles played by LCs.

Monopoly efforts vs. real competition

If the quality of LCs is improved and prices fall under the monopoly provision of infrastructure, what are advantages of introducing competition? Would improvements in LC services justify the monopoly provision of LCs?

The fact that service improvements are achieved by a monopoly provider is an evidence of pressure from possible competition. Users' request for LCs in terms of quality and prices have been the same since the time when the monopoly provision of infrastructure was taken for granted in Member countries. The possibility of introducing competition in domestic markets and *de facto* competition with foreign-based PTOs in international markets required PTOs to enhance their customer-oriented attitude. Those corporate users that need to tailor the telecommunications system to fit their own needs have commonly experienced that PTOs started to listen to them much more after the introduction of competition. The real competition, not merely an imaginary threat, may lead to even more service improvements.

Monopoly provision of infrastructure is likely to have adverse effects on service competition. Asymmetry in the access to information needed for service provision between the monopoly provider (PTO) and other service providers may remain. Such information would include network specifications, customer information, etc. Some crucial information for the provision of competitive service by a service provider may have to be passed on to a PTO for the network interconnection.

In short, as long as the possibility of asymmetrical distribution of information exists between the monopoly provider of infrastructure and its users, service competition is not fair or viable. The standardisation of user-network interface, such as Open Network Provision (ONP) and the institutional separation between policy-makers and the service operator may help to alleviate the asymmetry. They, however, do not entirely remove the problem as long as monopoly provision of infrastructure continues.

Which LC usages should remain regulated?

The difficulty in defining seemingly similar but slightly different terms of LC usage was observed in LC regulations in Member countries. Given a rapid increase in variation of usage, defining what is allowed and what is not is no longer a feasible solution.[12] A definition which is too precise results in too much rigidity in the regulation, or, at best, only makes the regulation rapidly outdated.

Policy makers may wish to consider setting the rules the other way round, i.e., to open the LC usage free with negative lists, rather than to list what is permitted. Increasing the level of competition in overall telecommunications will inevitably accelerate liberalisation in LC usage. A variation of uses allowed will outnumber the uses prohibited. The telecommunications chapter in NAFTA provides an example of the adoption of such a way of rule-setting.

Summary and Conclusion

The introduction of competition in the supply of LCs and usage liberalisation have to be undertaken simultaneously. LCs are not merely a flat rate service that may take communications traffic away from PSTN, or a special service for exceptional demand. Policy-makers should place importance on the vanguard nature of LC users as forerunners of advanced telecommunications use that may be made available for a wider public in the near future.

The dynamism of market liberalisation will continue to work to bring economic benefits to liberalised countries by taking potential benefits away from less liberalised ones. It is widely observed that the level of liberalisation is a substantially large factor in the decision-making of corporate private network users in the selection of location of communications hubs. In other words, one country's liberalisation directly affects other country's interests. While such a phenomenon is observed in both the PSTN and LCs, it will be more pronounced in LCs.

A means of enabling the unrestricted flow of information over the world needs to be developed in order for telecommunications policy to keep pace with the globalizing economy. This requires telecommunications policy to be examined in the wider context of information policy. The rules for LC usage certainly form a part of such a reviewing process in the sense that these rules matter for access to information. Such a review would promote the unanimous but harmonious liberalisation of LC provision and usage between Member Countries.

A major challenge for national policy-makers is that they have to formulate rules for the national telecommunications sector in an environment where a mismatch exists between economic and political logic; the former develops beyond national boundaries (economic globalization) while the latter stays within the boundaries (the national policy). As LCs play a strong economic role in corporate activities, this mismatch is clearer than the one for the PSTN. What is desirable for one country may not be so at the international level. Individual countries have to continue to seek a way in which the collective movement for market liberalisation in telecommunications leads to a globally desirable solution.

Glossary

Resale
The subsequent sale or lease of telecommunications services leased from a Telecommunication Operator (TO) with or without adding value on a commercial basis.

Capacity resale
The subsequent sale or lease of channels of LC leased from a TO with or without adding value on a commercial basis.

Simple resale
The subsequent sale or lease of channels of LC leased from a TO without adding value on a commercial basis.

Private network
A network dedicated to one user or a group of users for the transmission of message (such as simple voice and fax, data, video, etc.). Such a network may or may not be interconnected with the PSTN. Such a network may be composed of LCs and self-provided switches (PBX) or the PSTN (such as Virtual Private Network).

Shared use
The use of telecommunications services leased from a TO by two or more parties on a noncommercial basis. Contract of the lease with the operator are shared between the users.

Third party traffic
The use of telecommunications services, on a noncommercial basis, leased from a TO by a party and used by a third party who does not lease from the TO.

Value-added networks (VAN)
A communications network providing such additional features as message switching, on-line data processing, on-line database, electronic data interchange and protocol conversion. The provision of these services may or may not imply the ownership of transmission facilities (lines and switches) by the service provider.

Value added services (VAS)
A communications service using existing common carrier networks (including LCs) and providing such additional features as message switching, on-line data processing, on-line database, electronic data interchange and protocol conversion.

Virtual Private Networks (VPN)
Use of the PSTN in a way that simulates a separate system as if it were a private network composed of LCs.

Notes

1. Those telecommunications service providers who do not own facilities.
2. CRTC Decision 85-19.
3. Those telecommunications service providers who own their own facilities.
4. Those corporate users centred in International Telecommunications Users Group (INTUG) and the Commission on Computing, Telecommunications and Information Policies of the International Chamber of Commerce (ICC) are examples of such users.
5. Source: US Department of Commerce (1993), *US Industrial Outlook 1993*, Washington, DC.
6. Source: Hayashi, K., 1992, Review and future perspectives of the development of LC services towards one million lines (Original in Japanese), in *NTT Technical Journal*, Vol. 4, No. 3, Tokyo:

NTT.

7. Source: AT&T, 1974, *Event in Telephone History*, New York City.
8. Source: Reuters, 1993, *A written contribution on the 1992 Telecommunications Service Review by CEC*, London.
9. For example, IBM's VAN service, IN, includes voice telephony services for users who are confined to a closed user group. EDS, a VAN provider, provides the similar service.
10. A streamlined, fast packed switching technology made possible by the error free communications of optical fibre and improved transmission system. The introduction of this system started in the early 1990s, first by VAN providers who have traditionally offered low bit-rate packet switching service based on the X.25 standard. An increasing number of PTOs also started to introduce this technology.
11. A communication technology based on cell relay technology. SMDS is expected to be used to implement B-ISDN.
12. For example, should an information service provided by a bank using its financial information to its corporate customers be regarded as a part of banking business or commercial information service? The classification of such financial information services varies between countries. Furthermore, different regulations may apply.

Beyond Competition: The Future of Telecommunications.
Edited by D. Lamberton.

CHAPTER 6

From data to wisdom: does IT contribute to the Gross National Happiness?

Michel J. Menou
CIDEGI, Gentilly, France

The advent of "homocompunicans"

A new age?

For more than 20 years we have been exposed to millenarian descriptions of the last revolution in the history of human societies: the advent of an information society, a key feature of which would be that information, or knowledge, is substituting material wealth as the major basis of power. There are in fact a variety of concepts of the information economy or society. The pragmatic approach taken by Miles (1988) may provide a suitable alternative. He contends that the information society concepts, whatever long-term challenges they pose, refer in the first place to the "complex of changes associated with – not caused by in a simple deterministic sense – information technology (IT) and its diffusion through the formal economy and society at large". The key distinction in his view should be how far any activity is, or will become, IT intensive.

IT is neither a clear-cut or well-established concept. It is, however, mostly acknowledged that its core lies in the convergence of computers and telecommunications, which were tentatively coined as "compunication". From this point of view it is clear that this revolution is quite similar to the earlier technological revolutions in that it is driven by the availability and use of machines. Except perhaps for the fact that these machines are themselves undergoing changes at an unprecedented pace. And most of the excitement and fascination goes to the machines themselves, as was the case with the first chipped flint implements. Rather than an information revolution, we have a microelectronics revolution. There are nevertheless in this process changes of nature and scale which may be more far reaching. We will come back to them later. Unfortunately human reasoning often does not pay attention to the qualitative changes induced by quantitative jumps.

Since microelectronics-based devices are increasingly incorporated into all sorts of products for sensing, control and regulation, not to mention robots, IT may indeed

Address for correspondence: Michel J. Menou, CIDEGI, 129 Av. P. Vaillant-Couturier, F-94250 Gentilly, France.

become quite pervasive. Perhaps the "intelligent company" of the next century will be a fully automated organisation with no humans, except a few shareholders who will cash the dividends, and possibly a few controllers checking workflows at distance. This is more so since most failures in information flows are likely to be attributable more to human than material causes. We should concentrate our attention on the combination of computers and telecommunications, of which, even though it is not mentioned (perhaps a premonition), the human being is still a significant component. We are concerned with this complex in that it serves for human beings to create, record, store, share and use information. IT does not exist, as yet, for its own sake, but because human beings need to communicate among themselves and have something to say to their fellows, or at least they think they do.

IT applications and effects

Ang and Pavri (1994) offered a penetrating review of the IT applications and their impact on society, organizations and the individuals. They also stressed the piecemeal and often contradictory nature of the analyses and interpretations currently available. Trying to devise more structured typologies may be in time. Taylor (1986) rightly insisted that "If we accept the notion that science is "observation organized", then there are some very basic needs for the field of information. It has been argued here that, at this time, the basic needs of the information field are rigor in definition and systematic typologies of the entities and processes we deal with. We need to learn how to observe, how to organize those observations, how to apply the insights gained to operating systems, and how those observations can inform theoretical constructs".

The following parameters may be considered to build a typology of IT applications:

— Scale: compunication occurs at the global (entire world), international, national, organizational, home (family) and individual levels.
— Nature of the activity which is the primary focus of the IT application, e.g., economic, social, domestic, recreational, educational, cultural, etc.
— Nature of the function affected by IT, e.g., governance, process, interaction, structure (or organizations, irrespective of their size), learning, memory, creation, emotion, values.

A major difficulty in both observation and interpretation lies in the pervasiveness and versatility of information under all its forms (Miles, 1992). If one is concerned with the overall benefits which can be derived from information, the reductionism which commonly restricts peoples' roles in the information lifecycle to only one, e.g., producer, publisher, records manager, librarian, system's manager, intermediary, user, etc., is inappropriate since most people play all these roles, possibly at the same time. The diffusion of information technology makes this confusion even deeper. Neither is it appropriate to reduce information use to distinct situations. Information gained for recreational purposes may well prove critical in exploiting information for professional purposes and vice versa. Transactions data generated by automatic tellers are not only useful for book-keeping but may also be turned into criminal evidence.

An information item heard at a time it was felt of no value, e.g., a traffic jam, when one does not plan to leave home, may later become essential if one is called for an emergency. In traditional societies, drama, story-telling and trivial conversation, which could be categorized as primarily recreational, are all privileged means for sharing information to be used in economic or social activities. Information channels and tools may be used for any purpose: private, social, political, professional, educational, recreational. As time goes by the nature, value and possible applications of information change. Having experienced the effectiveness of alternative bulk mail distribution services prompts not only better planning of the dispatching of mail but also its preparation, while it may become the basis of fun stories as well as new commercial practices. All the cells in the matrix resulting from the above suggested typology are likely to constantly interact.

Expansion and integration

There is clearly an immense potential for installing IT in most parts of the world which are still underserved. If at least the technology can be made more suitable for the local conditions. Probably a third of the tape recorders and transistor radios available in Africa remain unused for lack of batteries, either because they are not in stock or, more often, because people or organizations cannot afford to buy them. Obviously more attention will be paid to the equipment of the more advanced countries, or wealthy segments of the developing countries, whose potential is attractive enough for many years. Especially since technological advances allow for steady cost reduction and demand seems to be primarily a function of cost according to de Sola Pool et al. (1984). The demand is also a result of the failure of prior, less IT-intensive systems from a cost-effectiveness and delivery point of view, as well as with regard to their adaptation to the workflows in organizations. Facsimile would be far less popular if mail had remained what it used to be and if telephone calls could reach the interlocutor more often without delay rather than an answering machine saying he/she is in a meeting. Many modern technologies can be seen as a mere palliative to the chaos generated by ever increasing spatial concentration, specialization, complexity forced on people and organizations whose characteristics have not so much changed over the centuries. Miles (1988) noted that: "the rapid diffusion of IT is unlikely to be matched by an equally rapid modification of organizational practices". In fact many organizations, especially in the public sector, are much like what they were a century ago, if not more. Except that, in accordance with Parkinson's law, the number of supervisors has increased in pace with the reduction of line staff.

Supply is also pushed by some strange factors. Fashion and the sensitivity to status symbols is one of them. Some years ago executives had to have their cars equipped with a telephone. Now no officer of supposed importance can walk in the street without showing his/her criticality to the organization and society by talking on a radiotelephone. Though this may often be for serious reasons, it is also frequently for trivial conversations. The same drive may in part explain the installation of costly color TV systems in many developing countries whose people suffer extreme poverty in their vast majority. Of course they will not see it on TV since most programmes are imported and the coverage hardly goes beyond the capital city. What really matters is that the local VIPs and those who visit them from abroad feel they are in a "civilized place" since they get the scores of the "world series" live. Even though the propagation of IT is obviously a must, given the current developments, it often displays a worrying similarity with the propagation of drugs among kids, through the imitation of their peers, duly encouraged by their status-prone parents, the media, advertising and even the educational system. So they do not count but use a pocket calculator, do not copy pieces of literature but make photocopies, as they do for geographic maps instead of drawing them, do not read plays but see the film on TV, etc. This remark will sound extremely reactionary, but are we so sure that these IT-assisted activities are not eliminating the basic exercises that the human brain needs in order to perform? Shall all human beings in the next century be unable to independently think and be fully dependant from IT and artificial stores of information? Both skills are certainly needed, but one may wonder if the balance is properly secured. It is even more questionable if the skills required in order to make sensible use of the information available from external sources, especially the media, are imparted. The coming integration and the increased reliance upon multimedia make the issue of literacy, with possibly various adjectives such as information or a computer (a very important one). It would perhaps be more appropriate, as suggested by Carbo Bearman (1992) to seriously consider the issue of what she called "mediacy", referring to the dominant exposure to fully integrated media of all kinds.

Integration is evidenced on the formal plane by the ever increasing capability of digital technologies to carry all kinds of information as objects (texts, still and animated images, sound, graphics) on a single medium. The original documents could thus be substituted by the digital storage units of their material contents, on the one hand, and the latter may be recombined at will into new information products, on the other hand. Integration is further manifested by the fact that all those pieces of information-as-object will be exchanged via a single channel of computers and high capacity telecommunications networks whose hybrid nature will remain transparent for the user. At least once standardization and user friendly and smart interfaces will have freed him/her from the present day burdens. This will allow for an almost comprehensive communication which will not be limited to the one-to-one interaction, which is still much the norm, but will enable collaborations to link anybody, anywhere, anytime as Strassmann (1993) stressed. Instead of the next generation of compunication facilities to serve as a mere substitute to the previous one, one will reach a stage where it will offer a reliable alternative for "multicultural cooperation and knowledge sharing" (Strassmann, 1993). At least when the systems will have been freed from the dictatorship of ASCII.

Integration may, however, also be witnessed in the concentration of major companies within the broad information industry. Although one of the principles, if we are not mistaken, behind divestiture in the telecommunications sector was the separation of the various functions in the information cycle, in fact a sort of separation of powers, we can now see that producers, distributors and carriers of all kinds of information are forming multipurpose conglomerates. They may soon establish their domination over the world's communication. Just like the motor highways, tomorrow's electronic super highways may interconnect quite a small number of high-tech spots and leave huge areas to progressive desertification. This concern is well presented in Kapor's (1994) remarks about the National Information Infrastructure when he says "The technical design of the NII will determine more about its public usefulness than anything else. We have a choice to make the NII open to a diversity of applications, information sources and services, or to keep it closed to all but those who own and operate the networks.....The NII may be a platform for the rich varieties of individual expression, for the transaction of commerce, and for exchange of ideas, or it may be nothing but 500 channels of least common denominator programming entertainment. We must take steps to ensure that the NII is more than just a repetition of the failures and shortcomings of mass media today". Let us hope that the pattern of traffic distribution shown in Fig. 1 is only a nightmare and not a premonition.

To some extent a significant change in the super highways is likely to bring to the media the ability to turn themselves into a point-to-point communication. The data from the USA-Japan information census (de Sola Pool, 1984) showed that the supply of information for entertainment far outweighed information for working or for living in the USA, even though the latter was more important in Japan. Since there were already signs of saturation in the USA, it may be tempting to find new markets. One would of course object that the highways are going to operate in a free market where "a sovereign consumer is able to freely choose" as Mattelart (1994) mentioned, emphasizing that this neoliberal vision of a "market democracy" is somewhat biased

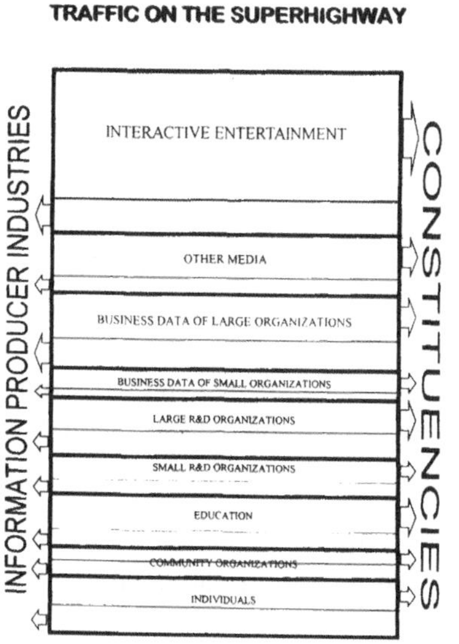

Fig. 1. Traffic on the Superhighway (©Michel J. Menou, CIDEGI, 1994).

and mainly serves as a "central piece in the legitimation process". Although the new IT has already demonstrated its unique capability for providing communication facilities to grass roots groups, nongovernmental organizations and the individuals at large, including as a means to resist oppression, e.g., their role in the recent political turmoils in the then USSR (Travica, 1992), its promises for freedom may not last longer than the dreams of free flying revealed by Blériot.

IT and information benefits: taking the substance for the shadow

The states of information

A lasting cause of confusion in all the debates about information is that all its different forms and the activities, trades and professions tend to present their part as the whole. Since telecommunications are a major channel of information, all the realm of information is thus at stake with them. Even if there is some truth in this, it would be better to leave with Ceasar what belongs to him. Compunications are a means for information. But what is information?

Information, like energy, can be viewed as a fundamental phenomenon whose attributes are experienced in various forms (Otten and Debons, 1970). It is often argued that because of this immediate experience, and the fuzziness of the concept of information, trying to define it would be a futile exercise. The present vagueness, to be polite, is becoming, however, a serious drawback. The various appearances of information, of which Belkin (1978) offered a most useful presentation, may be tentatively organized in a typology based upon what we would call "states", as follows:

a) information-as-object;
b) information-as-product (or services based there upon);
c) information-as-channel;
d) information-as-process;
e) information-as-resource; and
f) information-as-content (what is "said").

Information-as-resource may be differentiated according to its source, or store, as suggested by Ruben (1992). We would rather consider four instead of his three orders:

1. The hidden information existing in nature but yet to be discovered.
2. The information existing in the environment, consisting of messages or stimuli which are intelligible but waiting to be perceived and interpreted (also called raw data).
3. The information configured by individuals.
4. The information configured by groups or social systems.

There might soon be a need for a fifth order – the artificial information contained in computer systems for collective use.

All states constantly interact and are interdependent. But they are distinct and do not have the same "value". Information when communicated is at the same time product, channel and contents. Unfortunately, attention is usually focused on states b, c and d. Or, in other words, on the crafted material aspects, to the extent that both channels and processes are more often than not considered in conjunction with the products they circulate. This materialist perspective results also in the fact that information is somehow equated to formal information, i.e., the one whose states, sources and carriers are evidenced in material objects. For a variety of reasons informal information plays an important role and may well be the only form of information for many constituencies in developing countries. All well-considered, informal information is not less important in the more "advanced" societies, as many users studies demonstrated. When formal information sources, especially information systems and services, are not used, it does not mean that people live without information but simply that they rely upon other channels. The bushmen do not need satellites to find out where they are. For all people, irrespective of position, education and culture, asking a peer or a neighbour is the most natural way to obtain information. Many failures, especially at the highest levels of responsibility, are more attributable to an inability to ask people for, listen to, and appreciate their advice, than to a lack of formal information. Trying to assess the impact of information without paying equal attention to informal information and to formal systems and services would be a gross mistake.

Information is the life blood of action

Information as a process is the result of what Belkin and Oddy (1978) called an "Anomalous state of knowledge", i.e., a need for enhancing, thus changing, the current knowledge base for a new one. The recognition of the inadequacy of the knowledge base and the need for information stem themselves from the occurrence of a problem, or problematic situation, according to Wersig (1979). Even in the most simple situation of this kind, any human being, or adaptive system, is in fact exposed to, or using, a plurality of sources and channels conveying a plurality of messages coming from a variety of formal and informal information sources. The model in Fig. 2 attempts to show some of the key features of this situation. In first place the steps in the problem-solving process primarily rely upon the internal knowledge base to which external resources have to be combined. The internal knowledge base, in second place, operates under the influence of a series of factors, or motors, identified for the time being as personality, culture, emotion, logic (reasoning process) and intelligence (quality of judgement). The latter also continuously interact with one another. The distinction between logic and intelligence may find a basis in Brookes' observation (1980) that "Problems can be solved by objective calculation or by making a judgement. As sound judgements are of great social importance, it surprises me that so much effort is devoted to the teaching of calculation and so little to the teaching of judgement".

The vast majority of so-called users when asked about which information was useful to them, or that they valued most, etc., hardly understand what the question

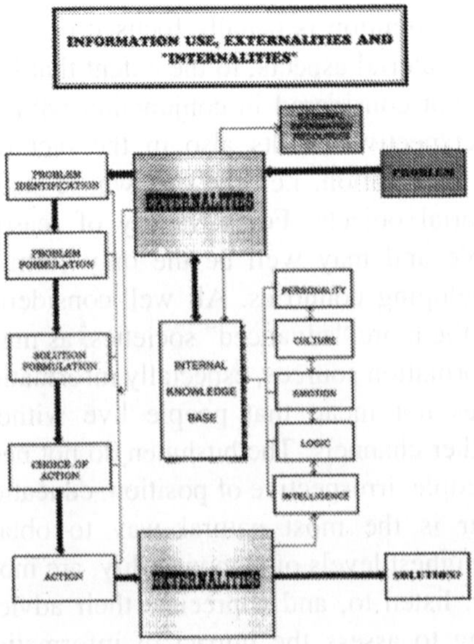

Fig. 2. Information use, externalities and "internalities" (©Michel J. Menou, CIDEGI, 1993).

is about. In any case, information is so much interwoven with the considered issues or behaviors that any attempt at isolating it would lead to a failure. People do not use information, they just take care of their business. The underlying question in any information impact assessment may therefore be "What, if any, is/was the contribution of information to the effective solution of a problem X?". This is not a simple question. Considering for instance a noteworthy reduction in the incidence of a disease, a number of tangible inputs, e.g., preventive and curative health care, delivery of vaccine, enforcement of legislation, more effective administrative procedures, etc., all play their part with public information campaigns and availability of appropriate information services to all concerned groups, health care personnel, families, etc. The effectiveness of some of the tangible inputs is itself partly relying upon effective information. Any achievement or goal should thus be decomposed in a complex series of inputs, outputs and associated supportive or inhibitive factors. They will form a nested hierarchy in which, at all levels it would become possible to identify the information activities which may have a significant role. The components of each block in the nested hierarchy would have to be selected on the basis of their playing a critical role in the achievement of the particular subobjective. Suitable procedures for constructing the nested hierarchies in the context of information impact assessment, and for discriminating information, information-dependant and information-independent factors are in our view a key topic for future research.

Even a simple problem, which primarily concerns a well delineated and small community, involves a number of distinct groups of stakeholders, especially if one considers the intermediate levels in the problem-solving process. In addition, each group is likely to have specific visions of information benefits and specific rationales for possibly changing the conditions of the problem-solving process or its information components. Furthermore, benefits for one group may well correspond to cost or losses for another group, short-term benefits to long-term losses. It is thus of paramount importance to consider all groups of stakeholders and take into account the benefits each of them considers as important. This may be of utmost importance in relation to decision-makers who may have dual, ambiguous or tacit agendas. Most politicians in particular do not consider only the intrinsic merits of any proposal but also, some say in the first place, what they may pay in return for their career, which is not always a subject for public disclosure. As Broadbent and Lofgren (1993) emphasized "a significant flaw in the evaluation approaches used in many previous information systems and services areas has been that only some of the stakeholders have been included in the evaluation process.... the perspective of senior managers has often been missing". Without this interaction one is at great risk of imposing preconceptions of benefits which have little or no bearing for them. For instance, time savings are often refereed to as benefits, without realizing the cultural bias associated with this kind of statement. Time is not the same for top management and line workers, nor for Britons and Mexicans. It is further necessary to distinguish among the various benefit areas and their mix, as well as the categories of benefits. Some are direct – in straight relation to the problem, some are only indirect, some are immediate, some delayed and some conditional – dependant on other critical factors.

Most assessments of, or predictions about, the effect of the on-going revolution have been primarily focused on short-term economic and social consequences, such as productivity, competitiveness, shifts in employment and the like. This is both short-sighted and culturally biased. Cherry (1977) already warned us that "these Western images of technologies such as television, telephones, books and newspapers are largely domestic ones. Their immense powers for "organizing" are perhaps inadequately emphasized". The growth of objective knowledge, the increasing complexity of society and social division of labour have resulted in the fact that individual and institutional memories contain less and less of the possibly useful information. The dependence from external, artificial, mechanized and collective information resources is thus expanding; and probably creating more imbalances and frustrations. But these resources, whatever their sophistication, e.g., expert systems, will remain only potential knowledge bases until they are matched and appropriated by the knowledge base of an adaptive system working on a problem. The IT revolution differs in this respect completely from the previous technological revolutions to the extent it no longer deals with material or physical attributes such as energy or motion. It deals with thinking, thus with the essence of human beings, further to Descartes' statement "Cogito, ergo sum".

So what?

Even if the effects of IT dependance have been identified and possibly classified along the above-suggested typology, the vexing questions will still remain to be answered: Do they correspond to an improvement in society's effectiveness? Do they improve people's lot? The all economic perspective of development has been increasingly criticized, with good reasons, but we are still far from a commonly agreed definition. The counterpoint between Gross Domestic Happiness and Gross Domestic Product precisely points to the core of the debate, which is whether the definition of what constitutes development is left to the people who are experiencing the socioeconomic transformation, or it is decided by the rulers on the basis of their idiosyncrasies, or those they borrow from the dominant nations or technostructures. We feel that it might be wiser to leave with the concerned people the responsibility to decide whether or not they are, or will be, better off once, or if, the transformation has been achieved, whether or not what happens can be called development and which elements in the profit and loss balance sheet characterize it.

What really matters

Signs vs. sense

Debons (1992) suggested that there are three basic types of information measures: psychological, linguistic and logistical. The latter are still dominating most of the informetrics and the physical information theory, grounded in the work of Shannon and Weaver (1949). Langlois (1982) stressed that the engineering perspective which commands this form of information theory prevents the required distinction between "the concepts of information itself and amount of information". Many efforts are directed at refining Shannon and Weaver's theory, e.g., with the concept of modulation proposed by Heilprin (1989), and incorporating into it some additional parameters which, by way of simplification, may be coined as "intelligence factors", such as Hayes' introduction of a syntactic measure (Hayes, 1993). But these models are mainly geared to the analysis of discrete transactions, either single or iterative, of single messages exchanged between a single pair of sender and recipient. They cannot apprehend the complexity of communication among adaptive systems, especially living ones.

"There is no information outside the subjects. There are only signals." This statement by Shpackov (1992) is an elegant way of stressing that neither IT or information really matters. Sperber and Wilson (1986) noted that "a fact is manifest to a person at a given time if, and only if, he can understand it at that time, and accept its representation as true or probably true". Information operates only when it "makes sense", i.e., fit with pre-existing meanings or can be integrated into them, and possibly transform them. Thus, information as contents may only matter. But only to the extent it has the capability of turning an initial knowledge base into a new more effective knowledge base. Machlup (1980) already alerted us that "information as the act of informing is designed to produce a state of knowing". The impact of information is to be seen in the improvement of a new state of knowledge over the previous one as a result of information-as-contents, i.e., in first place the reduction of the anomaly.

In search for the ore of sense

Information-as-contents is itself made up of three layers: semantic, syntaxic and paradigmatic. The same can be found in the knowledge base. In the absence of a supportive paradigm, a message recognized only at the semantic and syntaxic level will have no effect. Telling someone that he/she is in great danger because his/her blood pressure is 190/120 when that person does not know what this measurement means, and believes that sickness is a message from God, does not help much. Strong (1982) also stressed the importance of "another kind of information to be considered in behaving systems which is not directly treated in classical information theory – that deals with pattern, and thus with meaning in relation to goal-directed behavior". The reduction of the anomaly does not only add contents to the state of knowledge, it also transforms it, in its superficial contents and organization (semantic and syntactic levels) and possibly its structure (paradigmatic level). Brookes (1980) had already suggested that "information is that which adds to or modifies a knowledge structure". He also mentioned "yet the subjective provides the metacognitive basis – the overview, the perspective, the judgements – which gives purpose and direction to all objective thought". Assuming that the process of information reduction by which significance is established in terms of meeting goals and needs is primarily a pattern-based selection, as suggested by Strong (1982), the principal benefit of information is to be looked for in the enhancement of the paradigmatic structure of the knowledge base. Brookes' "metacognitive basis" is what we call the paradigmatic structure, for lack of a better expression. The "fundamental equation" proposed by Brookes (1980) is not radically altered in this statement, to the extent that knowledge structure is defined as being the paradigmatic structure. It can be reformulated as follows:

> The effect of information is equal to the new knowledge structure less the sum of the previous knowledge structure and the increment of information. However, one may wish to use two parallel equations of the same form, one being used for the superficial knowledge structure (semantic and syntactic) and one for the deep structure (paradigmatic). One may further consider that the ultimate benefit of information is a combination of the two.

The above implies that information as contents should present attributes that meet the requirements imposed by the paradigmatic structure of the knowledge base in order to be understandable and assimilable, and also that it allows a transformation of the knowledge base in such a way that it becomes more able to cope with the kind of problem at hand, or is perceived as such. The stock of paradigms is not static, even though it may evolve only slowly. It is also not isolated but permanently confronted with incoming information-as-contents it has to process and organize. The more information-as-contents carries patterns that can be absorbed, the higher its value. The more diversified, comprehensive and flexible the stock of paradigms, or if one prefers the stock of "cognitive maps", available to individuals and groups, the more they are able to interpret new information or to generate appropriate meanings, hopefully leading to adapted actions.

Assessing the information core

Should this premise be accepted, then comes the question of suitable ways to identify and characterize a stock of paradigms and their evolution. They are discussed in another paper (Menou, 1995) and can only be summarized below. Some methods are available for surveying the semantic and syntactic contents of knowledge bases, such as the grids for mental constructs developed by Kelly (1955), or the "informs" and "knowings" suggested by Debons (1992), which may provide a useful starting point. We do not know yet of any tool for digging out the paradigmatic structures, except perhaps those which were used by ethnologists for the structural analysis of tales, beliefs and the like and of course psychoanalytic techniques, which may be difficult to apply in the context we are considering. Trying to impose preconceived yardsticks may be fraught with ethnocentric biases. It seems more appropriate to derive the measures from the properties which will emerge from actual observations and be recognized as significant and discriminative by the various groups being investigated.

However, it may be helpful to refer from the inception to a number of dimensions which a priori appear to be useful pointers to the characteristics of a given stock of paradigms. We have tentatively identified seven dimensions on the basis of those particular aspects of appropriate information which are often mentioned in the literature, for instance by Neelameghan (1981) and/or those we had the opportunity to observe:

- Informal:Formal, which refers to the existence, or not, of an identified formal source for the paradigms.
- Endogenous:Exogenous, which refers to the fact that the paradigms originated, or not, from within the group.
- Resident:Circulating, which refers to the fact that the paradigms remain unsaid or are exposed to open discussion.
- Unconscious:Conscious, which refers to the degree to which the paradigms are grounded in structured justifications.
- Ancient:Recent, which refers to the lasting of the paradigms.
- Stable:Changing, which refers to the possibility for the paradigms to be adapted, or not, as a result of experience.
- Multiple purposes:Single purpose, which refers to the range of events that can be supported by the paradigms.

These dimensions should be seen as axes along which the paradigms can be positioned. Although it is possible that some paradigms fully fit at any one end of an axis, the more likely situation is that they take an intermediate position, being more formal than informal, more circulating than resident, etc. A "traditional" stock of paradigms could hypothetically be characterized as being fully informal, endogenous, resident, unconscious, ancient, stable and multipurpose. This implies no value judgement. Conversely, a flexible and responsive paradigmatic stock may be a priori depicted as one which would concentrate in the area along the axes which is more congenial to the problem at hand, or in general terms evenly distributed around the median position. Using these dimensions one may be able to draw maps similar to the chart of a field of vision. The differences between the paradigmatic stocks of two groups confronted with a particular problem, or the evolution of the paradigmatic stock of a single group over time, in conjunction with an improved supply of information for instance, could then be visualized and assessed. A theoretical example is shown in Fig. 3 where the paradigmatic stock of two groups, e.g., farmers and extension workers, involved in a given problem such as the preservation of water resources in a Sahel village are tentatively mapped. Empirical research is required in order to determine whether the model can indeed be applied, objective, or nonbiased, measures of these dimensions can be performed and the resulting maps allow for meaningful interpretations in the eyes of both the groups under study and the investigator.

The human dimensions

The ability of information as contents to enhance the knowledge base does not depend only on its attributes and their matching of those in the knowledge base. It also, and perhaps more importantly, depends on the characteristics of the adaptive system itself. Two traits may be regarded as especially important according to Lamberton (1993). Readiness, which can be described as the willingness and ability to act upon the problem, and capability which is the talent to turn an action into effective results. We would add a third one "appropriation" which is the sense of ownership of the means and ends, or more simply the perception that what is at stake is one's own business. Kuhlthau (1988) noted for instance that her subjects "revealed a sense of ownership on an internal, intellectual and private level. There was an

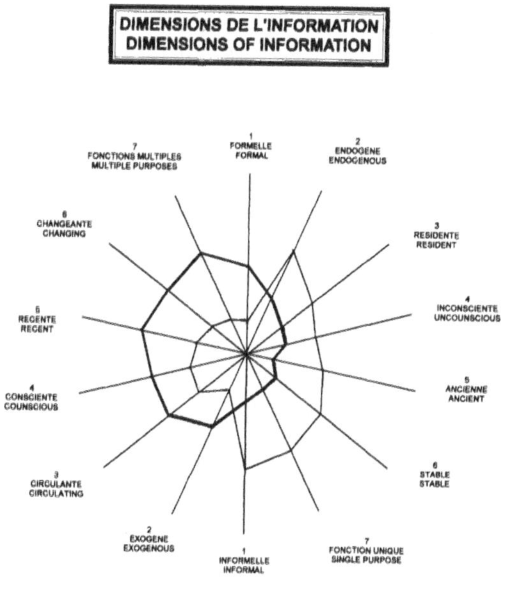

Fig. 3. Mapping the dimensions of information (©Michel J. Menou, CIDEGI, 1993).

element of possession in the process as well as the product". Appropriation may also be expressed through the belonging to the group, or relationship with the master, so essential in the initiation process, which is, as we once stressed (Menou, 1969) a basic information-transfer pattern. A fourth trait, which could be called "gestuality", may be of particular importance. It refers to the fact that the application of paradigms is evidenced in behavior. Strong (1982) mentioned the possibility "that pattern is constructed via behavior representations – as the "glue" relating one state to another. This particular form of organization of information may be a result of the social adaptation of primates whereby behavioral invariance is important in the ability to pass along group adaptations. This view would lead to a new basis for theories of the social construction of reality, in that it is much easier for organisms to copy one another's behavior than share one another's perception". The importance of tales in the transmission of knowledge, stressed by Bruner (1991) could be considered as an indication of the soundness of this assumption. Tales convey knowledge through the description of behaviors. It would not be surprising that what we call "innate" is nothing but paradigms of the knowledge base which were glued over so many thousands of years of behavior that they became part of our heritage, e.g., the fascination for fire we referred to in the above-quoted paper, or the standard male chauvinism. How could this be demonstrated is another story, worth attention. These traits are obviously dependant on the complex interaction of a wealth of factors. But they also suppose the existence of paradigms which can support them. In the absence of those traits, the availability of information-as-contents would make no difference, as we have plenty of evidence with regard to the protection of the environment, prevention of AIDS and the like. The timely provision of accurate and relevant information to the highly educated professionals who flourish in today's large and complex organizations seems on the contrary to breed more and more failures.

Any group of stakeholders cannot also be reduced to a few general attributes, as is often the case in, for example, occupations, level of responsibility, seniority, etc. The analysis which is outlined above can only be successful if the investigators have in the first place a fair understanding of what Taylor (1991) called the Information Use Environments (IUEs) of each group of stakeholders involved in the considered problem. The sets of parameters which characterize an IUE include the people, the settings, culture, the recognized problem, the expected solutions to the problems, the process of problem solving and the available information sources. Since Taylor's original work considered only American constituencies, he did not include culture as a standing feature of his model, though he stressed its importance and suggested one may have to. We have incorporated culture since we feel it should be part of the investigation in any case. Groups which seem culturally homogeneous may have significant differences among their segments. A description of this factor is also required for the interpretation of IUEs across cultures. The last element, information sources may need to be expanded into a particular study. We feel that a comprehensive description of the structure and operation of the total information lifecycle, both formal and informal, in which the constituencies are involved, is required. Because on the one hand one cannot understand the limitations of the final sources without knowing what is happening in the previous steps. And because on the other hand the information lifecycle is not a single spiral but a combination of spirals where the entire cycle is reproduced, under adapted forms, at each step. The basic structure of the IUEs and the specific data which are best suited for the description of IUEs to provide a meaningful picture have still to be ascertained, with due consideration of cross-cultural perspectives. Taylor (1992) insisted that the description of IUEs does not require a profound anthropological study. Rosenbaum (1993) has suggested a number of research issues associated with IUEs:

— What are their origins and how do they persist over time?
— What is their relationship with information behavior?
— What is their influence over the user's rationality?

The literature often refers to "rational behavior" or logic in conjunction with problem solving, decision making and thus information seeking and use. Lack of, or inappropriate, use of available, mostly formal, information resources is described as "irrational", e.g., consulting the witch doctor instead of a medical doctor. This is again a most annoying cultural bias. Not everybody operates like a graduate of the Harvard Business School, with all due respect. And even these people may, according to the circumstances, if not concurrently, rely upon different "logics", including the one underlying power struggle as opposed to straight economic rationality. Maruyama (1974) stressed that "There is not one but many logics. Aristotelian logic is not *the* scientific logic". One may add that although people in a given constituency, or IUE, may have a favourite form of logic it does not preclude their use of alternative ones when appropriate. There may be, and indeed are, different logics for presenting and justifying one's behavior to the public, or different segments of the public, which may themselves differ from those actually used. When investigating the impact of information care should be taken not only to recognize these different logics, but to systematically search for them. Benefits of information may relate to diverse logics, including for instance the satisfaction of ego.

Information richness vs. wisdom

Aspects of information richness

The link between information and its benefits is certainly more easy to trace when considering the information that is available and actually used. One may thus be tempted to concentrate on situations of information on demand. This could not do justice to the undetermined range of secondary and delayed effects. The difference between two constituencies regarding their information situation and their prospective benefits may indeed rely in the first place to the relative information richness of their milieu, as Mackenzie Owen (1992) stressed. Except for counting its material ingredients, e.g., number of documents, number of records, pieces of equipment, size and technological sophistication of the infrastructure, etc., we do not know what constitutes an information-rich environment. Identifying its components and their attributes and articulating them into some sort of general model is an exciting topic for future research.

One aspect we would love to see clarified is the contradiction we suspect between the breadth of the information available to privileged constituencies, in part as a result of the diffusion of new media, and the depth of the more limited information available to "deprived" constituencies. The richer may not be those we use to consider so. Another aspect of the same problem can be found in the fact that almost anyone over 10 years old in the North now has a pocket calculator, yet more and more people are unable to make a simple addition without the machine. IT addicts say they have shifted their skills to other and more rewarding areas. Maybe. It would be more convincing if we knew which ones and how more rewarding, and if we knew for sure nobody would ever need to make an addition. The same may well become true of writing when advanced wordprocessing packages will do it all, once prompted with the appropriate subject and a few specific data.

It should also be stressed that material counts of wealth indulge in some travesty of reality. When each individual in the privileged countries will have two micro-computers, five videophone sets, three HDTV sets, access to 500 hybrid networks, etc., as he/she seems to now accumulate other types of appliances, the actual use of those facilities will not increase in proportion. The value of information will also be much less. de Sola Pool et al. (1984) already stressed the discrepancy between the growth of supply and the growth of final consumption. Even if we assume that all the active time will be spent using IT, it will remain a maximum of about 16 hours a day and no more than a few applications, at best, at the same time. Unless the Big Brother on duty requires some of them to be kept permanently turned on. It does not seem that the concern for information overload has had any effect on the suppliers' strategies, except for product diversification. Neither does the growth in facilities and the slighter one in use, imply a better quality or adequation of the information, nor its more effective use.

"Nur Kultur and Barbarei Dinge zur Bedeutung sind" – Goethe – Gespräche mit Eckermann

More information and more IT may lead to higher chances for individual knowledge to be enhanced and for the collective virtual knowledge to be improved. Information becomes knowledge when it has been understood and absorbed or better appropriated. Machlup (1980) told us that "from various points of view, "living knowledge", or what living people know, may be the relevant stock of knowledge in society". An important consequence, especially for the developing countries, is that indigenous knowledge rather than the "world's stores of information" and access to them should be regarded as the main source of potential benefits. A rich knowledge base does not, however, guarantee that it will ever be used in a considerate fashion so that the reactions, actions or adaptations it will support will result in a well-balanced, more satisfactory, reversible and, so to speak, better situation. To that end, adaptive systems need to turn knowledge into wisdom.

"What is the process of learning, of growth in knowledge, growth in understanding and growth in wisdom?" were among the fundamental questions in information science research mentioned by Kochen (1984). Wisdom may be said of knowledge bases which have been so organized and factored that their use offers minimal risks of inappropriate outcomes and maximizes the comfort of the user(s). This may in part be dependant on the ability of the paradigmatic structure of the knowledge base. It is also dependant on the above discussed users' attributes. Criteria for wisdom are certainly not universal and may not apply beyond individual ideology. Wisdom can perhaps be witnessed only in retrospect or in comparison. The seven principles set forth by the Foundation for human progress in its manifesto "For a responsible and jointly liable world" (Fondation.., 1994) may offer a first approximation of wisdom. They are: safeguard, humanity, responsibility, moderation, caution, diversity and citizenship. Doris Schoenhoff (1993) reminds us that "In many Third World cultures, where formal education is still the exception, there is a greater emphasis upon wisdom than upon knowledge, a greater respect for the wise judge than for the scholar". This is also part of the cultural heritage of the First and Second Worlds, as evidenced by many proverbs. By any yardstick, most contemporary societies have a long way to go before they achieve it. Societies which are better off with regard to their material wealth, technological capacity and richness of their knowledge base can be predatory and ultimately self-destructive, as evidenced by so many examples in history and contemporary affairs. Is it wise, for example, to turn thousands of acres of Amazonian forests into grasslands for future hamburgers? The conditions for the transformation of knowledge into wisdom and the attributes of wisdom are subjects which deserve more attention. The spiral tentatively depicted in Fig. 4 may remind

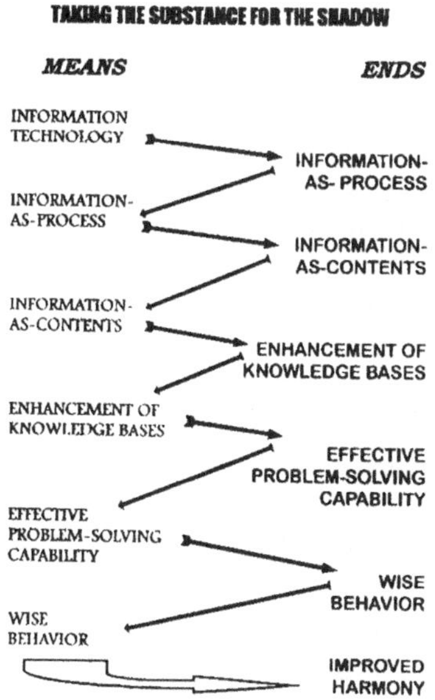

Fig. 4. Taking the substance for the shadow (©Michel J. Menou, CIDEGI, 1994).

us of the trade-offs between means and ends, and the long way to true goals. Investments in IT or expansion of telecommunications traffic fall short of the latter.

Investigating the impact of information: toward a cooperative international effort

The issue

International conferences, international information programs and national information policy documents steadily repeat the assumption that information is an essential resource which "constitutes a major factor for accelerating the development process" (UNESCO, 1979), which deserves special attention and high priority. It sounds like a lip service when considering the level of priority and support actually granted to information systems and services in most developing countries. Being "convinced that access to information is a basic human right" (NACOLADS, 1985) is of little help when many other, and more obvious, rights are still to be satisfied. At a time when competition for scarce resources is exacerbated, more solid arguments are indeed required.

Since the 1950s the efforts by institutions of the developing countries and technical cooperation agencies have achieved significant results in terms of number of services and systems established, increased access to information worldwide, increased indigenous capacities in information management, application of modern information technologies and strengthening of information infrastructures. The evaluation of information systems and services and of international projects to support them has provided over the years some evidence of their performance and effectiveness, of their cost-effectiveness in some instances and, to a much lesser extent, of their cost-benefit, mostly restricted to short-term immediate returns in organizations. Information specialists, and all those in the information business at large, have a firm belief that information and information systems and services are indeed relevant to decision making and problem solving, and consequently to the development process. This belief is supported by anecdotal cases. Systematic and/or quantitative evidence of the contribution of information activities, whatever their nature, to the achievement of the individual, institutional and social goals and the improvement of people's condition remains, however, to be ascertained. Due consideration and support of information activities can only be expected if it can be demonstrated that the provision and use of information make in itself a significant difference.

The challenge is even greater since from all corners, in both the South and the North, the relevance and appropriateness of the information systems and services is questioned (e.g., Eres, 1983; Heitzman, 1990; Sturges et al. 1991). They are said, not without reason, to serve only the elite, especially government bureaucrats, to be concentrated in the capital cities, to focus on international information at the expense of the indigenous one, to concentrate on technical processes at the expense of dissemination, to rely upon inappropriate formats for presenting information to the users. The use of modern information technology is often, and in our view not properly, judged prematurely or imposes excessive constraints. Or it is conversely presented as a panacea, as if CD-Roms and access to Internet were the answers to all problems. Many accounts of the failure or disappearance of information services are given. These criticisms, as those related to many other aspects of the development scene, are not without prejudice since similar stories can be told in all countries, especially with regard to equality of access (Doctor, 1992). The sustainability of information systems in developing countries has also become a serious concern. Agha's study (1992) singled out the lack of financial resource as the major drawback. It can in turn be attributed to the lack of recognition of the impact of information by policy and decision makers.

The project

The International Development Research Center (IDRC), Canada, launched in 1992 an international research project called "Impact of information on development". This project aims at articulating the basic concepts related to the benefits developing countries derive from information activities and formulating workable approaches toward their assessment (first phase), testing the proposed methods (second phase), consolidating the methods and assembling evidence of the role of information (third phase). Although the focus of the project is on developing countries, its findings are obviously relevant for all societies. From the inception the complexity and difficulty of the subject was fully realized. No quick and easy results were anticipated. On the contrary, the project was considered as a first step in a long-term research effort.

The first phase of the project consisted of an international computer conference, the first of its kind in this field, which lasted from April to November of 1992, and of a postconference workshop held in March 1993 in Nairobi, Kenya. Its objectives were to:

1. Examine as thoroughly as possible the relationship between information and development, identify possible benefits and suggest appropriate approaches to their assessment.
2. Propose a preliminary framework to be used in related empirical studies in the developing countries.
3. Provide a basis for further research.

More than 40 information specialists from 18 countries participated in one form or another during this phase. The proceedings and results of the first phase of the project are documented in a report which is now publicly available (Menou, 1993). The unpublished documents produced by the project could also be made available by IDRC to interested researchers upon request.

The second phase of the project immediately followed towards the middle of 1993, even before the full completion of the report of the first one; another distinctive feature in this kind of endeavour worth stressing. In the 3 years to come it will cover a series of interrelated activities:

- Raising awareness about the need for and feasibility of assessing the impact of

information on development, through publications, conference presentation and meetings.
- Testing the preliminary framework in the field, which is the major objective; five such exercises are already in progress in the Caribbean, Latin America, Africa and Asia in the framework of new information projects supported by IDRC; a number of other possibilities are contemplated in various regions and it is hoped that other organizations will also consider using this framework for conducting impact assessments within their own projects.
- Consolidation of some results of this and the first phase, even in this preliminary stage, e.g., guidelines on applying cost-benefit analysis to information projects have been prepared by Dr F.W. Horton Jr and will be published this year. They may significantly help the managers of information systems and services in developing countries when formulating their plans.
- Establishing mechanisms for ensuring a steady interaction among those involved in related activities.
- Creation of a decentralized research network linking individuals and organizations in the information science community and related areas which will choose to explore the many theoretical and applied research ideas encapsulated in the report.

The need for international cooperation

The complexity of the issues and the requirement for a bottom-up and case-specific approach will result in a long time lag before reliable results have been gathered. It may be reduced if many institutions and individual researchers, in the information science community and possibly related fields, join their efforts, form a critical mass and spotlight the impact of information as one of the key areas in the research front. Interest for the fundamental issues embodied in the impact problem has up to now been scattered and episodic. Cyclically, authors point to "the need to think more carefully about the economics of information and knowledge" (Langlois, 1982. Yet very little happens till the next call after some 10 years. It is hoped that a decentralized international research network could be formed to that end. The structure, governance and activities of the network would have to be defined by its members. The network could be decentralized and there is in principle no need for costly joint services and operations, as soon as effective communication mechanisms are available, on the one hand, and, more importantly, there exists a common will and commitment, on the other hand. The potential role of an international network relying as far as possible upon electronic communications was described by Dosa and Katzer (1991). IDRC will consider participating in its operation. Up to now the research capacity in information has been heavily concentrated in the North. But, for some years, suitable, though still limited, human and institutional resources have emerged in the developing countries. They could pool their efforts, as illustrated for instance by the recent establishment of a consortium among four library and information science schools in Africa, and cooperate with the concerned institutions in the industrialized countries, especially those which are active on the international scene and receive a significant number of international students.

So far, for a number of reasons, including IDRC's own mandate, the focus of the project has been on developing countries, which now include Central and Eastern Europe. This does not preclude the recognition that most of the related phenomena and issues also exist in all other countries, though under different conditions. This is obviously true when considering the information available to disadvantaged groups or minorities, as Cochrane and Atherton (1980) showed many years ago. It is even more true when considering the implications of the trend toward multiculturalism in many industrialized countries. Not only are there pockets of "underdevelopment" in all high income countries, but all countries may be regarded as developing countries, when it comes to the advent of an "information society" and even more a "knowledge society". Theoretical and methodological developments are needed which will improve investigation in any environment and have an universal bearing. A more accurate and reliable understanding of the role of information and its impact has clear implications for all aspects of the field, including the design of new technology applications (as opposed to the common repackaging of the same good old products in new media). Therefore research on this topic which would be directed at constituencies in the industrialized countries, or of a theoretical nature, are certainly part of the universe and should be included in the cooperative program. Comparative research should further be emphasized as a most promising avenue in this area since parallel investigation of constituencies sharing some basic characteristics, but operating in different environments, is likely to greatly help in the identification of key factors as well as commonalities and specificities between different sociocultural environments. The above-suggested research network would thus not need to restrict itself to developing countries but may possibly be far more extensive in its geographic and subject coverage.

We believe that a solid starting block has been provided. A comprehensive review of the nature of information and its role in the development process has been offered. More importantly, key aspects of assessment exercises have been outlined. They should be problem-center and beneficiary driven. They should avoid any preconception of benefits. The latter, and the parameters or indicators which may depict them, should be defined in a steady and interactive process involving representatives of all groups of stakeholders. An appropriate familiarity with the various attributes of IUEs and the total information lifecycle associated with the considered problems is a prerequisite for identifying benefits. Costs related to the provision and use of information should always be taken into account in order not to establish information as a good in itself. The diversity of information activities and resources should be considered, in particular the role of informal information resources and channels, which is too often overshadowed by formal products. The diversity of, and contradictions among, the values, logic and expectations of the various stakeholders should also be recognized.

We hope that this project will indeed serve as a catalyst for a major stream of research to emerge.

References

Agha, S.S., 1992. *Sustainability of information systems in developing countries: an appraisal and suggested courses of action.* Ottawa, Canada: International Development Research Centre.

Ang, J., Pavri, F., 1994. A survey and critique of the impacts of information technology. *International Journal of Information Management*, 14(2):122–133.

Belkin, N.J., 1978. Information concepts for information science. *Journal of Documentation*, 34(1):55–85.

Belkin, N.J., Oddy, R.N., 1978. *Design for an interactive information retrieval system based on an anomalous state of knowledge hypothesis.* Quoted in Wersig (1979 op. cit).

Broadbent, M., Lofgren, H., 1993. Information delivery: Identifying priorities, performance, and value. *Information Processing and Management*, 29(6):683–703.

Brookes, B.C., 1980. Measurement in information science: objective and subjective metrical space. *Journal of the American Society for Information Science*, 31(4):248–255.

Bruner, J., 1991. *Car la culture donne forme à l'esprit. De la révolution cognitive à la psychologie culturelle.* Paris, Eshel. (Translation of Acts of meaning.) Boston: Harvard University Press, 1990.

Carbo Bearman, T., 1992. Information transforming society: challenges for the year 2000. *Information Services and Use*, 12:217–223.

Cherry, C., 1977. *World Communication: Threat or Promise? A Socio-Technical Approach*, 2nd Edn. Chichester, UK: John Wiley and Sons.

Cochrane, G., Atherton, P., 1980. The cultural appraisal of efforts to alleviate information inequity. *Journal of the American Society for Information Science*, 31(4):283–292.

Debons, A., 1992. The measurement of knowledge. In: Shaw, D. (ed) *Information Management on the Move. Proceedings of the 55th Annual Meeting of the American Society of Information Science*, Vol. 29, (pp. 212–215). Medford, NJ: Learned Information.

Doctor, R.D., 1992. Social equity and information technologies: moving toward information democracy. In: Williams, M.E. (ed) *Annual Review of Information Science and Technology*, Vol. 27, (pp. 43–96). Medford, NJ: Learned Information.

Dosa, M., Katzer, J., 1991. Electronic networking in support of South-South cooperation. *Journal of Education for Library and Information Science*, 32(1/2):84–96.

Eres, B.K., 1985. Socioeconomic conditions related to information activity in less developed countries. *Journal of the American Society for Information Science*, 36(3):213–219.

Fondation pour le progrès humain, 1994. Pour un monde responsable et solidaire. *Le Monde Diplomatique*, April:16–17.

Hayes, R.M., 1993. Measurement of information. *Information Processing and Management*, 29(1):1–11.

Heilprin, L.B., 1989. Foundations of information science reexamined. In: Williams, M.E. (ed) *Annual Review of Information Science and Technology*, Vol. 24 (pp. 343–372). Amsterdam: Elsevier.

Heitzman, J., 1990. Information systems and development in the Third World. *Information Processing and Management*, 26(4):489–502.

Kapor, M., 1994. Quoted in: NII Priorities: a Sampling of Views from the NII Advisory Council. *Information Industry Association – Information Policy Online*, 1(3) (May 1994).

Kelly, G.A., 1955. *The Psychology of Personal Constructs*, Vol. 1. A theory of personality. New York: W.W. Norton.

Kochen, M., 1984. Information science research: the search for the nature of information. *Journal of the American Society for Information Science*, 35(3):194– 199.

Kuhlthau, C.C., 1988. Longitudinal case studies of the information search process of users in libraries. *Library and Information Science Research*, 10:257–304.

Lamberton, D.M., 1993. *Private communication*. (See also his contribution – Information and organization: questions and clues. In: Feeney, M., Grieves, M., (eds) *Changing Information Technologies: Research Challenges in the Economics of Information. Proceedings of the 3rd International Information Research Conference, Poigny la Forêt, France, 11–13 July 1993.* London: Bowker & Saur, 1994.

Langlois, R.N., 1982. Systems theory and the meaning of information. *Journal of the American Society for Information Science*, 33(6):395–399.

Machlup, F., 1980. Knowledge: its creation, distribution and significance. Vol. 1. *Knowledge and Knowledge Production*, Princeton, NJ: Princeton University Press.

Mackenzie-Owen, J.S., 1992. *Waarde-aspekten van informatie-verzorging (Aspects of the value of information).* The Hague, The Netherlands: RABIN.

Maruyama, M., 1974. Endogenous research vs. delusions of relevance and expertise among exogenous academics. *Human Organization*, 33:318–322.

Mattelart, A., 1994. Comment résister à la colonisation des esprits. *Le Monde Diplomatique*, April:28–29.

Menou, M.J., 1969. Knowledge transfer from autoconsumption to mass production. In: Mikhailov, A.I., Chernyi, A.I., Gilyarevskii, R.S., Afremov, V.Y. (eds) *Theoretical Problems of Informatics*, (pp. 41–63). Moscow: Viniti, FID 435.

Menou, M.J., 1993. *Measuring the Impact of Information on Development*. Ottawa, Canada: International Development Research Centre.

Menou, M.J., 1995. The Impact of information (Part 1): toward a research agenda for its definition and measurement. Trends in....The impact of information (Part 2): concepts of information and its value. *Information Processing and Management*, (accepted for publication).

Miles, I., 1988. *Mapping and Measuring the Information Economy*. Brighton, UK: Center for Information and Communication Technologies, Science Policy Research Unit, University of Sussex.

Miles, I., 1992. *Program of Research on Engineering Science and Technology*, University of Manchester. Private communication.

NACOLADS, 1985. *Seminar on the Evaluation and Analysis of National Information Policy. Final Report*, (p. 8). Kingston, Jamaica: Office of the Prime Minister, National Council on Libraries, Archives and Documentation Services.

Neelameghan, A., 1981. Some issues in information transfer: a Third World perspective. *IFLA Journal*, 7(1):8–18.

Ostrom, E., 1990. *Governing the Commons: the Evolution of Institutions for Collective Action*. Cambridge, UK: Cambridge University Press.

Otten, K., Debons, A., 1970. Towards a metascience of information: informatology. *Journal of the American Society for Information Science*, 21(1):89–94.

Rosenbaum, H., 1993. Information use environments and structuration: towards an integration of Taylor and Giddens. In: Bonzi, S. (ed) *Integrating Technologies, Converging Professions. Proceedings of the 56th ASIS Annual Meeting*, Vol. 30, (pp. 235–245). Medford, NJ: Learned Information.

Ruben, B.D., 1992. The communication-information relationship in system-theoric perspective. *Journal of the American Society for Information Science*, 43(1):15–27.

Shannon, C., Weaver, W., 1949. *The Mathematical Theory of Communication*. Urbana, IL: University of Illinois Press.

Shpackov, A.A., 1992. The nature and boundaries of information science(s). Letter to the editor. *Journal of the American Society for Information Science*, 43(10): 678–668.

Schoenhoff, D.M., 1993. *The Barefoot Expert. The Interface of Computerized Knowledge Systems and Indigenous Knowledge Systems*. Westport, CT: Greenwood Press.

de Sola Pool, I., Inose, H., Takasaki, N., Hurwitz, R., 1984. *Communication flows. A census in the United States and Japan*. Tokyo: University of Tokyo Press.

Sperber, D., Wilson, D., 1986. *Relevance: Communication and cognition*. Cambridge, MA: Harvard University Press. (As summarized in (p. 604) of Harter, S.P. Psychological relevance and information science. *Journal of the American Society for Information Science*, 43(9):602–615.)

Strong, G.W., 1982. Adaptive systems: the study of information, pattern and behavior. *Journal of the American Society for Information Science*, 33(6):400–406.

Strassmann, P., 1993. *Systems Integration: the Way of Information Management in Year 2000*. Keynote presentation at the 56th Annual Meeting of the American Society for Information Science, Colombus, OH, October 25 1993.

Sturges, P., Mchombu, K., Neill, R., 1991. Indigenous knowledge base in African development. *Social Intelligence*, 1(3):213–237.

Taylor, R.S., 1986. *Value-Added Processes in Information Systems*. Norwood, NJ: Ablex Publishing.

Taylor, R.S., 1991. Information use environments. In: Dervin, B., Voigt, M.J. (eds) *Progress in Communication Sciences*, Vol. 10, (pp. 217–255). Norwood, NJ: Ablex Publishing.

Taylor, R.S., 1992. *Private communications*.

Travica, B., Hogan, M., 1992. Computer networks in the xUSSR: technology, uses and social effects. In: Shaw, D. (ed) *Information Management on the Move. Proceedings of the 55th Annual Meeting of the American Society of Information Science*, Vol. 29, (pp. 120–135). Medford, NJ: Learned Information.

UNESCO, 1979. *Intergovernmental Conference on Scientific and Techological Information For Development. UNISIST II*. Paris, 28 May – 1 June 1979. Final report (p. 18). Paris: UNESCO, General Information Programme.

Wersig, G., 1979. The problematic situation as a basic concept of information science in the framework of the social sciences: a reply to N. Belkin. In: *FID, Theoretical problems of informatics: New trends in information and its terminology*, (pp. 48–57). Moscow: VINITI.

Part II: Management and policy

Beyond Competition: The Future of Telecommunications.
Edited by D. Lamberton.

CHAPTER 7

The impact of structural changes in telephony prices

David Cracknell
Telephony Modelling and Forecasting, BT, London, UK

Introduction

Over the past 2 years, BT has completed a fundamental review of the structure of its telephony prices, and the findings have been reflected in call price changes made in the first year of RPI-7.5 price control. The work was initiated because BT's call pricing structure had remained effectively unchanged since 1958, and a review was appropriate given BT's commitment to putting the customer first and policy of continuous improvement. At the same time tighter regulatory controls increased the extent to which BT will reduce telephony prices over the next 4 years, and competition is steadily developing, particularly in the Personal and Small Business markets.

Regulatory background

The regulatory background against which BT currently makes pricing decisions is the need to make reductions to a basket of main telephony prices of 7.5 points below the rate of inflation for each year between 1993 and 1997. Inflation is currently running at a low rate in the UK – BT's reference point for the 1993/1994 pricing year was the June on June general price increase of 1.2% – so in 1993/1994 BT was required to reduce the basket of telephony prices by 6.3%. Within the basket BT's practice in recent years has been to increase exchange line rentals by 2% more than general inflation, and such increases were again applied in 1993/1994. A reduction in call prices of over 9% was therefore needed to achieve the required overall price cut, together with enhancements to maintain the value of BT's portfolio of focussed pricing packages.

Faced with the prospect of revenue reduction from regulation and with the growing impact of competition, it has become increasingly important for BT to grow the calls market, and improve the value of customer response to price reductions, as measured by price elasticity.

Oftel (1992) reports full details of the BT retail price cap.

Address for correspondence: David Cracknell, Manager Telephony Modelling and Forecasting, BT, Room 3074, 2–12 Gresham Street, London EC2V 7AG, UK. Tel.: +44-171-356-7621. Fax: +44-171-356-7051.

Price setting and network capacity

The theory of peak load pricing is widely used in industries where output cannot easily be stored and where demand fluctuates in a predictable and cyclical way. Given some simplifying assumptions, the theory indicates that the time of day pricing structure should be such that the contributions to the capacity costs at each time of day equals the total capacity cost. Park and Mitchell (1987), fully explore the economic case for peak load pricing of telephone calls. BT's policy in the 1970s and 1980s was to broadly follow this model as, in an analogue network environment, the additional transmission and switching capacity required to handle increases in calling rates was expensive to provide. The analogue network also had a large number of small routes which all had to be dimensioned to the peaks for each individual route. Network resilience was also a problem if peaks in demand exceeded capacity, and thus it was important to provide sufficient capacity to cater for even localised peaks. As a result, network costs were very sensitive to local traffic peaks. As a result, BT developed three time-of-day prices which operated as shown below, in terms of calling time purchased per 4.2p unit for Local calls, and BT's then three long-distance rates (a, b1, b).

		Seconds per 4.2p unit			
		Local calls	Long distance		
			a	b1	b
08.00–09.00	Standard rate	80.00	36.15	32.00	25.60
09.00–13.00	Peak rate	57.50	27.00	23.95	19.20
13.00–18.00	Standard rate	80.00	36.15	32.00	25.60
18.00–08.00	Cheap rate	220.00	80.80	50.35	37.95

The morning peak rate was introduced in 1970 and applied between the hours of 09.00 and 12.00 during weekdays. The peak period was extended by 1 hour in 1975. The introduction of the pricing differential had the anticipated effect in that peak rate traffic grew less quickly than other traffic in the following years. However, the morning peak rate persisted for Local Calls until, in 1986, further rebalancing took place.

Figure 1 illustrates the time of day traffic profile which arises when such a structure has been in operation for some years. Three peaks are evident, with the highest often occurring in the evening (when prices are lowest). The size of the morning and afternoon peaks are very similar.

Consumer research

Consumer research on attitudes towards prices has been a key input into the price

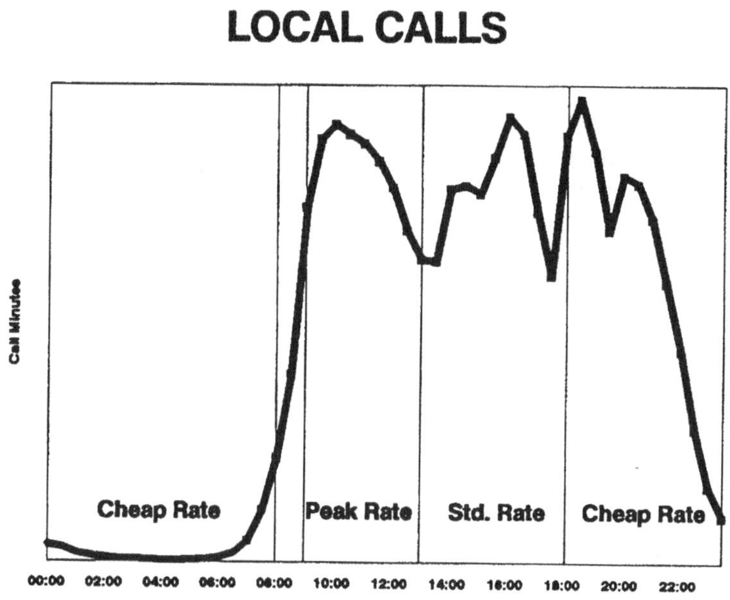

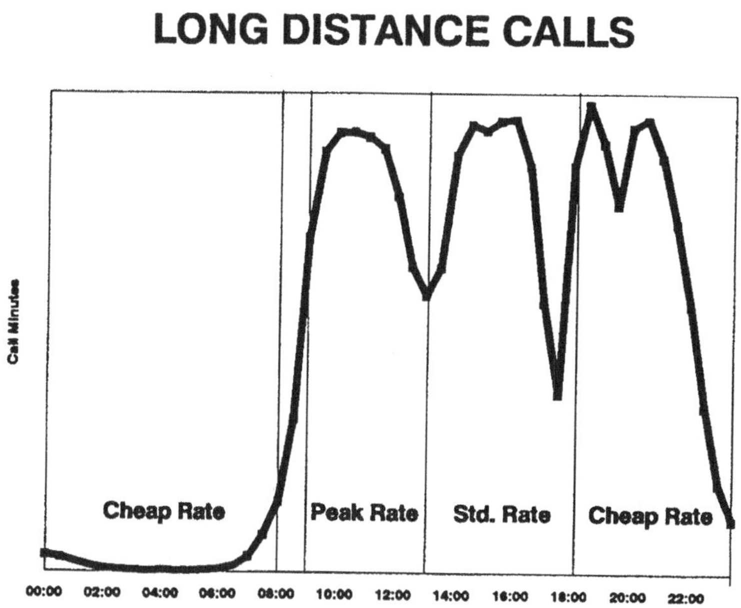

Fig. 1. Illustrative Weekday Call profile.

review carried out over the last 2 years, and this was investigated using both qualitative and quantitative market research surveys.

Existing market research studies indicated that BT's call prices were the main reason for customer dissatisfaction, with overestimation of both prices and the differential between BT and competitors, and dislike of BT's unit based charging system. Customer response to call price changes, as measured by price elasticity, had been stable and predictable, but lower than observed in Europe and North America. The new qualitative research results confirmed the received wisdom, showing that customers had a very poor understanding of the existing structure which led to confusion and dissatisfaction. They indicated a key requirement as being reduced prices with a simple and fair pricing structure.

These conclusions were also borne out by the results of quantitative studies. Knowledge of the pricing structure was low, with awareness of the terms 'Peak', 'Standard' and 'Cheap' being less than 40% in all cases. Perception of call prices was significantly higher than reality. The average response showed prices to be between 2 and 3 times greater than actual; for a 5-minute Cheap Rate Local Call the average answer was 4 times the actual price. Customers with a good knowledge of the time of day structure felt that it imposed restrictions on the way they made use of the network. There was a strong preference for the removal of Peak Rate, simpler distance bands and time of day structures.

As a result of this research, and the change in the cost structure which arises with the move from analogue to digital technology, BT concluded that there was a need to simplify the call pricing structure. This is also important in the application of Peak Load pricing theory, where efficiency arguments are based on the assumption that customers know what prices apply at different times. It is not enough simply to set the prices and then rely on customers to act accordingly. For example, a multiplicity of prices, especially if they were changing very frequently, would be likely to confuse customers about the levels of prices, rather than serve the function of informing them about underlying network costs. Such prices would then not influence consumption patterns in the way required.

Testing the market – special pricing offers

Over the period November 1992 to July 1993, BT ran a number of special pricing offers with the objective (amongst other things) of finding out more about price elasticities for different call types at different times of the week, and the extent to which they could be influenced by effective customer communication. The offers included Sunday Special (UK long-distance calls priced at Local rates on Sunday afternoons), European Special (Calls to EU countries on weekday evenings and weekends), Down Under Special (Calls to Australia and New Zealand on Saturdays), Local Special (Local calls on weekday evenings and weekends) and North American Special (Calls to Canada and USA on Saturdays and Sundays). They thus covered a wide range of call types, and times of the week, but concentrated on the Personal market (rather than Business).

Historically, like most other telephone companies, BT's traditional source of price elasticity information was time series econometric modelling. The time series models included price among the explanatory variables, and hence yielded estimates of elasticity. In a pricing environment characterised by lack of customer specific options and special offers, time series model results provided managers with the elasticity information they needed to make pricing decisions. They could be used to predict the market and market share response of a given change, and hence, pricing policy could be developed to meet BT's strategic objectives.

The need, however, to model telephone demand in a dynamic framework is well recognised in the literature. Taylor (1980) explores this in some detail, and explains that longer-run elasticities are nearly always larger than short-run estimates. The solution to this problem adopted in North America was to obtain elasticities using models based on a pooled time-series cross-sectional estimation. It was argued that such models captured longer-run elasticities as they used data reflecting underlying market dynamics, such as telecomms intensiveness, and that such elasticities were of most relevance when considering structural tariff changes.

The pooled time-series cross-sectional approach cannot be as readily adopted in the UK as in North America, as BT has only three Long-Distance tariff rates, but modelling work on UK data gave an estimate of the long-run Daytime Long-Distance call elasticity, which was 4.5 times greater than that obtained from time series modelling, as predicted by economic theory. A similar study on Intercontinental calls (from the UK to countries outside Europe) gave a long-run price elasticity which was 4.9 times greater than the time series estimate.

The detailed work on monitoring and evaluating the impact of the special pricing offers is reported in Cracknell and Knott (1995). The key conclusions were that:

- The calls market is potentially much more price elastic than would be indicated from time series based econometric modelling. Special price offers, coupled with effective customer communication, are a means of stimulating the call market, albeit on a temporary basis, immediately a price change is implemented.
- Personal customer price elasticities appear to be higher at the weekends, than during the week, as more time is available for discretionary communication.
- Price elasticities on individual International routes are very variable. They are strongest where there is a clear community of interest between the distant country and the UK.
- Effective customer communication is an essential ingredient in achieving the full potential traffic stimulation from a price change. This is an important message when handling revisions to standard prices.
- The offers can have a strong impact on perceived Value for Money, and can therefore assist the longer-term strategy of raising customer calling rates.

Figure 2 shows a comparison of price elasticity results obtained from time-series (short run), cross-sectional (long run) and special-offer evaluation studies. Also shown are representative elasticities from other countries, culled from past International Telecommunications Society and other conference papers. The conclusions which can

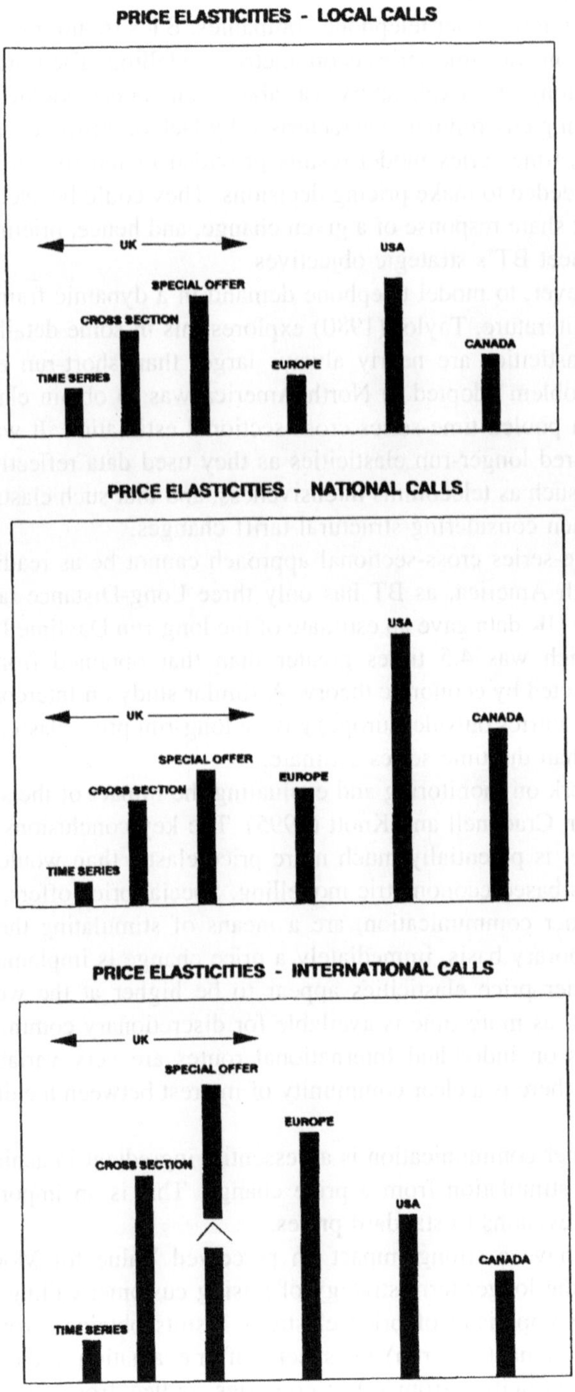

Fig. 2. Price elasticity international comparisons.

be drawn are that:

- Time-series model based estimates of the Local Call price elasticity are similar to those in Canada and Europe. Elasticities similar to those in the USA were measured from the special offer.
- Time-series model based estimates of the National Call price elasticity are very low in the UK. The special offer led to a price response similar to that expected in Europe or Canada, but below the USA.
- Customer response to International call price changes is particularly stimulated by a special offer. The elasticity for the period of the Australian and New Zealand special offer was much stronger than had been observed from previous BT International call price changes.

The special-offers programme also presented an opportunity to make a detailed study of the relationship between advertising and customer response — outline results are reported in Cracknell and Knott (1995), and indicate that timing is crucial to maximising the short-term call stimulation arising from a price change.

Modelling structural price changes

The consumer research indicated that a key area of concern to customers was the time-of-day call pricing structure, and as a result, this has been a major area of modelling work over the last year. To meet the need for increased flexibility in pricing, and to test the effect of increasing the number of charge periods, or varying the time of day for which they apply, a methodology was needed for calculating price impacts on traffic on a half-hourly basis, for both weekdays and weekends. The BT approach was based on that adopted by AT&T for International calls, and reported in Natarajan and Hunger (1986). The methodology presented in their paper has been adapted to reflect:

- The overall call price elasticity, estimated using the short run, long run or special-offer approach.
- Information on own price and cross-elasticities by charge band, derived from the special-offer analysis.
- The observed tendency of price elasticities in the UK to increase in proportion to the real tariff. This is reflected by relating the log change in demand to the absolute change in price.

The methodology was programmed into a Lotus spreadsheet so it could be updated and adapted as necessary for practical implementation. This was named TANDOORI (Time-of-Day Analysis using Natarajan-Hunger methodology and Outturn data for Optimisation and Redistribution of Income).

The mathematical details of the AT&T algorithm are fully reported in Natarajan and Hunger (1986). This shows how a set of time-of-day price elasticities, consistent with an overall total, can be derived. The properties of the elasticities are:

- Peak-period demand is relatively price inelastic when compared with off-peak demand. This result was untested by the BT special offers which had applied to calls in the evenings and weekends.

– Empirically, the ratios between the highest and lowest elasticities were reported to be consistent with other AT&T market studies.
– The direct dependence on the demand profile yielded results which AT&T reported as robust when tested on differing traffic streams.

Figure 1 illustrates the type of traffic data used to calibrate the model. Separate call minute streams were available for Business and Personal Calls for Local and BT's three Long-Distance charge steps, by half-hour period for weekdays, Saturdays and Sundays. These results were then stacked so that the modified AT&T algorithm could be directly applied. Factors derived from BT's internal estimates of own and cross-price elasticities were applied to the volume-based factors from the AT&T algorithm to give a set of own price and cross-elasticities for each half-hour period.

BT's 1993/1994 pricing programme

As a result of the consumer research and the elasticity evidence from the special pricing offers, the two key structural pricing changes made by BT in 1993/1994 are:

– The announcement of a move to elapsed time, rather than unit charging in 1995.
– National Weekend Rate, introduced in December 1993. This applies to Long-Distance calls on Saturdays and Sundays. The three previous Long-Distance charge rates were merged into a single rate with a clearly defined price message (10p for 3 minutes – including 17.5% Value Added Tax) to assist customer communication. The price reductions ranged from 9% for the cheapest to 55% for the most expensive Long-Distance calls.
– Abolition of the Peak Rate which had applied between 09.00 and 13.00 in March 1994. All calls between 08.00 and 18.00 are now charged at the previous Standard Rate (renamed Daytime Rate).

In a further price change, implemented in September 1994, BT continued its price simplification programme by merging its two long-distance tariffs ('b' and 'b1') to create a single price band for calls of more than 56 km.

The impact of these price changes on call demand is being tracked using the same methodology applied to the special offers in 1992 and 1993, reported in Cracknell and Knott (1995). The results can be compared with the expected outcome given the three types of elasticity, and provide a further check on the empirical accuracy of the Time-of-Day algorithm. They also provide further information on the dynamics of price elasticity development, and the importance of customer communication in the response to price changes.

National Weekend Rate

The logic behind the introduction of National Weekend Rate was to offer a price reduction at a time of the week when special-offer analysis indicated that price elasticities were highest. Weekend traffic rates are lower than weekday, and the capacity of the BT network to handle large increases in weekend traffic had been confirmed during the Sunday Special offer in November and December 1992.

Figure 3 shows the customer response to National Weekend Rate when compared with the model prediction, over the first year of its operation. It shows that Saturday performance started very strongly, although this was in part due to the influence of Christmas Day. Growth fluctuates from month to month, but is consistently outperforming special-offer elasticitiy expectations. Sunday growth performance started slowly (although above the Long-Run elasticity), but now consistently exceeds that expected from the special-offer elasticity.

These results are particularly encouraging because they show that:

- Growth is well above that which would have been traditionally expected from a generic price change.
- Special-offer price elasticities can be achieved from a permanent price change if it is well communicated and understood by customers. The short-term price elasticity would only apply if these conditions are not met.
- The selection of a clearly communicable price point, and development of a marketing campaign to accompany a price revision contribute towards the achievement of a strong price response. The studies of the links between advertising and call growth established during the programme of special offers were used in deciding the timing of the marketing activity.

The time of day modelling approach has been broadly validated by the results to date.

Daytime Rate

The Daytime Rate marks a change from BT's pricing approach of the 1970s and 1980s. It has been introduced because of the views of customers, as expressed in the pricing research study. BT's trunk network has now been completely modernised, and in March 1994 97% of customer lines were served by modern exchanges. With a digital network, many of the factors which led BT to introduce the Peak Rate in 1970 have diminished significance because localised peak capacity costs are no longer of such importance when:

- Digital technology makes larger switching and transmission modules economic, and the network has fewer sites connected by much larger transmission links.
- Modern network management techniques allow alternative routings to be used to spread any localised traffic peaks which do arise.

Under these circumstances it becomes more important to consider the impact of price changes on the overall time-of-day profile. As price elasticities vary at different times of day, this is likely to adjust to any price revision. The impact of the removal of the Peak Rate on the network was therefore assessed against the impact of an alternative price revision which would have been needed to comply with BT's price-cap regulation.

Figure 4 shows the customer response to Daytime Rate when compared with the

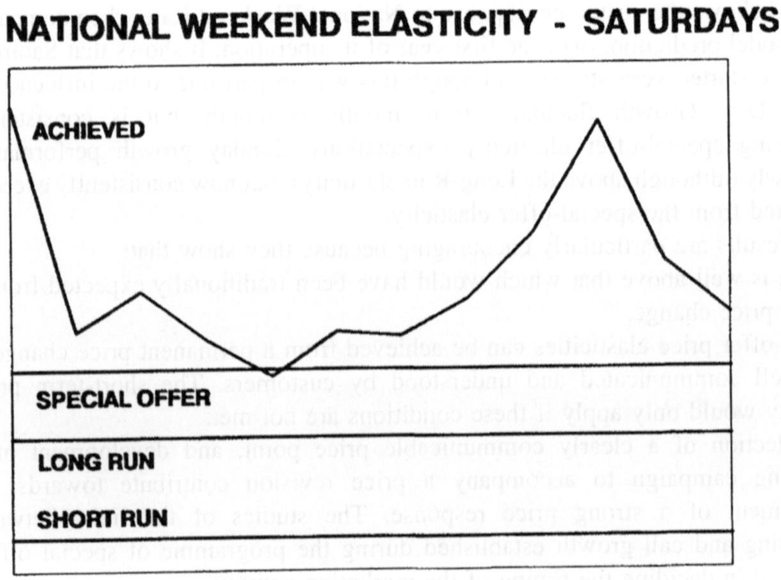

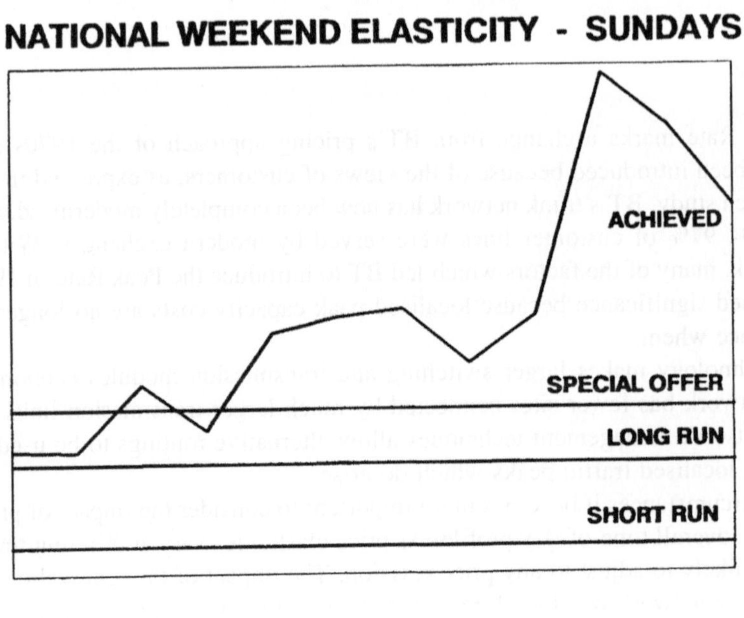

Fig. 3. Price elasticities – National Weekend Rate.

DAYTIME RATE ELASTICITY - LOCAL CALLS

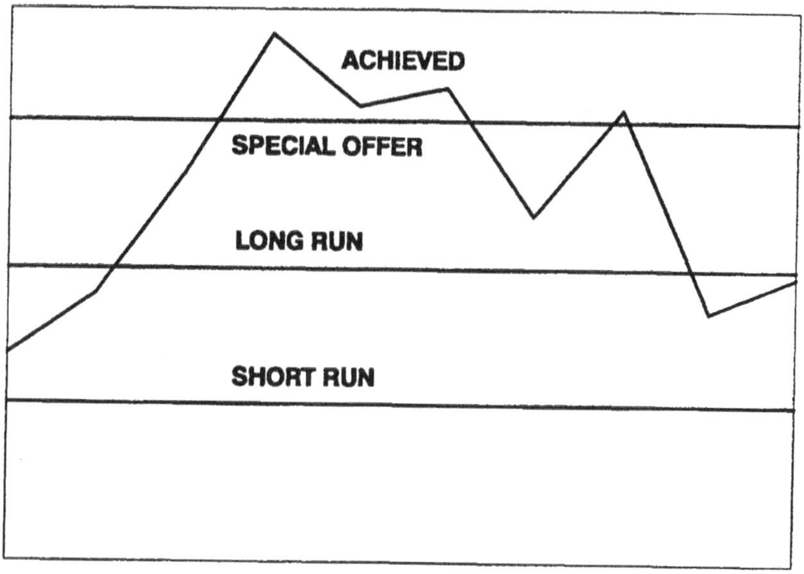

DAYTIME RATE ELASTICITY - NATIONAL CALLS

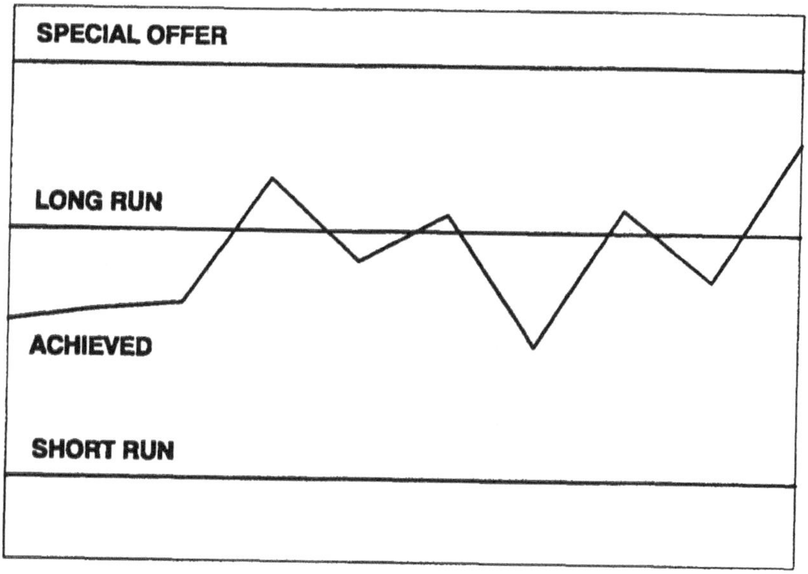

Fig. 4. Price elasticities – Daytime Rate.

Table 1. Price structures in other European countries.

Country	Morning peak	Evening reduced	Further reduced
Austria	No	Yes	–
Belgium	No	Yes	–
Czech Rep./Slovakia	No	No	
Denmark	No	Yes	–
Finland	No	Yes	Late evening/Night
France	No	Yes	Late evening/Night
Germany	No	Yes	–
Greece	No	Yes	
Hungary	No	Yes	–
Ireland	No	Yes	–
Italy	Yes	Yes	Night
Luxembourg	No	Yes	–
Netherlands	No	Yes	–
Norway	No	Yes	–
Portugal	No	Yes	–
Romania	No	Yes	–
Spain	No	Yes	
Sweden	No	Yes	–
Switzerland	No	Yes	
Turkey	No	Yes	–
UK	No	Yes	Weekends

model prediction, over the first ten months of its operation. They confirm the prediction from the special offers that weekday elasticities are less strong than those at the weekend, but show the impact steadily building (with some fluctuation on a month-by-month basis), and traffic stimulation around the long-run elasticity point by the end of the period. The impact of the change continues to be monitored on a weekly basis.

European comparisons

Table 1 shows how the UK compares with other European countries in terms of tariff structure, with afternoon prices taken as the norm. It shows that:

- The only European country which currently has a morning Peak Rate is Italy.
- Other countries have low off-peak tariffs in the late evening and night time, but the UK is unique in Europe with a low weekend tariff.

On this basis the BT price action in 1993/1994 might be seen as bringing the UK in line with the European practice on morning Peak Rate, while introducing an innovative pricing structure at weekends which is not currently matched elsewhere in Europe.

Future developments

The objectives of making call prices simpler, fairer and better value will set the agenda for BT tariff changes over the next 5 years, with the merging of the two long-distance tariff bands, and the move to per second charging in 1995 as the next steps. The modelling work for the 1993/1994 price changes described in this paper will provide BT's basis for the commercial evaluation of customer reaction to alternative pricing proposals.

Acknowledgements

The author gratefully acknowledges the assistance of the BT Telephony Pricing and Regulatory Affairs units in the preparation of this paper.

References

Cracknell, D.R., Knott, M., 1995. The measurement of price elasticities, the BT experience. *International Journal of Forecasting*, forthcoming.

Natarajan, M., Hunger, G., 1986. *Hybrid Forecasting Models.* Paper presented to International Telecommunications Society conference, Tokyo, December.

Oftel, 1992. *Future Controls on British Telecom's Prices*, statement by the UK Director General of Telecommunications.

Park, R.E., Mitchell, B.M., 1987. *Optimal Peak-Load Pricing for Local Telephone Calls.* Rand paper.

Taylor, L.D., 1980. *Telecommunications Demand, a Survey and Critique.* Cambridge, MA: Ballinger.

Beyond Competition: The Future of Telecommunications.
Edited by D. Lamberton.

CHAPTER 8

Network management and the interconnection of networks. From the exogenous to the endogenous: the strategic dynamics of interconnection policy

Howard Williams and John Taylor*
Network and Resource Management Centre, University of Strathclyde, Glasgow, UK

Introduction

In recent years, the development of new network-management technologies has been one of the primary foci of research and development by the telecommunications service providers. Yet the implications of such technological developments continue to remain outwith the main focus of public debate and regulatory developments. It is issues such as licensing and interconnection that have dominated the regulatory debate, particularly in the UK, in the postduopoly review. The implications of the diffusion of network-management technologies as a response to regulatory change on the structure of the industry as well as the regulatory debates in themselves have been to a large extent ignored. However, it is the central thesis of this paper that these new technologies portend a major restructuring of the industry in terms of both service provision and its structure. The argument advanced in this paper is that the major telecommunications operators are seeking to reshape their businesses not around, that hitherto central axiom, the *ownership* of network elements but around the manipulation and control of network-management technologies, and in particular the new information flows these technologies support. Therefore, the implication is that to focus regulatory policy on issues such as equal access and the pricing of interconnection is to address partially the issues surrounding the development of a competitive market and the evolving industrial structure. In essence, there is a tension between the policy environment that is encouraging the development of competition and, therefore, increased uncertainty for firms, and the strategic behaviour of firms designed to reduce the uncertainties they confront. Almost inevitably, a regulatory policy that focuses on licenses, equal access and interconnection, will involve a response by firms, particularly the incumbent(s), in areas through which they might regain relative certainties. It is the contention of this paper that the implementation of network-management technologies is one such response by the major telecommunications companies.

Address for correspondence: Howard Williams, Network and Resource Management Centre, University of Strathclyde, Glasgow G1 1XH, UK. Fax: +44-41-552-6686.
*John Taylor is Professor of Management at Glasgow Caledonian University, Glasgow, UK.

The evolution of network-management technologies

Network-management technologies are a recent phenomenon; their existence is contingent on digital technologies that allow for new ways to collect, manipulate, store and communicate information. Therefore, the diffusion of new network-management technologies and their use in reshaping organisational structures, processes and decision making in individual telephone companies has been closely aligned with the processes of network modernisation which were initiated in the early 1980s.

The development of the telecommunications infrastructure around analogue technologies, step-by-step switching and transmission occurred alongside organisational structures, decision-making processes and resource allocations, particularly that of the workforce, centred upon individual network elements. Furthermore, given the inescapably geographical structure of the network, not only were resources focused on particular network elements but on network elements in specific localities. Thus, surrounding the development of the network grew organisational structures that were not only technocratic but also topocratic. That is, these organisational structures were secured and legitimated both by technical competencies and the specificity of *place*. In effect, the management of the network was devolved and placed in the hands of what might be seen as "Baronies" or independent islands of managerial behaviour. As a result, network management tended to be manual, ad hoc and nonoptimal.

This legacy of network development in terms of organisational structure and decision-making processes meant that the initial development of network-management technologies tended to be incremental and orientated towards those areas most resonant with the primary concerns of atomised element management, namely diagnostic systems. This pattern of incremental development is well documented in a recent survey by Goyal (1991). The most common network-management applications within the diagnostics domain are monitoring, troubleshooting and fault diagnosis. This bias towards diagnostic systems can be seen to reflect the pre-existing focus on network *element* management and the desire by management in particular organisational units to ensure the reliability and integrity of the network elements in their "jurisdiction". Moreover, these differentiated behaviours were reinforced by quality of service measures which were interpreted locally so that, for example, the Northeast region focused on labour productivity and Northwest region on customer service.

However, this incremental development has brought in its wake some deficiencies. Many of the deficiencies of existing network-management systems stem from a lack of integration, or at least the inability to integrate such systems and include:

- *Lack of cooperation:* related management applications which are developed independently tend to lack the basic ability to cooperate and communicate effectively.
- *Lack of flow-through:* where information is required to pass transparently from management application to management application, this cannot be achieved efficiently unless considered at the initial design stages.
- *Duplication of functionality:* independently developed management applications often incorporate similar or identical functionality. For example, user-interface

services, which are common to many management applications, are embedded within each, rather than being available as a common service to all.

- *Representation inconsistency:* independently developed management applications are unlikely to have common representation techniques. They will have been developed to suit local knowledge and needs resulting in inconsistent user interfaces. They are also unlikely to have a common input from appropriate interface experts.
- *Communications inefficiency:* where automatic communication requirements have not been designed into the management applications, attempts to do so result in efficiency limitations.
- *Abstraction inefficiency:* management functionality is often concentrated at lower levels than is desirable, resulting in a lack of control at the appropriate higher level. For example, tools exist to manage elements of the network but not to manage a network of elements.
- *Multiple input and output devices:* operators have to cope with a multiplicity of terminals and interfaces.
- *Lack of system security:* with the responsibility of management-application security residing with the application itself, there is no control upon the overall system security.
- *Inconsistent use of terms and concepts:* independently developed management applications are unlikely to have used common terms or concepts, which leads to the inability to use common information bases.

These deficiencies have been thrown into stark relief as changes in regulatory frameworks and the emergence of increasingly competitive markets have given rise to the need to manage a network as a single, albeit distributed, entity. This new emphasis has given impetus to the design and development of *integrated* network-management technologies.

Integrated network management

The basic concept behind integrated network management is to provide a coherent structure to achieve interconnection not only between network elements and their management but also between various network elements and, increasingly, various networks. Therefore, for a service provider, integrated network-management technologies can be seen to provide the main focus of service provision, new service development, quality of service, and the control of costs (through enhanced and integrated management of numerous network elements and scheduling resources).

From this perspective, integrated network management can be seen to embrace a number of functional areas, namely:

- *Design:* design encompasses the engineering and documentation of network elements, customer equipment and service provider's equipment based on specified requirements. This functional area can be decomposed into service design, network design and element design.

- *Planning:* planning activity is concerned with the organisation and control of the implementation of enterprise or business layer decisions for the introduction and management of communications systems. Its role is to bring about the introduction of services in a cost-effective manner while being cognisant of the requirements of the customer and operator.
- *Installation*: the placement of network elements and the physical and electrical (or optical as appropriate) interconnection of the network and supporting elements is the function of installation.
- *Provisioning*: provisioning functions are responsible for activating network resources that will supply the customer with their requirements. It is the first of the in-service functions of the lifecycle and, as such, is responsible for the activation of all other functional areas in that domain. The relationship between customer and provider is established at this point of contact and set out in contractual terms. Activities dealt with in provisioning are customer request for service, resource assignment, contractual negotiation and service activation. Modification to and deletion of services are also specified. This functional area also provides an interface between the customer and other functional areas, such as maintenance and accounting for the purposes of dealing with customer difficulties.
- *Maintenance*: the commonly accepted interpretation of maintenance is usually on-line corrective fault management. This concept has been expanded, with particular consideration being given to preventive and surveillance-based maintenance, e.g., as described in Comité Consultatif International de Télégraphique et de Téléphonique (CCITT) Recommendation M.20. Scheduling of maintenance tasks and the logistics of workforce management are of central importance here.
- *Performance management*: performance management is employed to regulate the network in order to maintain the quality of service levels agreed with customers. This is an increasingly important area as new network configurations, such as broadband and ATM, offer the potential for multiple services and quality levels. The cost-effective optimisation of resources is also an objective for this functional area. The activities undertaken in performance management are monitoring, analysis and problem alerting, diagnosis, optimisation and control.
- *Security management*: security management is defined as being the provision of security services for customers, such as authentication services, nonrepudiation, integrity and confidentiality services. Authentication services involve the transmission of given criteria that can be identified as valid by cooperating entities; password recognition is one such service. Nonrepudiation services are concerned with providing users with proof of the origin or delivery of data or its contents. Integrity services allow the originator of the data to provide confidence to the recipient that the data has not been modified. Confidentiality services provide an assurance that data transmitted in a particular instance of communication has not been modified, which also includes the data not having been made available to unintended recipients.
- *Accounting management*: the use of network resources and services is measured

by accounting management functions for each customer and the customer billed accordingly. Such use-records and tariffs are used in the calculation of charges to customers who are presented with invoices at predetermined intervals, or in terms of predetermined credit limits. Billing intervals and payment methods are agreed at the contractual phase of the service-provisioning activity. Cost accounting is carried out to control costs, revenues and tariffs. Inter-operator accounting is also part of this functional area.

– *Customer query and control*: the facility to allow large and sophisticated customers access to particular areas of network-management functionality is expected in advanced network environments, such as an integrated broadband network. Facilities envisaged are reconfiguration of the customer access (by the customer), queries to accounting and other access similar to those available to the operator. The security aspects related to this facility tend to be considered as being of sufficiently high importance as to warrant a separate functional area.

The development of a particular integrated network-management system is well illustrated by the case of British Telecommunications (BT). Building from a variety of sources of work (some of which are set out in Appendix 1), BT has sought to create a framework for integrated network management. This framework is illustrated in Fig. 1 and is referred to as a cooperative networking architecture management (CNA-M). One of the fundamental principles embodied within CNA-M is the partitioning of management functionality into distinct roles and the ability to conceptualise service delivery separately from the ownership of network elements. A summary of the CNA-M model is provided by Strang and Troullides (1991):

> "BT's own CNA-M provides useful suggestions as to how the anticipated complexity of functions could best be structured. One of the fundamental principles embodied within CNA-M is the partitioning of management functionality into distinct roles. These roles correspond to the various layers of the CNA-M model. These layers are represented in a cone structure....

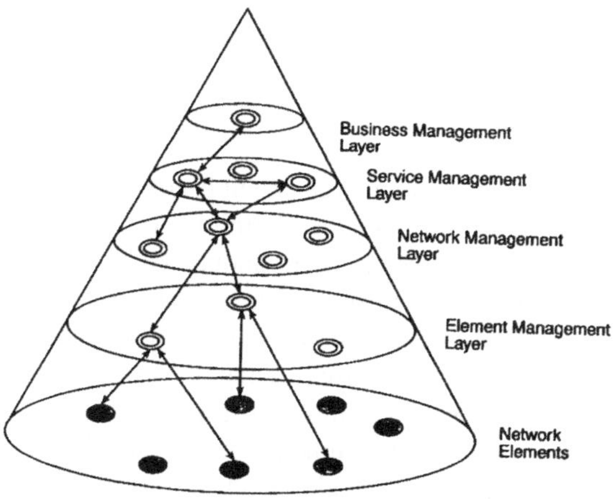

Fig. 1. Cooperative Networking Architecture Management.

> Working from the bottom of the cone, the network elements are the physical equipment which the management system manages. The network element management layer is then concerned with those processes which have a network-wide view of this system.
>
> Within the service management layer the management of networks is split from the management of the services which they support. This division assumes that services are provided across networks and that these networks can, and should, be managed independently from the services which they supply. Therefore service management builds upon the underlying support provided by the network and network element management and provides a technology-independent view.
>
> The most strategic type of management decisions are addressed by the business management layer. These include decisions about the levels of service provision, priority and so on. These are best separated and classified as aspects of management which are dedicated to a particular business organisation.
>
> These management layers lie stacked one upon another, and help to partition responsibilities and provide a further classification which aids the construction of operations systems for management systems." (pp. 70–71).

The development of integrated network-management technologies offers the potential for the structural separation of network elements from network management and service delivery. This separation is well illustrated in the case of CNA-M. Importantly, from a service delivery perspective, integrated network-management technologies facilitate the interlinking and management of service delivery across separate networks. It is these key characteristics of integrated network-management technologies that challenge both the existing organisational frameworks for service provision and industrial structure. Moreover, it is in this context of the development of integrated management technologies, processes of organisational change and the broad reshaping of the industrial structure of the telecommunications sector that the debate over interconnection can be located.

Interconnection

The current policy debate over interconnection has to a large extent become colonised by accountancy; the issues have become centred on the definition of costs, the allocation of costs by service and the way in which costs should be distributed between different network operators. Complex in themselves, these issues are further complicated by the legacy of a pricing structure in which costs and tariffs are out of balance on a service-by-service basis, and asymmetric regulation which places particular obligations on the incumbent, most notably universal service and quality of service standards. The way in which costs of the incumbent operator are allocated and its tariffs restructured is a critical issue in the development of a multivendor, multiservice environment. As evident from the experience in the UK, new market opportunities for entrants are to an important extent based on their ability to arbitrage between the prices of the incumbent and their own costs. It is in this context that interconnection rates become a critical issue. The argument of the regulator, and that of new entrants, is that interconnection, and in effect equal access, should allow new entrants access to network elements on the same terms and conditions as those which the incumbent provides services to itself, and thereby enjoy the benefits of economies of scale and scope generated by the incumbent's network. Underlying this position is the implicit argument that the advantages of economies of scale and scope that the incumbent enjoys are the consequence of previous public policy decisions and should therefore be equally available to new entrants.

Inevitably, there is a rich debate as to how to define the framework for establishing interconnection charges. If interconnection prices are set too high, then incentives to invest in new networks will be relatively greater, and if they are set too low, the incentive to invest will be diminished. Given a liberalised and competitive environment which is largely predicated on facilities-based competition (such as in the UK, USA and Japan) then the incentives to invest are critical to the achievement of broader policy objectives. If the investment incentive structure, reflected in interconnection prices, is distorted then the possibility of "market failure" is considerable. In evidence to the House of Commons Select Committee on Trade and Industry on 13 April 1994, Lord Young commented that the current framework for interconnection charges in the UK is giving the wrong price signals, and as a consequence Cable & Wireless (Mercury) have been "encouraged" to invest at least £100 million more than it would otherwise have done in building its core network in order to overcome "excessive" interconnection prices. If this critique of the interconnection pricing regime can be sustained, then questions must exist over the economic viability of the current investment commitments (a total of some £4–5 billion) being made by new entrants into the UK telecommunications market, most notably by the cable television franchises.

A number of approaches to defining a framework for interconnection charges have been taken. One approach is to define the problem in terms of incremental costs of supplying an individual access link into the network. Another approach is to use public finance theory to develop taxes on interconnection in order to subsidise universal service. Alternatively, some argue that the problem of pricing interconnection should be decided within a framework of intellectual property rights. A pragmatic approach can also be used to establish interconnection prices, whereby prices in the first period reflect the desire to establish a competitive market and therefore would tend to be below cost, and that over time or as market shares increased so interconnection prices would be increased to reflect true incremental costs.

Defining the interconnection debate, and those other policy decisions which are linked, such as equal access and funding universal service obligations, largely in terms of the accountancy paradigm, fails to consider the dynamics of the industry and the inter-relationships between structure-conduct-performance and processes of technical change.

In a game-theory approach the current definition of the interconnection policy debate can be seen as analogous to those situations where it is assumed that the costs of new entrants are largely exogenous to the firm, i.e., the set-up costs are associated with building a network to deliver a homogeneous product. Moreover, there is an assumption that for any firm, given the same technical choices, the costs of the building network would be the same, *ceteris paribus*, equal. In effect, the assumption made is that all new entrants face the same costs and that economies of scale are unimportant as a constraint on the equilibrium structure of the industry. In some respects the issues surrounding the nature of economies of scale in the telecommunications industry remain contentious but there is common agreement that the existence and nature of economies of scale is not such that they sustain the argument for natural monopoly. Contemporary theoretical and empirical investigations as well as changes in the nature of telecommunications technologies have tended to question the level of economies of scale in telecommunications networks. Such debates have been mirrored at the political level where the arguments of a natural monopoly, sustained by economies of scale, have been undermined, thereby sustaining, in part, the development of competition policies in the telecommunications sector. Thus, the assumption of the relatively diminished importance of economies of scale in defining the equilibrium industrial structure may not be unreasonable in the context of the current telecommunications industry.

Where costs of entry can be seen to be exogenous, then as the market expands so too will the equilibrium number of firms increase and so industrial concentration will decline indefinitely. Thus, such a theoretical model offers the possibility of transforming highly concentrated (*in extremis* monopolistic) markets into increasingly competitive markets. This transformation of industrial concentration is perhaps even more marked in those situations where the monopolistic industrial structure was initially enshrined in legislation, as is the case in the telecommunications sector. In a game-theory perspective, where costs are assumed to be exogenous, such an outcome is inevitable because in round one the behaviours of the incumbent cannot change the distribution of profits achieved in round two. In effect, the escalation of outlays by firms in round one in order to gain competitive advantage in round two are relatively futile.

An elaboration of this relatively simple model, and one which may well parallel the telecommunications industry, is where there is product differentiation or consumers are spread geographically or both, their consumption is, to some extent, a function of this geography (cf. a locational model of the Hotelling kind). In such cases, consumers can still be seen to be seeking lowest costs, but here costs include not only the price paid to the firm but also 'transport' costs, i.e., those psychic and actual transport costs involved in purchasing from distant suppliers. Under such circumstances multiple equilibria are endemic. In general, for any given market size, a fragmented equilibrium tends to exist in which a large number of firms sell in *one* location and a small number of firms sell in *many* locations. Such a description appears to characterise the current UK telecommunications market; relatively numerous new network operators providing services in particular locations and relatively few who are providing facilities-based services in numerous locations.

While recognising that regulatory and licensing conditions are also shaping this outcome, it can be argued that the current industrial structure is perhaps one that will survive the easing of these legislative conditions. Through product differentiation, enhanced quality of service and competitive pricing structures the cable television companies have been able to exploit specific territorial markets. For example, product differentiation, in particular the bundling of TV and telephony into a single service offering, has been used as a market entry strategy. In some circumstances this market entry strategy has meant that initial access to the telephone network has been "subsidised" by the operator. Evidence from the Office for Telecommunications (Oftel) suggests that as a result of this product differentiation there are now some 377,000 new telephone subscribers, one in seven of whom were previously "untelephoned". (As of April 1994 there were 642,000 subscribers to cable networks in the UK.)

While these two outcomes are predicated on the assumption of exogenous costs, more complex and unpredictable outcomes emerge if the costs faced by firms are assumed to be endogenous, i.e., related to activities such as R&D and advertising. In effect, the assumption of endogenous costs recognises the existence of information asymmetries between firms and that the availability and nature of information is integral to the behaviour of the firm, its products, prices and quality. Under such circumstances, the behaviours of individual firms, in response to changing market characteristics, are inevitably both different and differentiated. As a result the performance of individual firms is contingent on the way in which information asymmetries manifest themselves in differentiated firm strategies and behaviours in the market place. Therefore investments, an escalation of outlays, by an individual firm in one period can produce an uneven distribution of profits in subsequent periods. If costs are seen to be endogenous then the behaviour of individual firms, strategy development, investment in novel technologies as well as the nature of intra- and inter-organisational behaviour begins to influence the evolution of the industrial structure.

The definition of costs as being largely endogenous to specific firms opens up a complex interconnection debate as the regulatory decisions are inevitably interacting with the development of firm-specific strategies and the dynamics of the market place in order to change the distribution of profits in subsequent periods. What makes the situation more complex is that while information asymmetries can be seen to exist between firms, especially between the dominant firms and new entrants, regulators also confront considerable information asymmetries. Under such circumstances, where costs are defined endogenously to individual firms, regulatory decisions can only be seen as part of a political process where the policy objectives are shaped and reshaped by inter-dependent relationships between actors in the policy community.

The policy implications from this approach of understanding changing patterns of industrial organisation, and in the context of the telecommunications sector the interconnection debate, rests on whether costs are exogenous or endogenous to the firm, particularly for new entrants. It can be argued that costs in the telecommunications sector are moving from being defined exogenously to being endogenous.

This shift in the nature of costs is occurring in the telecommunications industry as operators seek to tailor their services to the needs of individual customers and the nature of products and services moves away from being based on standardised hardware to bespoke software services. Such is the change that is occurring that, for example, Cable & Wireless expects half of its turnover to come from new software products by 1999. Moreover, related to this growing predominance of software products and services is the reduction in the period of product lifecycles from decades to at most several years.

Therefore decisions over the nature of the interconnection arrangements between networks inevitably raise important questions over the development of the industry and not just about the appropriate allocation of costs. Such a perspective opens up a way to review the nature of interconnection, the behaviours of firms and the resulting changes in the industrial structure.

Network management, service provision and industrial structure

Previous sections of this paper have set out some of the characteristics of integrated network-management technologies and a conceptual understanding of the nature of the regulatory game, in particular the way in which the distribution of profits in periods subsequent to round one can only be influenced if costs are assumed to be predominantly endogenous. The purpose of this section is to explore the ways in which, in a competitive environment, integrated network-management technologies are transforming the nature of costs incurred by any one firm away from the exogenous to the endogenous and how incumbents may develop strategies to reduce regulatory and competitive uncertainties. In so doing, this section points out the complexities of the debate over interconnection and, by implication, the relative narrowness within which the contemporary debate has been cast, i.e., largely within accountancy frameworks.

Multiple services from a multitude of service providers is one of the key characteristics of the contemporary telecommunications environment. Multiple vendors is the primary policy objective in some countries and, for example, in the UK the choice between at least three suppliers is becoming the foundation stone of a new universal service policy. However, running contrary to this supply-driven proliferation of services and service providers is a desire, on behalf of consumers, to see the integration of service provision and, therefore, a reduction in the transaction costs associated with exercising competitive choices. In part, the development of integrated network-management technologies is a response to this newly competitive market and perceived consumer demand. The development of integrated network-management technologies allows for the provision of multiple services across different networks and such service offerings can be managed and controlled independently from the network elements on which they depend. Therefore, the provision of multiple services to any one customer is capable of being delivered independently of the ownership and control of network elements by the service provider. Thus, a focus on market development has begun to move the emphasis of the PTOs away from the construction of networks, in particular geographically defined markets, to a focus on the needs of particular customers and markets defined in terms of their communications needs, e.g., the oil and gas industry, and money-transmission services for retailers.

The transformation of telecommunications networks from a basic, largely analogue, infrastructure only providing voice services to a "nervous system" based on new digital technologies providing an increasing array of services critical to the economic development and success of firms, regions and countries has also brought about fundamental changes in the organisation and management structures of the telephone companies. Thus, while technical changes have brought about a switch from analogue to digital technologies, organisational changes have sought to reorientate inward-looking bureaucratic hierarchies towards more customer-centred and outward-looking cultures. In this process there has been a shift away from the control and management of costs, largely set by the capital costs of equipment and nationally negotiated wage rates, to a focus on revenues, customer needs and demands. In this change process there is an inevitable rebalancing of costs from the exogenous to the endogenous as the understanding and interpretation of customer needs as a basis of business performance supplants the basic need to provide dialtone.

Using BT as an exemplar, its current phase of organisation restructuring, based upon Project Sovereign, can be seen as an attempt to move the company away from its historical *technocratic* and *topocratic* structures towards an organisation focused upon customers and markets. Thus, as illustrated in Fig. 2, the organisational structure of BT is moving towards a "customer-driven" approach where the traditional pyramids of separate administrative and technical hierarchies are positioned in support of customer-facing activities rather than dominating the behaviour of the organisation. This layered model of the company can be seen to be resonant of the conceptual model set out in the CNA-M scheme (Fig. 1). The changes that this new organisational form are bringing about are profound and manifest themselves throughout the company. At a general level the company has embarked on a process of "business process re-engineering". This process has been closely aligned with attempts to reshape the organisation and to eliminate organisational structures, decision-making processes and activities which have left a legacy of relatively high costs. This process of organisational transformation has been underpinned by the development of new information systems of which integrated network-management systems are playing an increasingly important role. (For a detailed study of "business process re-engineering" in BT see Taylor and Williams (1994).)

In the case of integrated network-management technologies, BT has through their implementation been able to break free of the pre-existing organisational structures centred on management of network elements which have hitherto imposed substantial costs on the company, especially labour costs. Given that headcount (staff-in-post) is one of the most powerful and easiest measures of efficient performance, and that it is a measure widely used by the investment community, any reduction in headcount benefits the company in a variety of ways. For example, reducing headcount helps

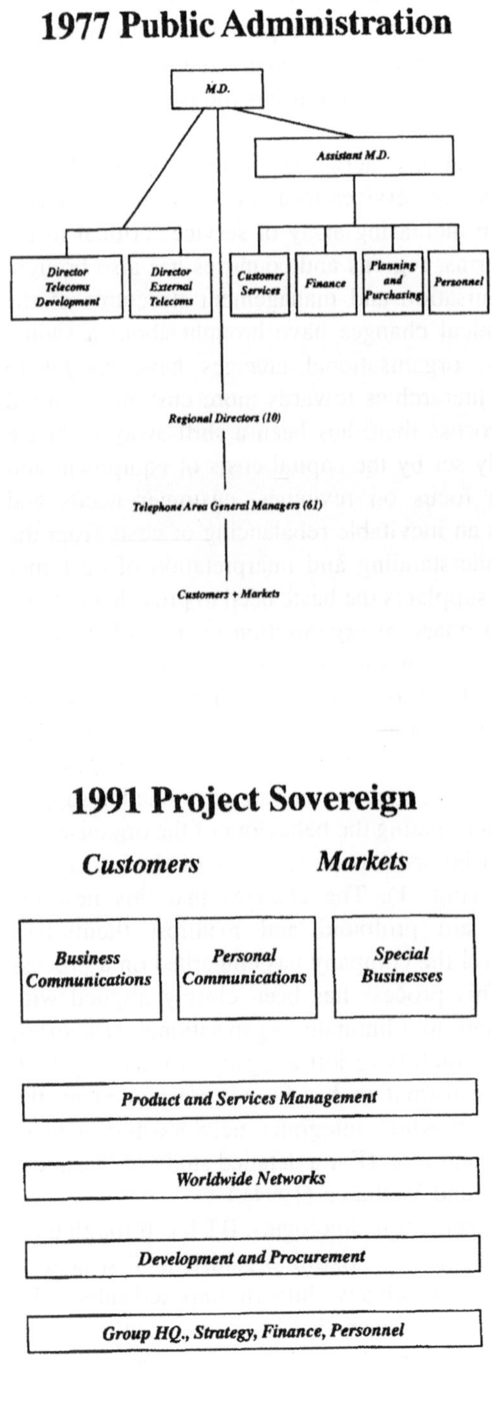

Fig. 2. Design archetypes in British Telecommunications in 1977 and 1991.

to create a virtuous (financial) circle where lower overhead costs lead to increased margins and profitability, which in turn leads to lower costs of capital which in itself feeds into better financial performance. This strategy is clearly in place and given that a substantial proportion of the current staff with BT are devoted to the management of network elements, the automation and integration of these functions will bring about substantial reductions in staff. Recent evidence suggests that BT is anticipating job reductions of some 55,000 over the next 5 years with an estimated annual saving in overhead cost of £1.1 billion (*The Guardian*, 10 June 1992). In the 10 years since privatisation, some 70,000 jobs have been shed by BT. In their 1993 annual report, BT gives details of the numbers of staff and costs involved in restructuring to date, as well as signalling the need for a further 15,000 jobs to go, particularly amongst the senior management cohort. Moreover, unofficial commentary suggests that BT is intent on reducing its staff numbers to 100,000. The organisational changes brought about through the introduction of integrated network-management technologies does not necessarily result in a changing balance between endogenously and exogenously defined costs, but it does undoubtedly bring the costs of BT into close alignment of those of its competitors who, in many cases, have exploited the advantages of being a new entrant and established lower costs than BT.

Furthermore, organisational changes have resulted in the strategic refocusing of business units. For example, the remit of the research laboratories at Martlesham Heath has evolved from "national" research laboratory focused on fundamental research through the exploration of novel technologies to the current emphasis on "Realising the Potential" (sic) of new techniques. In effect the focus of the laboratories has been reorientated from *research* into network elements and new technologies to a focus on customer needs and the array of *techniques* available to satisfy these needs in novel services and products. Within this new focus on techniques there is also a concern to use research and development to create regulatory uncertainties, for example, through a "regulatory bypass" strategy such as that embodied in "video-on-demand". The objective has become the close alignment of research with new revenue streams or reducing costs or both, as well as to destabilise the regulatory regime. In the context of the arguments in this paper, this alignment of research and development expenditures with the business needs of the company changes the characteristics of such costs, increasingly with the proportion than can be seen to be defined endogenously. Moreover, this shift in the definition of costs away from the exogenous can be seen to be occurring throughout the company as processes of organisational change seek to reshape the business around understanding consumer demand rather than the provision of dialtone.

The diffusion of network-management technologies is also laying the foundations for a radical restructuring of the industry in that it allows for a focus on service delivery and network management to be divested from the ownership and management of network elements. In such a situation, the interconnection debate becomes one not merely concerned with the manner and terms with which new entrants interconnect with the dominant incumbent service provider but a policy debate focused on the nature and structure of the industry itself. As integrated network-management technologies become established as the platform from which to deliver services, then the interconnection of networks becomes dynamic and key issues arise over the flows of information which are central to such novel technologies. The growth of integrated network-management systems is resulting in the flows of information across the network becoming the new foundations of market power and not the ownership of network elements. In effect, it is the software, new flows of information and its analysis that are critical, not the ownership of assets.

Such a perspective raises interesting and perhaps speculative questions over the future of the telecommunications industry. If network-management technologies do actually result in the transformation of service delivery then the requirement of operators to own and manage network elements is diminished. In such circumstances the issues become centred on access to bandwidth and the quality of service provision. Thus, the argument can be advanced that through network-management technologies the emergence, or perhaps maintenance, of a dominant carrier becomes highly likely. However, the dominant carrier may not necessarily be the owner of substantial network elements. Perhaps a precursor to this outcome is the current calls by some of the new entrants/operators in the UK to link regulatory freedom for BT, particularly lifting cable-television restrictions, in exchange for interconnection arrangements which see a further development of the division between the wholesaling and retailing activities of BT. The wholesaler would in effect become the national bit-hauling carrier; the retailing division having to defend its markets in a newly and considerable intensified competitive arena.

Related to this potential for divestment of network elements is a strategy to move away from the ownership of network elements in specific localities yet retain effective control of service delivery through integrated network-management technologies. In many geographically defined markets the costs of operating the network within the structure of a large telecommunications service provider and the relatively low levels of revenue, from say residential users, sit uncomfortably alongside the opportunities in other geographically or service-defined markets. Moreover, these more problematic markets which lack competition because of their structure sustain the interest and concern of the regulator in the entire operations of the dominant provider. Thus, the apparent surrender of these areas to other companies through the transfer of network *elements* can be seen to create regulatory slack for the dominant supplier, while at the same time, through the use of network-management technologies, maintaining their dominant position.

In short, taken together these threads can be seen to help BT and for that matter other telecommunications service providers in escaping from the autocracy of the telecommunications regulators. The development of integrated network-management technologies is allowing the appearance of new telecommunications competition, fuelling the policy objective of "competition and choice" yet allowing the established telecommunications operators to re-establish their dominant position.

Appendix 1

The development of integrated network-management technologies has been of interest to a number of research institutions and projects as well as standard-setting bodies. While all have tended to accept the basic functional areas set out above, there are a number of different approaches.

At an international level both CCITT and the European Commission, under the auspices of RACE, are supporting the development network-management standards. The CCITT view is embodied within recommendation M30 (Principles for Telecommunications Management Network). While this recommendation is seen as providing details on the general principles for planning, operating and maintaining a network-management system it does have its limitations, in particular, the lack of a defined structure within the operations system and therefore the need for other sources of such information.

Within RACE a number of different but related approaches have been adopted. This development of network management is centred upon several projects, such as NETMAN, ADVANCE, AIM, NETSYS. A detailed discussion of these approaches to network management can be found in Mamdani (1991).

Within the UK, the Alvey project ANSA sought to organise the network-management functionality based on distributed systems which behaved as a unified entity. The ANSA project developed four representations of the problem domain, namely:

- *The enterprise model* which specifies the overall objectives of the system in terms of activities, people and policies of the organisation of which it is a part.
- *The information model* which describes the elements, rules and relationships between the information elements of the system. No discrimination is made, at this level, between those parts which may be automated and those which are human.
- The modelling of *the operations for information transfer*, storage and retrieval for an automated system are described in the computational model.
- *The engineering and technology models* which describe the physical distribution and realisation of the computational model.

References

Bernstein, S., 1987. DesignNet: an intelligent system for network design and modelling. In: Sassa, D.J. (ed) *International Communications Conference*. New York: Institute of Electrical and Electronic Engineers.

CCITT, 1988. Blue Book M30. *General Principles for Telecommunications Management Network*. Comité Consultatif International de Télégraphique et de Téléphonique, Geneva.

Del Monte, A., 1992. *Recent Developments in the Theory of Industrial Organisation*. London: Macmillan.

Goyal, S.K., 1991. Knowledge technologies for evolving networks. In: Krishnan, I., Zimmer, W. (eds) *Integrated Network Management II*, (pp. 439–461). Amsterdam: North Holland.

Guattery, S., Villarreal, F., 1985. NEMESYS: an expert system for fighting congestion in the long distance network. *Expert Systems in Government*. Institute of Electrical and Electronic Engineers, October.

Mamdani, E.H., 1991. Guideline – RACE Project R1003, AIP and Standards for TMN. European Commission, 03/BTR/712/DS/B/009/b1, Brussels.

Mantleman, L., 1986. AI carves inroads: network design, testing, and management. *Data Communications*, July:106–123.

Prerau, D., Gunderson, A.S., Reinke, R.E., Goyal, S., 1985. The Compass Expert System: verification, technology transfer, and expansion. In: *Second Conference on Artificial Intelligence Applications*, (pp. 597–602). Washington, DC: Institute of Electrical and Electronic Engineers.

Strang, C.J., Troullides, Y., 1991. Development of network and customer administration systems. *BT Technology*, 9(3):68–79.

Sutton, J., 1992. Implementing game-theoretic models in industrial economics: levels of attack. In: Del Monte, A. (ed) Recent Developments in the Theory of Industrial Organisation. London: Macmillan.

Taylor, J., Williams, H., 1994. The transformation game: information systems and process innovation in organisations. *New Technology, Work and Employment*, 9(1):54–65.

Thandasseri, M., 1986. Expert systems application for TXE4A exchanges. *Electrical Communication*, 60(2).

Vesonder, G. et al., 1983. ACE: an expert system for telephone cable maintenance. In: *8th International Joint Conference on Artificial Intelligence*, (pp. 116–121). Menlo Park, CA: American Association for Artificial Intelligence.

Walles, A., 1991. Functional descriptions of network management. *BT Technology*, 9(3):9–17.

Beyond Competition: The Future of Telecommunications.
Edited by D. Lamberton.

CHAPTER 9

Network evolution and prospects for switched video service in UK subscriber networks

Yogesh Sharma
Department of Economics, Brunel University, Uxbridge, Middlesex, UK

1. Introduction

In September 1993 British Telecom (BT) announced its intention to offer interactive services over its existing copper subscriber network. *Asymmetric digital subscriber line* (ADSL) is a compression technology allowing a bandwidth of up to 2 Mbit/s over an average length of standard copper pair. Having successfully conducted technical trials of *video-on-demand* (VOD) to the homes of 60 BT engineers, BT intends to offer an extended commercial trial to 25,000 homes by the end of 1995 offering access to banking, retail, education and movies-on-demand.

BT is also experimenting with ATM switching over fibre to provide a range of integrated voice, video and data services. Therefore the promotion of multimedia services over ADSL-enhanced metallic subscriber networks may represent a transitional approach to enhancing local networks since in the long term, switched (interactive) services could equally be proposed in conjunction with a range of *fibre-in-the-loop* (FITL) schemes, and offering significantly more bandwidth than ADSL-copper.

This raises a number of issues from the viewpoint of operators attempting to offer new services and of governments charged with regulating markets which independently generate new technical alternatives. What is the best approach to modernizing subscriber networks given the current scale of interest in broadband technologies and integrated digital networks? Should greater functionality or bandwidth be offered in the context of slightly improved existing infrastructure or radically different FITL programs? Will the market respond sufficiently to justify the financial implications of offering interactive services using either ADSL or FITL technologies?

Section 2 describes general developments in UK subscriber networks. Section 3 identifies alternative networks based on ADSL, hybrid and all-fibre technologies. Section 4 discusses ADSL technology and how UK regulation has responded to it. Section 5 analyses general prospects for interactive service provision followed by conclusions in the last section.

Address for correspondence: Yogesh Sharma, Department of Economics, Brunel University, Uxbridge, Middlesex UB8 3PH, UK. Tel.: +44-895-274000. Fax: +44-895-203384.
e-mail: Yogesh.Sharma@Brunel.ac.UK.

2. UK subscriber networks

Network architectures in telecommunications represent a traditional engineering compromise between transmission and switching. The design of local loop lengths are limited to a few kilometers due to the inherent problems of signal distortion in copper wire. The development of amplification techniques still preserve the need to balance transmission and switching but enable longer line links to manage higher load factors (more bandwidth) than were previously possible in older networks. Overall, network operating costs have both declined and expanded in relation to lower unit costs of transmission and higher costs associated with switching and multiplexing carrier signals.

Network modernization

In the UK, British Telecoms's (BT) local network is comprised of about 7,500 semielectronic analog switches and fully electronic digital switches, which may be either stand-alone or remote concentrator units. Small groups of *local exchange* (LX) sites are linked by regional *digital cell center exchanges* (DCCE). DCCEs further connect to 64 *digital main switching units* (DMSU) which effectively constitutes the national trunk network connecting major cities.[1] The majority of DCCEs and all DMSUs are electronic and use *System X* SPC switches. Table 1 describes the composition of full switches, concentrators or remote units in BT's network. Table 2 describes the percentage of BT subscriber connections by exchange type.

Since the mid-1980s BT has replaced the vast majority of its older LXs by modern *stored program control* (SPC) systems. It is anticipated that 2.7 million access lines will be connected to some sort of fibre by 1997 suggesting only a gradual migration towards fibre access (Payne, 1993). This further suggests that fibre deployment plans will largely be made on a cost justifiable basis and mainly for large business users as optical component and system costs continue to fall.

Therefore despite advancements in switching, control and transmission, copper-enhancing technologies like ADSL will continue to ensure copper will remain a predominant access technology in subscriber networks over the next 10 years. One

Table 1. Composition of UK local exchanges.

Digital exchanges
3600 System X (550 processors, 3050 concentrators)
1000 System Y (150 processors, 850 concentrators)
Semielectronic and analog switches
800 Strowger
500 TXE4s
1600 UXD5s (small rural switches for 30–40 lines)

Source: Industry source

Table 2. BT subscriber connections by type of exchange (%) (Williams and Taylor, 1994).

Exchange type	1985	1987	1989	1991	1993
Digital	1	2	23	47	64
Electronic	33	41	41	37	32
Crossbar	–	17	15	–	–
Strowger	–	31	21	–	–

estimate suggests that the use of copper pair transmission media will continue to account for a significant share of worldwide *new* subscriber connections, from 66% in 1993, falling to 8% by approximately 2003 (Matthews and Darabi, 1994).

While fibre deployment continues to remain a costly proposition, BT is still deploying fully managed fibre networks primarily, although not exclusively, for large business users. The plan involves installing *Flexible Access Systems* (FAS) by first laying fibre rings in urban centers (such as the City Fibre Network in London's financial district) in order to establish a reliable primary network. This is followed by connecting users via a secondary distribution network of "blown" fibre. A service access link provides the final broadband connection between the subscriber and the distribution network. FAS networks optimize the design period since users may be connected "just-in-time" thereby enabling the deferral of critical (and costly) capital expenditure. Alternatively, either wireline or wireless technologies may be used to connect subscribers to the point at which high capacity fibre ends, usually at a primary distribution point. However, since BT is prohibited from entering the mobile business, it is unable to use either CT2 or DECT standard wireless technologies.

The progression from narrowband to broadband service provision represents a "design period" in which the telecommunication network becomes a communication network based on optical fibre architecture. However, in the UK the notion of local fibre loops remains rather more conceptual than practical, except in the trunk network which is entirely replaced by fibre. Over a period of 10 years since 1984, BT has increased the content of fibre in its networks from under 500,000 km to some 2.4 million km, with an estimated 300,000 km in the local network.[2] In June 1992, Oftel proposed that BT should install around 3.5 million km of optical fibre in its network by 1997.

And despite recent interest in announcements BT made intending to invest some Ecu20 billion (approx $US23 billion) in constructing local fibre loops covering the whole of its access network, such enthusiasm could be moderated by regulatory restrictions currently preventing it from delivering entertainment services (see section 4). This view suggests that unless the Government overturns an entertainment-delivery ban on BT sooner than the first proposed review in 1998, the intervening period will see only large businesses benefiting from fibre technology. Yet despite the restriction on delivering certain services, the ban does not necessarily rule out the provision of interactive services on either currently available ADSL technology or future fibre-based access networks.

BT services

The range of existing and future service applications proposed by BT are consistent with a broadband platform in the context of providing integrated voice, data, video and related high-bandwidth applications on demand. Equivalently, access to a number of "treatment facilities" within the network ensure stable and error-free transmission paths between terminal nodes. Table 3 provides a list of current and proposed narrowband and broadband ISDN (NISDN and BISDN) telecommunications-based applications offered by BT, along with associated network-based treatment facilities.

However, despite predicted deployment rates of new enabling technologies employed by BT and other network operators, opinion still suggests that copper will continue to serve as the main access medium in the drop loop at least until early into the next century (Allen and Marshall, 1993; Matthews and Darabi, 1994; Rowbotham, 1991; Valdar et al., 1992). Yet as a consequence of unifying switching nodes in the primary network of LX sites and secondary exchanges (DMSUs) the opportunity to deploy fibre in the loop becomes evermore realizable.

Table 3. BISDN telecommunications-based applications and network-based treatment facilities (Allen and Marshall (1993)).

Applications	L	M	S
Megastream	/	/	
ISDN30	/	/	
Fractional rate ISDN (part ISDN30)	/	/	/
Fractional rate ISDN (n*64)	/	/	/
ISDN2	/	/	/
Flexible bandwidth service	/	/	
Featurenet1000	/	/	/
Featurenet5000	/	/	
SHDS (2Mb/s)	/	/	
SHDS (34Mb/s)	/		
Frame relay	/	/	/
Kilostream	/	/	/
Analogue PC	/		/
Analogue PSTN	/	/	/
Callstream	/	/	/
Linkline	/	/	/
International private leased circuit	/	/	/
Managed links service	/	/	/
Treatment facilities			
Digital exchanges			
Network control facilities for private circuits			
Managed bandwidth service multiplexors			

Note: L = large site >34 Mbit/s; M = medium site 2–34 Mbit/s; S = small site <2 Mbit/s.

Telephone networks

The local loop analogy offers a stylized representation of a local telephone network. From an historical perspective, early telephone engineers conceived this part of the network as a *loop* of metallic wire connecting the LX to each subscriber and back again thereby establishing an electrical closed circuit (Calhoun, 1992). Figure 1 presents a simple local loop topology indicating the feeder network, the distribution network and the service access network. Traditionally, knowledge of switching and transmission costs have guided network planners in constructing the most appropriate network topology for narrowband (telephony) and broadband (cable television) service delivery. Common physical layouts used in constructing new telecommunication networks are *star, double-star, bus* (or *tree-and-branch*) and *ring* topologies.

A star network (Fig. 2) is characterized by offering a dedicated link connecting every subscriber to the LX. The presence of dedicated links ensures an efficient delivery of switched services when transmission costs are low. The star network also ensures maximum privacy since the delivery of bandwidth capacity is not shared among subscribers. The double-star (Fig. 3) network economizes on the amount of cable used compared to the star network by sharing the feeder part among different subscribers connected to the same exchange. The feeder segment consists of the line links between the LX and a remote distribution point such as a remote distribution unit (RDU).

The concept of shared capacity in the feeder is achieved by the presence of electronics in the LX and at the RDU which multiplex separate signals onto one channel, ultimately resulting in a reduction in the amount of cable needed in the feeder network. Beyond the RDU flexibility point, the network resembles the star network providing each subscriber with a dedicated link to their specific RDU. Despite the expense attached to installing RDUs, double-star networks still retain efficient transmission economies and can handle both switched (telephony) and distributive one-way services like cable television.

The bus network (Fig. 4) achieves yet further plant sharing compared to the double star configuration by linking the RDUs in series. In the case of three connected RDUs, the third and furthest RDU can only switch signals if a given transmission cable can provide sufficient capacity to carry unique channels to each subscriber. Economies of plant sharing are achieved by invoking *pair-gain* techniques that

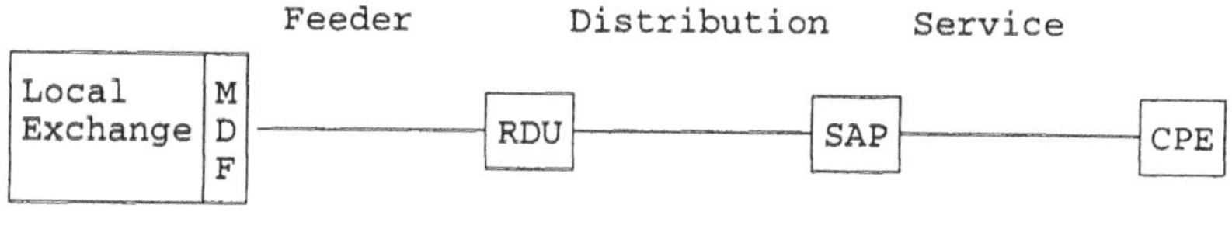

MDF Main Distribution Frame
RDU Remote Distribution Unit
SAP Service Access Point
CPE Customer Premises Equipment

Fig. 1. Local access network topology.

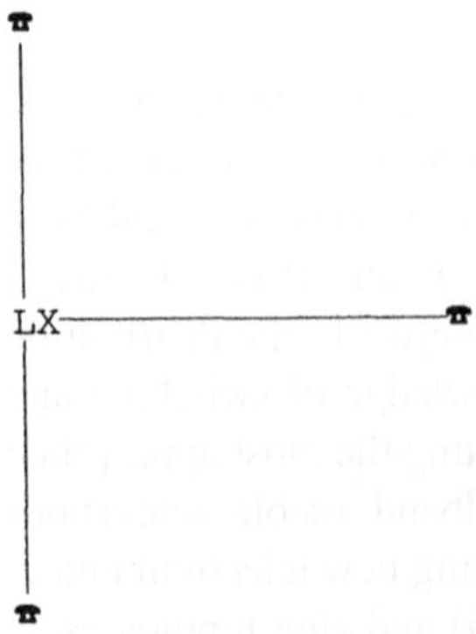

Fig. 2. Star network.

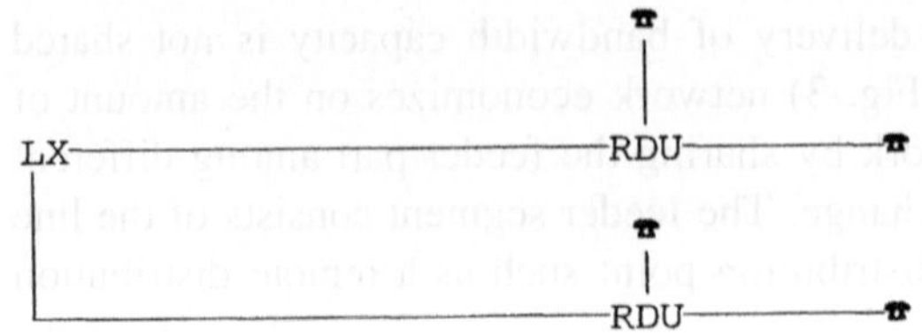

Fig. 3. Double star network.

multiplex signals in the feeder part of the network. The ring layout (Fig. 5) provides a closed loop configuration through which each subscriber is connected to the LX.

Star networks are more applicable in cases where signal transmission is switched or communicative (meaning two-way) like standard voice telephony.[3] Therefore in star-type telecommunication networks dedicated to voice-type service provision, the switching function within the network takes a relative priority over transmission. This explains the tendency for telephone networks to employ a relatively low-value medium in copper wire for signal transmission, but using relatively high-value switching equipment in the LX. Star networks are also suited in circumstances where subscribers are located close to the LX (Reed, 1991).

Due to the nature of infrastructure sharing in ring and bus-type network topologies, these network are "more likely" to be used for *distributive* (meaning one

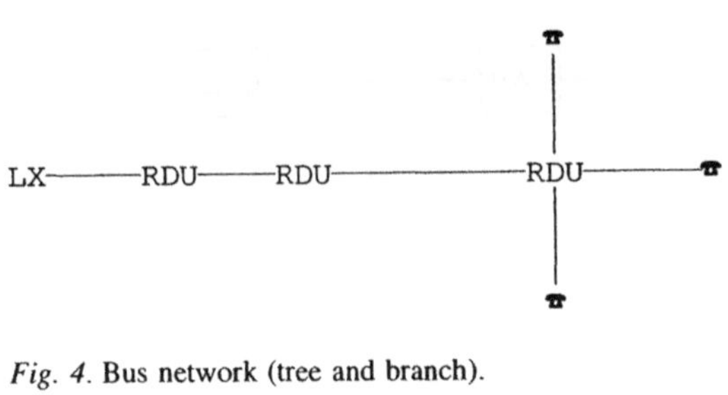

Fig. 4. Bus network (tree and branch).

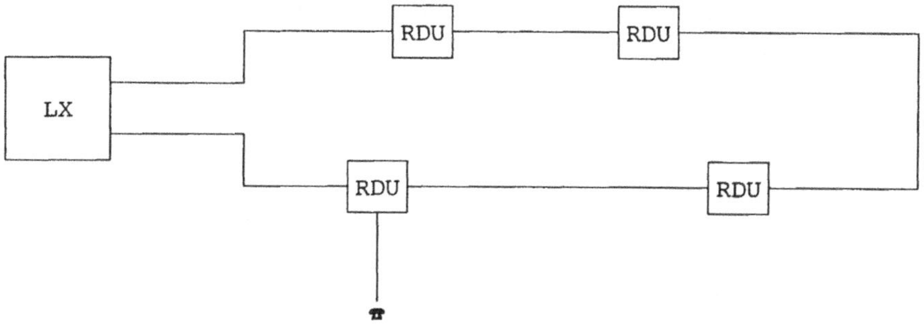

LX: Local Exchange
RDU: Remote Distribution Unit
☎: Household

Fig. 5. Ring network.

way) kinds of transmission like television signals. In this context, the transmission requirement takes a relative priority over switching thus facilitating an efficient delivery of simplex broadband television signals rather than narrowband telephony signals. On the downside a greater loss of privacy may result from more extensive use of remote nodes.

The cable television industry distributes *wideband* signals over tree-and-branch and fibre-ring topologies using relatively high-value transmission media such as coaxial cable and fibre. Identical one-way (hence distributive) signals transmitted from the video head end to the subscriber become progressively split during transmission. The ring layout has the advantage of minimizing disruption in the event of one of the cable links failing to operate. Tree-and-branch and ring networks also minimize the use of dedicated plant. In summary, low bandwidth services such as telephony favor star-type networks using mostly dedicated plant while high-bandwidth services such as video favor bus (tree-and-branch) in an attempt to minimize the use of dedicated plant.

The design period

The installation phase or "design period" in a telecommunications network makes available dedicated access facilities to households requiring telecommunication services now and at some period in the future. However, due to the uncertainty of future demand, it is normal practice for engineers to install an element of excess capacity in the access network in relation to the degree with which anticipated future demand for telecommunication services exceeds current demand. For example, the main cable in the feeder network connecting the LX to the distribution cable may be 500 pair in anticipation of future demand growing beyond the current community of, say, 250 households. This represents a "lumpy" investment proposition in local loop plant, whereby a proportion of installed asset remains idle until future demand requires it. In contrast, wireless communication systems overcome restrictions implied by wire links and in most circumstances can be more responsive to service requirements.

In the above example, if the capacity in the feeder distribution cable is completely utilized, new circuits may be appropriated via *electronic pair-gain* without having to physically install a new feeder cable. In some circumstances electronic pair-gain systems can offer only a temporary response to extra burdens being placed on the network. For instance, physical deterioration of the main feeder distribution network cannot support electronic pair-gain systems. Furthermore, if extra circuits are required in the *critical path* of the network where dedicated facilities reside (i.e., the drop loop to the customers' premises), this implies that further installation of distribution cables *may be* required and drop-line facilities *definitely* required. Hence expanding the network to cope with extra demand for basic services can only be accommodated in the long run as extra dedicated facilities are installed. This represents a significant part of the local loop plant cost structure.

Access economics and network economics

The reference to network designs above suggests that a group of subscribers are collectively served by the same LX. Each subscriber to the network is physically connected to their own LX switch by means of an appropriate access link consisting of a *feeder network*, a *distribution network* and a service access link or *drop loop*. In the majority of cases, the transmission link in the drop loop is twisted-pair copper wire.[4]

The drop loop masks a problem well known in network-based industries, that of asset specificity. For much of a typical 1,440-min day, dedicated facilities such as subscriber access lines and dedicated premises equipment may only be in daily use for a few minutes. As these facilities can remain totally idle for much of an average day, (the final tranche of) the local loop represents an inefficient investment proposition for telecommunication network planners.

This implies that local loop networks located in sparsely populated areas automatically take up a higher proportion of total investment costs in loop plant compared to installing dedicated access facilities in more densely populated areas. Therefore providing access through physical wireline media such as copper to rural users can become relatively uneconomic in relation to the quantities of access equipment required to connect them to the nearest switch, in comparison to centrally located subscribers in urban centers. Table 4 summarizes the physical and technological differences between network and access environments, representing long distance and local networks, respectively.

Table 4 suggests a real distortion in access economics compared to network economics. Whereas dedicated facilities in the access network exhibit extreme diseconomies, elements of the core network exhibit favorable scale economies since network facilities are more likely to be integrated allowing shared-usage among

Table 4. Network and access: basic comparisons (Calhoun, 1992).

Long distance network	Local access
Shared facilities	Dedicated facilities
Integrated	Fragmented, not integrated above feeder level
High usage	Low usage
Controlled environment	Field environment
Rapid technological evolution	Technological stagnation
End-to-end digital technology	Partly digital or semielectronic
Software dominated	Hardware dominated
Increasing returns to scale	Decreasing returns to scale

common subscribers. The rationale for building FITL networks immediately gives subscribers the access-capability to new (high bandwidth) tele-interactive services. If access is inefficient or prevented due to the use of outmoded technologies, a bottleneck is said to exist. Alternatively, a bottleneck may be also interpreted in terms of the scarcity of bandwidth in relation to the demand for new services by existing and future subscribers.

The limited bandwidth capacity of copper and its predominant use in local access represents an engineering *bottleneck* in service provision, despite technical developments such as digital compression and advanced modulation techniques. Compared to the capacity of BISDN optical networks, the nature of this bottleneck is still relatively significant.

This suggests that a number of new technologies and developments will succeed in extending the lifetime and functionality of existing telecommunications infrastructure, while over a longer period, technologies like fibre and wireless radio will (invariably) attempt to penetrate from the outer reaches of the core network, into the local access network, and eventually to subscribers' premises. In BT's case while fibre will support a number of BISDN applications, the design period also intends to optimize existing copper plant through digitisation and compression as a foundation for NISDN service applications. The management of different traffic requirements according to user needs will be based on four main transport mechanisms:

1. Point-to-point fibre systems (or active star network).
2. Point-to-multipoint fibre systems or *passive optical networks* (PONs).
3. *High Speed Digital Subscriber Loop* (HDSL) which provides up to 2 Mbit/s over two or three pairs of loop grade copper using *digital signal processing* (DSP)
4. Wireless radio.

By the late 1990s, it is anticipated that 1,000 business subscriber sites will require access to bandwidth exceeding 34 Mbit/s and at least 80,000 will require up 2 Mbit/s.[5] BT plans to deploy about 3 million fibre connections supported by a fully managed niche of fibre networks, HDSL networks and wireless radio networks by the year 2000 in anticipation of future bandwidth requirements for wideband and broadband applications (Payne, 1993). The long run strategy for network evolution is to connect 95% of all business users to fibre rings by the end of the decade.

Research indicates that large users are best served by single-star networks, while residential users and small-to-medium business users most suited for active double star and passive star/bus networks (Rowbotham, 1991).

3. Alternative networks

The background to fixed local subscriber networks above suggests an *evolutionary*, rather than *revolutionary*, approach to adopting advanced fibre transmission technology beginning with existing and enhanced (ADSL) copper solutions, and followed by fibre and copper hybrid and all-fibre solutions. The introduction of optical technology in local networks depends on views about the "correct" approach to network evolution with respect to current and future demand for bandwidth capacity. Table 5 presents a UK-specific framework to network evolution and level of service provision based on alternative local subscriber networks.

Existing networks

Telephone networks are essentially distribution systems over which circuit-switched signals (calls) are inter-connected between subscribers. Prior to multiplexing techniques, traditional star-type networks used dedicated copper pair throughout the local network for each subscriber. For longer line links, this necessitated a requirement to install repeater equipment in order to maintain signal quality. The advent of multiplexing resulted in economies of plant sharing by combining multiple carrier signals onto single copper pairs in the feeder and distribution network. For larger networks using a double-star topology, the plant sharing occurs between the

Table 5. Alternative fixed networks and service provision in the UK.

	Narrowband voice and nonvoice services[a]	Switched video (VOD)	BISDN services
Existing networks			
Copper local network (present)	*		
Copper local network (ADSL)	*	*	
Fibre and coaxial (FTTC)[b]	*	*	
Hybrid networks			
Digital Loop Carrier (DLC)	*	*	
Fibre and Copper (FTTC)	*	*	*[c]
All-fibre networks			
NISDN on Fibre	*	*[d]	
BISDN on Fibre	*	*	*

FTTC: fibre-to-the-curb; POT: plain old telephony; NISDN: narrowband integrated services digital network; BISDN: broadband integrated services digital network. [a]Services like call waiting, video phone and caller identification. [b]Based on telephony provided by the CATV networks. [c]Over 10 Mbit/s transmission over copper is possible for pedestals placed close (150–300 m) to the subscriber. [d]An ISDN chip allows the integration of voice and video on the same fibre.

MDF at the LX and the RDU. While upgrading existing local networks to ADSL represents a relatively uncomplicated procedure.

Hybrid networks

Multiplexing traffic results in cost economies since the overall use of cable and electronic plant is minimized compared to star networks. Growth in future traffic is also easily accommodated via electronic pair-gain techniques. In this context, pair gain is achieved by replacing electronic components at the LX and the RDU either side of the feeder segment (Reed, 1991). This system of digitizing and multiplexing signals over a single wire is called *digital loop carrier* (DLC).

Modern DLC networks use fibre optic cable in the feeder segment in order to overcome the need to install repeater equipment that is otherwise required in traditional *all-copper* DLC networks. In practice, DLC networks are hybrid networks since fibre is substituted for copper in the feeder segment. Signal transfer in the remaining distribution and drop segments is facilitated over standard copper pair. In the US, DLC networks have been shown to be cost-effective for local access lines exceeding 3 km (Jenson, 1993). Below this distance the network service access is best provided with dedicated links avoiding the need to install RDUs. By 1993 Ameritech, a US local telephone company, used DLC for 8% of its subscriber base representing 17 million lines. In aggregate about 10 million access lines in the US were supplied by DLC systems in the same period (Oakley et al., 1993).

Fibre-to-the-curb (FTTC) represents a more extensive fibre reach than DLC networks since fibre goes beyond the RDU to an additional flexibility point called a curb or pedestal (Fig. 6). According to the required level of service FTTC networks can be either fibre\copper or fibre\coaxial. The pedestal contains an optical network unit (ONU) which converts optical signals to electrical signals and demultiplexes them. All-fibre networks represents the most ubiquitous presence of fibre in the local loop.

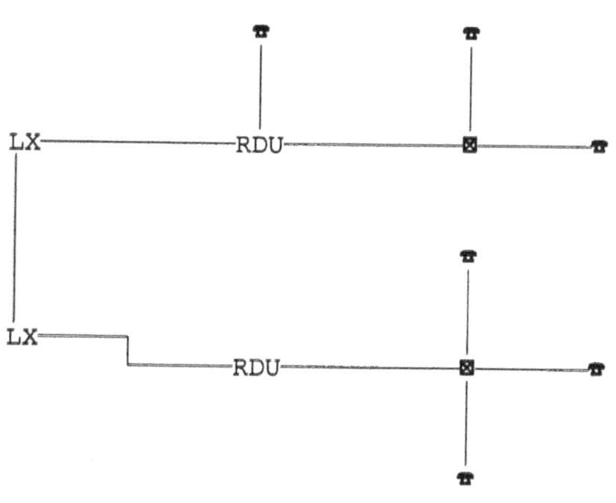

Fig. 6. Hybrid network (FTTC).

All-fibre networks

Fibre-to-the-home (FTTH) networks extend fibre connections all the way to subscribers' premises. This provides the potential to access broadband capacity in line with BISDN applications. FTTH can be either *point-to-point* or *point-to-multipoint*. Point-to-point FTTH networks resemble simple star structures (Fig. 2) providing a dedicated fibre link from the LX to the subscriber thus eliminating the need to split fibers or multiplex (and demultiplex) signals. Point-to-multipoint networks are *passive optical networks* (PON) and represent a variation on broadband network design.

All-fibre PON networks achieve a high degree of plant sharing in the feeder segment by using devices known as *passive optical splitters* which divide downstream signals in the feeder network into identical measures before delivery to a number of pedestals located beyond the RDU. For example, a splitter may disperse into four or eight pedestals each serving a collection of subscribers. Each signal distributed from the LX is received by every subscriber simultaneously. The appropriate channel for each subscriber is demultiplexed by the *customer premises equipment* (CPE).

Each distribution fibre emanating from the pedestal is connected to the splitter in the RDU. The use of passive electronic components rather than active electronics reduces costs due to the possibility of extensive fibre sharing in the feeder network permitting the apportionment of transmission costs over many subscribers, compared to dedicated fibre networks. For example a dedicated fibre network serving 20,000 subscribers which instead operated a 1:32 way split would only require 625 active fibers in the feeder network.

PON networks can be arranged as a double star, bus or ring. With PON Double Star optical signals travelling between the LX and the RDU are multiplexed on to one or more common fibers. Optical splitters in the RDU disperse equal amounts of optical energy to remote pedestals each connecting dedicated fibers to every subscriber. Since no demultiplexing occurs in the networks, CPE demultiplexes the appropriate channel relating to each subscriber.

PON bus or ring networks are similar to PON/double star since the primary ring network may be arranged as a logical star network. The secondary network resumes a tree-and-branch topology from each network node (the RDU). The required channel is accordingly "tapped off" and directed to the CPE. In both cases the LX transmits time *division multiplexed signals* downstream to the subscriber premises and uses *time division multiple access* for upstream signal transmission back to the exchange.

PON networks were originally constructed as a cost-effective network solution for telephony provision and low-bit-rate services. By 1990, BT's network provided *telephony on passive optical networks* (TPON) dispersing 20.48 Mbit/s transmission capacity through 128 fibre ends up to variously located network termination points, further subdividing into 294 separate 64 kbit/s channels (Rowbotham, 1991). BT currently intends to deploy business TPON as an approach to local fibre deployment aimed at small and medium-sized companies. One industry source suggests up to the end of 1993, the majority of new local fibre network trials in 26 European countries indicated a preference towards point-to-multipoint passive optical networks (73%), while a significantly smaller proportion of networks were based on point-to-point networks (8%). The remainder were based on variants of the two.

As an alternative to time-division multiplexing, *wavelength division multiplexing* techniques enable passive optical networks to upgrade from TPONs to *broadband PONs* (BPON) for high-bandwidth delivery such as switched video. With WDM, the multiplexing is applied directly to incoming and outgoing light signals. Each traffic channel is defined as a single optical wavelength and subsequently identified by an optical filter, diffraction grating or prism at the subscriber site. Additional capacity is easily served by using other wavelengths.

Hybrid networks such as DLC and FTTC represent cost effective interim solutions towards all-fibre, all-digital, completely integrated BISDN networks. The gradual introduction of fibre in local networks beginning with the feeder segment commensurately displaces narrowband copper in the remaining parts of the network. By confining an overall network's dependence on "bottleneck" segments, DLC and FTTC hybrid networks offer higher bit rates (bandwidth) than would otherwise be possible in a network based on purely narrowband transmission media. For example as fibre-reach penetrates closer to the subscriber, standard twisted pair can transmit in excess of 10 Mbit/s over distances between 150–300 m (Jenson, 1993).

It follows that fixed networks can adopt a number of different designs according to the relative penetration of fibre compared to other network topologies. Hybrid networks use a mix of delivery technologies depending on the relative proximity of fibre evolution to the subscriber. The gradual deployment of fibre progressing towards ubiquitous fibre networks results in a cost/benefit trade-off pitching network investment against ever closer subscriber access to end-to-end all-digital broadband capacity (Reed, 1991).

4. ADSL-enhanced telephone networks

Beyond the obvious differences in cost and performance between optical and electrical transmission systems, and despite the notion that copper wire is synonymous with analogue technology, its functionality has been greatly enhanced by enabling technologies responsible for allowing digital transmission and compression over metallic wire. Compression technology accounts for a reduction in the amount of bit rates needed to reproduce a given signal, while advanced modulation techniques such as digital transmission have increased the information throughput measured in bit rates over a given amount of bandwidth. The integration of advanced switching and transmission techniques has thus permitted a truly end-to-end digital transport capability. The application of ADSL further introduces the potential for higher bandwidth capacity by allowing the delivery of low definition television signals over existing copper lines.

The delivery of switched video service over the local telecommunications network contrasts with *distributive type* entertainment services offered by the cable companies, since films can be repeated over relatively short time intervals. VOD allow telephone subscribers the opportunity to make real time access of centralized libraries of viewing material stored on video servers.

The notion of a switched service such as VOD is analogous to telephone users choosing to *switch* to other telephone subscribers in standard voice communication. Users are able to select material independent of broadcasters' schedules in contrast to *near demand* services offered by pay-movie channels and video rental shops. The policy climate hinges on whether BT may, under current rules, be allowed to offer switched-video entertainment. This point highlights the potentially uncertain effects new interactive services like VOD may have on the financial stability of UK cable franchises delivering distributive entertainment services.

Regulation

Following the Government's review of UK telecommunications in 1991, fixed-link telephone operators were prohibited from delivering broadcast-entertainment services over their networks for a recommended period of 10 years subject to an interim review in 1998. However, BT's proposal to offer a nationwide video dial tone service appears to have escaped the original restriction on conveying entertainment services despite opposition from the cable industry.

With reference to the 1990 Broadcasting Act, the UK Independent Television Commission (ITC) and the Department of Trade and Industry (DTI) have suggested that VOD is a *mono*cast service since it is conveyed in a *point-to-point* context, such that the service provider responds to unique requests from individual subscribers. In contrast, services such as terrestrial television or CATV are *point-to-multipoint* thereby qualifying as a distributive *broad*cast service rather than a communicative monocast service like switched video.

Furthermore the 1990 Broadcasting Act explicitly states that no company is permitted to deliver entertainment services in an area already served by a *local delivery license* (LDL) for the explicit purposes of delivering entertainment. This then makes it impossible to grant additional LDLs in areas already franchised for cable television service, accounting for about 64% of UK television homes. Since VOD is currently regarded as a *mono*cast service it escapes current LDL legislation set up for *broad*cast entertainment. Consequently BT appears free to offer switched video using ADSL technology. However, interpretations of what exactly constitutes an entertainment service may well be challenged in the courts by the cable industry despite the protection *it* already has from the original entertainment-delivery ban imposed on BT.

In general the entertainment-delivery ban captured a view that BT would engage in unfair competition by letting its telephony business subsidize new entertainment services. More specifically the ban was imposed as a measure designed to restrict competition while the cable companies satisfied their network build obligations. A recently convened House of Commons trade and industry select committee investigating the development of optical fibre networks recommended a definite removal of the entertainment restriction by at least 2002 if not sooner. For BT, offering interactive services in the meantime still poses a dilemma of how costs and revenues are to be attributed between voice and video services.

ADSL technology

In engineering terms ADSL involves installing a decoder and receiver set-top unit at the customer site and an interface (the ADSL unit) in the LX. The ADSL-encoder combines video and telephony signals at point of concentration located beyond the LX and transmitted over copper to the home (Fig. 7). The video signals are generated by a video server which may be located independently of the local telephone exchange. The ADSL decoder at the subscriber site separates video and telephony signals. ADSL technology enhances the capacity of local copper networks up to a capacity of about 2.0 Mbits from the exchange with a 16 kbits control channel from the customer site, hence the term asymmetric. The return channel from the customer site is necessary for subscribers to communicate with the video server. Some versions in the US such as *ADSL3* can support up to 6 Mbit/s over 2 km distances.

ADSL, like ISDN (Integrated Services Digital Network) and HDSL (High Rate Digital Subscriber Line) are a class of local loop technologies which exploit the theoretical capacity of copper using frequencies outside the range used to carry voice signals. Pair-gain systems also fall within this technology group generically described as *digital data over voice* (DDOV) systems. The advantage of DDOV is seen in terms of the additional circuit capacity created thus avoiding the need to install new cable plant.

Although copper is technically outmoded compared to any number of FITL solutions, BT's vast network of copper lines are still undepreciated largely due to the extensive accounting life given to cable technology in general. It is therefore not surprising that BT is exploiting compression technology in conjunction with delivery of video over copper. With distances between 4.5–6 km defining a reasonable outer range for transmitting video over copper and average loop lengths varying between 2–2.5 km, roughly 90% of BT's residential customers are ADSL-compatible. But even this may be an optimistic estimate given the potential interference caused by co-existing high frequency network applications such as ISDN or Megastream services which BT offers to business subscribers.

Transporting video over telecommunication networks further relies on compressing a large library of viewing material on powerful computers acting as video servers.

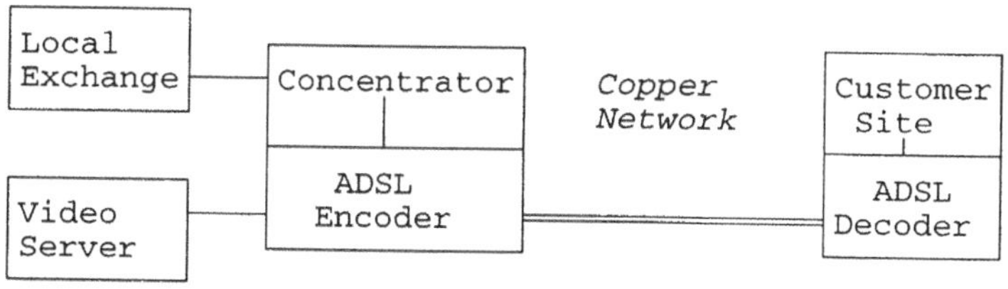

Fig. 7. ADSL-enhanced copper network.

These digital servers, through a system known as *massively parallel processing* are capable of handling tens of thousands of simultaneous video requests from anywhere in the country. These developments are analogous to trends witnessed in the computer industry as comparative value shifts from hardware to software. Traditional telecommunications networks (the hardware) are becoming transformed into intelligent networks driven by software innovations capable of delivering enhanced services.

Digital compression technology lies at the heart of these developments, succeeding in altering the cost-performance trade-off between copper and fibre by harnessing the potential bandwidth available from traditional copper pair. This suggests that the benefits of fibre weighted against its cost continue to cast some uncertainty over the eventual arrival of ubiquitous subscriber-based broadband networks delivering an array of distributive and interactive services.

5. Video distribution and interactive service delivery

Historically, the idea of service convergence through integration is not new. In the US during the 1970s, the use of coaxial cable by the cable industry for one-way broadcast service delivery was considered an ideal medium to explore the potential for two-way services such as telephony, data and video on demand. Although trials were conducted, most failed (Noll, 1991). Effectively the same debate has reemerged but centered on the potential that optical fibre has to offer. BISDN is suggested as the most appropriate platform for integrating nonstandard service applications through a single digital network. Without integration, it remains incumbent on separate communication networks to provide access to applications ranging from voice telephony, data services and video communication, to broadcast entertainment. Whereas previously coaxial cable was the domain of the cable industry, discussions centered around broadband networks anticipate developments to be led by the telecommunication industry.

However, due to uncertainty of future demand for interactive services combined with the cost constraints associated with deploying networks equipped for broadband access, medium term developments are only likely to address the immediate needs of a market which is increasingly perceived as becoming more segmented in terms of user requirements. Commercial success for interactive services like VOD will depend on both standard price considerations as well as their general appeal over existing mainly distributive entertainment services. Nevertheless, in the US, preliminary results of VOD and *near*-VOD trials already appear to show encouraging results compared to the more limited services offered by *pay-per-view* (PPV) premium CATV channels. Table 6 shows customer response to three US VOD trials.

In comparison with traditional PPV services, the buy rates in each trial, measured as movies purchased per month, have varied from twice (GTE) and eight times (Time Warner Cable) the industry average to more than 12 times (TCI). The trials have also reinforced some negative perceptions of traditional PPV service such as inconvenience, limited choice of titles and difficulty in ordering services (Tachau, 1994).

Table 6. Customer response to video-on-demand trials in the USA (Tachau, 1994).

Test	Test pool	Sponsor	Buy rate
Viewer controlled TV	150 Denver households	Tele-Communications Inc.	2.5 movies/month
Quantum system, New York	5000 Queens households	Time Warner Cable	1.5 movies/month
Cerritos CA test site	Cerritos CA households	GTE Corp.	2 movies/month (40% of participants)

Ultimately the long run competitive outcome between switched services and the combined efforts of "near" demand services (cable and video hire) will influence the correct approach to technology deployment in the local loop. Local telecommunications infrastructure in the UK is undergoing important changes given the activities of BT and the cable industry. The development of ADSL in the local loop will depend on the outcome of VOD trials and regulatory interventions determining the distinction between interaction and entertainment.

The difficulty arises in predicting the mass-market's response to new interactive services, given that satellite and cable delivery of pay-television services already reach nearly 20% of UK entertainment homes. The cable industry's nationwide build program, which currently represents around 70 operating franchises out of 130, has already passed in excess of 5 million households and continues to maintain an average penetration of around 22% of homes passed. Total build obligations run to nearly 15 million television households representing some 64% of UK television homes (ITC data).

If the cable industry maintains its share of television subscribers by the end of their build program, then by the end of the decade they alone will account for at least 22% of UK entertainment homes. Furthermore some cable franchises have already announced an interest in delivering interactive services on their own. Nynex CableComms has conducted tests exploring the interactive potential of broadband in its North-West of England franchise area. While Videotron already offers a "games and information" channel incorporating a number of low level interactive services.

These developments suggest a less than certain outcome for BT in its attempt to deploy a choice of networks based on ADSL-enhanced and different FITL solutions. Nevertheless ADSL appears to offer an interim solution to equipping communication networks with interactive capability while ultimately delaying the need to deploy extensive local fibre. A recent Bellcore study substantiates this point by stating that although in general, access costs rise linearly with subscriber connections in a wireline network, ADSL is less costly in comparison to FTTC when penetration rates are under 27% of households. However, the ADSL cost advantage yields to FTTC when penetration rates rise to and exceed 55% (Matthews and Darabi, 1994).

6. Conclusion

Proposed telecommunication services like video on demand, whilst enabling technology to adjust more directly to consumer taste compared to near-demand services, raises doubts as to whether and how many such "entertainment" networks and/or services can coexist in a single competitive environment. Ultimately telephone operators face the uncertainty of not knowing the extent to which switched video will displace other media such as terrestrial satellite, cable and time shift viewing. Also, building integrated networks for the purposes of multiservice delivery may not deliver the benefits of economies of scope under some circumstances, suggesting instead an evolution path in the direction of separate networks rather than one integrated network.

As the distinction between communication and information services continue to blur, it may be conceivable for video services to be offered by merged entities comprising local telephone companies and franchised cable operators serving their territory, given the scale of recent alliances in the US.

In the short term, multimedia prospects in the UK will witness continued efforts to enhance network capacities with BT attempting to offer a number of switched video services to residential users. Business users will continue to benefit from the steady roll out of point-to-point fibre and Passive Optical Networks to large customer sites requiring multiservice delivery. The prospect of multiservice delivery will also depend on the efforts of third party providers developing software capable of offering value-added services in conjunction with network enhancements. For example, insufficient complementarity between hardware and software is regarded as one of the shortcomings of ISDN, that is, in lacking sufficient third party actors.

ADSL technology provides telephone operators an interim solution to delivering a combination of switched services along traditional copper wire. The challenge of deploying fibre in the local loop remains a cost issue and may continue to do so until regulations currently restricting BT from delivering entertainment services until 2001 are removed. The current view from the Independent Television Commission suggests that video on demand does not violate the conditions set out in the 1990 Broadcasting Act. However, cable industry finds difficulty in locating a clear distinction between interaction and broadcasting. BT's interest in delivering interactive services represents an attempt to halt further decline in its market share as the combined effects of competition in the UK from the cable industry and other operators continues to develop.

Acknowledgements

The author thanks Martin Cave, Mark Shurmer and Keith Ward for comments made on an earlier version of the paper. Financial assistance from BT, ITS and the ESRC (grant no. L114-251028) is acknowledged.

Notes

1. In 1989 BT had 7,062 local exchange switches and 360 DMSUs.
2. Source: British Telecom.
3. Although signal transmission is two-way, the form of communication is strictly half duplex, and not usually full duplex, since there is a listening party and a talking party. Cable television networks are distributive networks since signal transmission is usually one-way.
4. Virtually every access link in any given local telecommunications network is twisted-pair copper wire, and has been the case since telephony was invented in the late 19th century. Physical installation "guiding" access links to network subscribers is usually either by overhead pole or underground conduits buried in shallow trenches.
5. BT's top 200 business subscribers have between them about 90,000 UK sites each using more than one line (Allen and Marshall, 1993).

References

Allen, P., Marshall, J., 1993. *Integrated Services Delivery to Major Customer Sites*, BT Working Paper.

Calhoun, G., 1992. *Wireless Access and the Local Telephone Network*. MA: Artech House.

Jenson, J.D., 1993. Rational and Economic Fibre Systems. In: Ritchie, W.K., Stern, J.R. (eds) *Telecommunications Local Networks*. London: Chapman and Hall.

Matthews, J., Darabi, F., 1994. *The Local Loop: Market, Technical and Regulatory Strategies*. London: Ovum Limited.

Noll, M., 1991. *Introduction to Telephones and Telephone Systems*, 2nd Edn. MA: Artech House.

Oakley, K.A., Guyon, R., Strern, J.R., 1993. Fibre in the access networks. In: Ritchie, W.K., Stern, J.R. (eds) *Telecommunications Local Networks*. London: Chapman and Hall.

Payne, D.B., 1993. Opportunities for advanced optical technology in local access networks. *BT Technology Journal*, 2(April);11–18.

Reed, D.P., 1991. *Residential Fibre Optic Networks: An Engineering and Economic Analysis*. MA: Artech House.

Rowbotham, T.R., 1991. Local loop developments in the UK. *IEEE Communications Magazine*, March:50–59.

Tachau, J.S., 1994. *Endangered Species? Prospects for Video Distribution in the United States and in Europe*. Paper presented at the 16th International IDATE '94 Conference, Montpellier, France, 16–18 November.

Valdar, A., Newman, D., Wood, R., Greenop, D., 1992. A vision of the future network. *BT Engineering*, 11(October);142–152.

Williams, H., Taylor, J., 1994. Competencies and diversification: the strategic management of BT since 1984. In: Bohlin, E., Granstrand, O. (eds) *The Race to European Eminence*, (pp. 67–77). Amsterdam: North-Holland.

Beyond Competition: The Future of Telecommunications.
Edited by D. Lamberton.

CHAPTER 10

The sustainability of the Broadband Network

Denis Joram
France Telecom/CNET, Issy-les-Moulineaux, France

Introduction

In the past, the universality of the network has justified protection or even institutionalisation of telephone monopolies. Because they were protected, the monopolies have been suspected of abusing this favourable position (mainly in the United States). Moreover, the subadditivity of the cost function was contested, econometrical studies on American data leading to the opposite results. The criticisms that were levelled at them were sharpened by the increasing weight of telecommunications in economic activities.

More and more, the telephone and connex services became an input in the production function. For households, the telephone takes an importance similar to the television, domestic appliances or the automobile.

In parallel, these increasing needs were satisfied by the implementation of new networks devoted to these new applications. Facing the inefficient proliferation of specialized networks, the technology can lead to a more efficient way to supply many services by integrating them all in the same network. That is the aim of the broadband network.

The idea of universality resurfaces but one does not push the idea of an universal service forward, rather the idea of an universal medium.

Beyond this idea of a universal infrastructure, there is the problem of the service and network dissociation.

The question is: if the broadband network is efficient, can it be sustainable?

In the second section, we propose a listing of the main phenomenons that could put a brake on the emergence of the broadband network.

In the third section, we compare the development of two networks which can be conceptually assimilated to multiservice networks, the videotex and the ISDN in France. The success of the former and the difficult start-up of the latter highlight the viability conditions of the broadband network which would be the ultimate focus: balance between cooperation and competition.

Address for correspondence: Denis Joram, France Telecom/CNET, 38–40, rue du général Leclerc F-92131 Issy-les-Moulineaux Cedex, France. Tel.: +33-1-46-29-51-86. Fax: +33-1-46-29-33-70.

The sustainability of the broadband network

Four main issues will be described in this section.

The cost problem or the efficiency threshold of the broadband network

The cost problem can be summarised by two considerations. Firstly, the subadditivity of the cost function has not yet been demonstrated on the whole set of possible services. Secondly, the existence of increasing returns does not imply today that the broadband network is the less expensive way to satisfy the demand. There probably exists an efficiency threshold, in terms of the type and of the number of services, that makes inefficient the development of new applications on specialized networks. The question is: Can this threshold be reached?

a. *The cost advantage of integration*

Henriet and Volle (1987) lays the foundations of the reflection to be done on this subject. Based on several assumptions on demand and the relative costs of service supply on a specialized or integrated network, the efficiency conditions of an integrated network are established (Fig. 1).

The cost to offer x services on an integrated network (hereafter IN) is supposed to be ax+b. The same supply on specialized network (hereafter SN) will cost cx.

b is the cost of integration and the marginal cost of an additional service on the IN is below the one on the SN, that is to say a<c.

At a given price, the level of demand is supposed to be higher in the IN case than in the SN case, that is to say Di(x)>Ds(x). Henriet and Volle (1987) attributes this difference to an "usage ergonomy gain" on the IN which can be assimilated with standardised access or usage procedures.

If the number of services is below X the services supply is efficient on SN at Xs.

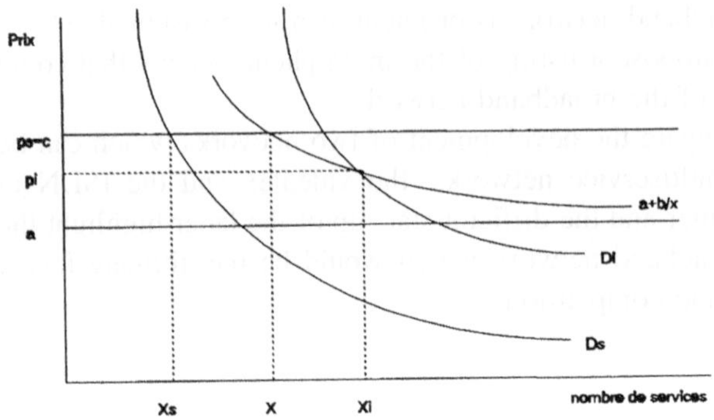

Fig. 1. Efficiency threshold of the integrated network.

The positive effect of integration on the level of demand gives, for the IN, an efficient location in Xi. Henriet and Volle (1987) concludes that the integrated supply is superior to the specialized supply on an welfare point of view because a higher demand is served at a lower price.

Some questions can then be addressed: What organization is more favourable to the differentiation of services (which is an efficiency condition) and what are the viability conditions of the IN?

b. *The organization of supply*

Three institutional configurations have been considered in Henriet and Volle (1987):
1. The specialized networks: each service needs one specific network.
2. A network monopoly and competition on the service supply side
3. An integrated monopoly:a unique enterprise furnishes the network and the services.

The main assumption is the existence of diseconomies of scope in the building of services. The argument used to justify this hypothesis is that the preponderance of specific competencies constitute the service itself: access to information and a very precise knowledge of the demand for certain services.

Specialized enterprises (banking, press, tourism, mail-order, property business,...) would then be able to propose, in their domain, a service with the lowest costs and the highest quality, but would not be able to diversify their activity efficiently in other domains. Then, the decentralized supply should be more efficient.

Based on a model of monopolistic competition, the authors show the superiority of the second solution, *"the network monopoly favours for all reservation prices the service differentiation (that is to say, for the same reservation price the network monopoly leads to a greater number of services, and consumers with low reservation prices who could not access the network in the SN case, can find services at a low price)".*

With the integrated monopoly, two opposite effects are perceptible: the network scale economies give an incentive to cover a wide market, the service diseconomies of scope give an incentive to reduce the range of services. *"When network and service supplies are bundled, it leads either to the competition between specialized networks or to a sustainable monopoly. If the competition is opened it is not sufficient to induce the emergence of the efficient structure".*

In fact the previous monopoly is sustainable if it reduces the number of services leading to a less efficient configuration than the one associating the network monopoly and services competition. What is called "network monopoly" is a technical monopoly that is to say a unique actor (or a coalition of actors) insures the consistency for the network (interfaces, network protocols,...) and its technological evolution. The technical monopoly coincides with the unicity of the network "manager".

Finally, this configuration will actually be efficient only if the network is organized as a market place which favours the differentiation of competing supplies by cutting network access charges and by widening the potential market. The best example of this configuration is probably the Minitel network.

The pricing problem or the flimsy sustainability of the broadband network

The flimsiness factors of monopolies are more virulent in the telephone case than in the broadband case.

In the telephone case, the universality leads to compromises which weaken the dominant position. The universal access leads to a uniform quality and the concern for equity and its corollary the balancing out pricing, constrains the operators to setting up a pricing structure with cross-subsidies.

These two factors (uniform quality, balancing out pricing structure) weaken the network coherence. Firstly, if the qualities needed by some agents are very heterogeneous, the price cut obtained by the economies of scale on the general network, may not be sufficient to avoid some users developing an interest in using a parallel network supplying a quality that better matches their own taste, eventhough an increase in prices (Henriet, 1986).

Secondly, competitors can choose to supply a service on gaps where the incumbent prices are not representative of the costs engaged (skimming strategies, Tirole (1988)).

In the broadband case, there is not only an arbitrage between the desired quality, the economies of scale and the price of the service, but an arbitrage between type and quality of services desired individually and economies of scale and scope generated upon the broadband network.

In this framework, a pricing structure which allows the network to be sustainable is more difficult to set up. Henriet and Volle (1987) show that cross-subsidies have to be used to attract enough customers necessary to the efficiency of the broadband network.

They show how a tariffication differentiation can limit the adverse selection phenomenon. Some consumers can have costless solutions out of the integrated network; in leaving the IN they induce a loss of efficiency (increase of the average cost on the IN) prejudicial to the collectivity. For example, if we consider a two service case, the user of service 1 pays more than the average cost but less than what he would pay if he was alone to use the IN; but the user of service 1 pays less than the average cost. The situation can then be the following: the user of service 2 can accept to pay more than the user of service 1 in order to attract the user 1 on the IN and then to take advantage of the cost cut resulting from scale economies.

If this tariffication scheme seems to be efficient, the heterogeneity of consumers and the variety of alternative technical solutions to the IN make the set-up of a pricing structure that can avoid adverse selection consequences tricky. (Moreover, the set up of this pricing structure needs to solve the joined costs to each service allocation's problem. It is well known that the analytical accounting is one of the network operators current priority).

The ONP process (which obliges the operators to rent the access to their infrastructure at the internal transfer price to competitors) can then be understood as an attempt to avoid redundant investments while favouring a complete cover of user needs at the best prices.

The replacement of the network technology

A wide integration of services will constrain the operators to replace the technology. Actually, the current technology is limited in transmission capacities as well as in the switching process of very heterogeneous signals. Additional problems arise. Besides its implementation financing problem, one can not avoid the transitory concomitance of several technologies in the network. This concomitance sets some pricing issues previously exposed (for example the problem of cross-subsidies between users of the new and the old technology). Here we want to describe the modal competition with alternative technologies particularly efficient to skim some lucrative gaps (as mobile services). This potential competition can change the operators' investment priorities. These operators would favour some strategic investments (e.g., in the mobile network) rather than efficient investments (long-term process innovation).

Finally, the regulation process can put a brake on the emergence of the broadband network. Actually, in the reformed framework of an open network architecture, the set up of an integrated network could not be used as an entry barrier by network operators, eventhough the constitution of entry barrier could have been precisely an argument for setting up an integrated network.

After a rapid presentation of fiber optic systems, we will show that the potential properties of the fiber system cost function takes, once again, into consideration the primary importance of the natural monopoly. We will then fix the main obstacles to the deployment of fiber in the local loop. We shall conclude on the emergence conditions of a universal network.

a. *The fiber optic technology*

The current technology already allows service integration but this integration is partial because of the limited performances of copper. The optical fiber pushes forward these limits. The monomode fiber needs impressive manufacturing performances: core diameter of only some microns, utilization of lasers for emission. Moreover, the fiber has some advantages compared with copper:

1. A broadband transmission capacity.
2. Insensibility to radioelectric parasites.
3. Excellent viability and longevity.
4. Low insertion loss which permits transmitting at long distances avoiding the amplification of signals.

The fiber optic system presents some inconveniences due to being a new technology:

1. The setting of sophisticated optoelectronic equipment as lasers or photodiodes.

2. Some functionalities located at the switch are located near the subscriber (subscriber line cards, etc.).
3. New powering system.
4. The training of a specialized workforce and the setting of new measures and installation equipment.

It is not surprising that fiber optic systems are already implemented in the core of the network (between switches or in transatlantic transmission) because high performance is required and because high costs are not prohibitive. But how could the fiber spray up to the local network (the most costly part of the network) where the bit rate is lower and where the willingness to pay is weaker?

Two responses are given today in which we find the traditional distinction between product innovation and process innovation. In the first case, one considers that fiber is more expensive but that it brings new services which allows the operator to increase the bill. In the second case, fiber is considered as a process innovation transparent to the user. Then we look for productivity gains or at least the cost parity with copper for POTS service. This is clearly a paradox because the main advantage of fiber would not yet be used. Between these two distinct positions a lot of compromises are conceivable.

b. *The cost function properties of fiber optic local network*

The scale economies. Firstly, in comparison with copper, we can observe additional economies of scale because the fiber can be shared between a lot of subscribers. Moreover, the economies of scale on civil works are stronger in absolute value, since the fiber optic cable volume is lower than copper cable volume.

The technical economies of scope. What are the marginal costs of the supply of an additional service on the fiber optic system? The transmission capacity of the fiber allows one to support simultaneously a great number of services (the nonduplication of the support permits a gain of about 10–20% for each service in comparison with the service necessitating its own network). Some equipment (like optical emission or reception cards) can be assigned to several services. Only very few additional equipment or functionalities are to be affected to a certain type of service (information treatment software, broadcast station for teledistribution, videotape providers for the pay-per-view,...).

It is highly probable that fiber based systems are subject to great technical scope economies in the field of service provider.

High scale economies, scope economies: the optical fiber leads to primary importance the contested natural monopoly (especially in the United States), starting in the long distance network 20 years ago, and going on today in the local network. It is a technical monopoly, as defined earlier (single actor or actors coordination insuring interfaces and transmission means efficiency, network protocols or its technical evolution). It is then noticeable that in the technological replacement question some similar problems to the ones discovered by telephone pioneers when they were about to develop their network remain.

We propose, therefore, a survey of the main obstacles against the introduction of optical fiber in the local loop.

c. *Obstacles to the deployment of optical fiber*

Since optical fibre cannot be deployed in the distribution network instantaneously, some criteria must be defined to determine the best way of introducing optical fiber in the local network.

The spatial criterium. The deployment of fiber in the local loop can be optimised by targeting deployment areas on a priority basis. This must take into account the differences in space of the distribution network:

1. The difference in space of consumer requirements: knowing where the demand is coming from can help to target potentially more profitable areas.
2. The differences in network topography: with fixed component cost, an area's density and the availability of the necessary infrastructures affect the total cost of equipping the area

By juxtaposing these two aspects we can prioritise the different areas to be connected (and also establish volumes because in certain areas, the introduction of optical fiber will not be justified) and we can also determine the limits of optimisation of optical fiber deployment.

If we assume a closed market because of the priority given to national producers by the operator or because of the specificity of the network (the operator has to develop a new system), the scale economies generated by the market of all these areas will determine the competitivity of the fiber. Actually, costs are dependent on produced quantities. In others words, the more the targeting of areas is precise in order to avoid an anti economical voluntarism, the less the scale economies play their part, therefore, the fiber based system remains less competitive.

In the French case, area targeting excludes cabling of a great number of areas thus generating a limitation of scale economies on fiber based systems which can be solved only with the scale economies generated in other countries more advanced in their spreading. Therefore we shall obtain two kinds of extreme deployments: the first one consisting of cabling only the most richer part of the customers (for instance the professionals) at high costs; and a great size deployment where scale economies pitch with network externalities to make the fiber network viable. The targeting deployment on the business fringe finds a strategic position in counterpart of high installation costs.

The dynamic criterium. Which is the optimal date of fiber introduction and which is the investment journal which insures an adequation between new services and new demand?

They are two main kinds of approach in the fiber based technology deployment:

1. A deployment "au fil de l'eau" that consists of cabling the areas where the

demand is obvious. This generates:

a) A weak initial investment, but a costly intervention each time a new customer asks to be part of the network.
b) A break for scale economies since the companies of equipment do not immediately benefit from a great market.
c) Some learning effects, e.g., while the equipments are installed.
d) A better prospect for demand. Therefore the forecasted profit of the project will eventually give the possibility of abandonment if the market shows signs of briskness.
e) A greater possibility of upgrading the network in case of process or product innovation.

2. The whole deployment of the network which anticipates the demand. This implies:
   a) A very high connecting fixed cost, but more adapted to satisfying the forecasted demand since the network access would only be needing limited intervention (consisting for instance in a simple plugging).
   b) A better competitiveness in terms of cost, generated by scale economies of the network components.
   c) Sunk investments facing an uncertain demand.

The optimal starting deployment date also generates the problem of cost decrease. Is it necessary to wait to benefit from decreasing costs (especially in the case where process innovations are foreseen), or the opposite, to install a new network as soon as possible in order to develop new services and attract the potential demand?

Meanwhile, the age of equipment and its life time leads to the question of optimal replacing date choice. Is it necessary, for instance, to wait for the complete economical obsolescence of copper to install fiber? Boiteux (1956) shows that the replacement must happen when the old system operating cost becomes greater than the new system cost (including the economical amortizing). The optimal replacement of a copper network must therefore be analysed considering their operating cost linked with their age. These results remain valid when the investment cost linked to the new system decreases continuously, depending upon time and technological innovation. Nevertheless, the link between cost and production scale is not taken into account. The opportunity of new services generating new takings is an essential element for introducing a new technology. At this level, the first conclusion to make is that the profits linked with the appearance of new services even on a medium term shall be key element in the decision making process in the favour of fiber.

The main obstacles to the fiber deployment in the distribution network are:

1. Regulation which can make it impossible to provide an integrated service on optical fiber, whereas it is integrated services that help to justify the use if this technology (generating additional incomes).

   By its properties, the new technology puts on the first level the cross-subsidization problem. They could take the following characteristics:

   a) Spatial. That is if the connecting of dense areas, probably presenting weak installation cost per user subsidizing for instance connecting in rural areas.
   b) Per type of users. If pioneer customers subsidize access to the others customers

assuming it is possible to discriminate users.

c) Per type of services. If some services with low profit are subsidized by juicy services.

d) Intertemporal. When the start up of a new service, before reaching its development critical mass, is temporarily subsidized by other services.

In the context where regulation evolves, and where competition becomes more important in some parts of the market, to limit, even to forbid, cross-subsidies would put a brake on the furnishing of the new technology to the whole population.

2. The problem of externalities which can take two forms:
   a) To propose a technology that is competitive in terms of cost, manufacturers must achieve large economies of scale, in other words a wide market. This market will only exist if operators install systems. Now, if the operators base their decision on the cost of the new technology, the market may well never exist. The dilemma can be solved and turned to advantage if the regulatory obstacles are lifted or more generally if elements outside the market (commitment from the public authorities or dominant operators) alter the operator's expectations of return on investment in an optical fiber system.
   b) Generating supplementary services assumes a customer base of connected lines. These customers will accept increased costs only if the services are already there. In other words, there is a "two-dimensional critical mass" (minimum range of services, minimum demand) which must be reached for a market to develop on a multiservices network.
3. We can not forget the modal competition. If new operators offer competing transmission media, it can change the traditional operator's equipment priorities (he then no longer has to manage only a range of services but also a range of media). If the operator's most profitable customers are threatened (as with mobiles which attract high revenue customers), capital resources can then be allocated on the basis of strategic rather than purely economic criteria. More precisely, as we have seen, although optical fiber must be initially planned on a targeted basis, it will eventually be a universal network available to most subscribers. The threat to the potential revenues of this network from techniques more able to creaming off the most profitable segments of the market, such as microwave radio distribution, are probably the main factor against the deployment of optical fiber.

d. *The concomitance of two technologies: some unsolved tariffication issues*

The replacement of the old technology will not be instantaneous. It supposes a temporary concomitance of technologies. Then what about the costs due to the juxtaposition of two technologies (additional maintenance, double training of employees, compatibility of equipment to continue the service delivering)?

Moreover, some tariffication problems are predictable. In some areas equipped with optical fiber, the new network shall absorb the old one by providing new services in addition with existing services. Which is the pricing structure to adopt for the users of a telephone service on optical fiber next to copper?.

If the pricing is oriented to the cost, for instance, imposed by the regulator, it seems obvious that if the new technology is weakly deployed the users of the "fiber" telephone will pay more than the users of the "copper" telephone. Nevertheless, once the fiber has become the majority, the operating cost of the old technology might become prohibitive and its users shall then pay a higher price.

There is, thus, an equity problem in the pricing between old and new technology. Pricing perequation should imply a pricing structure including cross-subsidies. The solution of this problem probably lies in the differentiation of existing services (between the new and the old technology) justifying a difference in prices, or simply in abandoning the pricing equity, at least for the "non public" services.

Mitchell et al. (1991) have already brought up this problem. "*In this case, high-use customers would be the first to join the broadband network, by-passing the old network and necessitating price increases for the remaining customers. On the other hand, with only the broadband network in place, incentives and opportunities may remain for some customers to by-pass the (broadband) network, by taking cheaper low-quality alternatives*".

Finally, the fiber based technology tends to considerably reduce the marginal cost of the use of the network. The type of service supported by the network is transparent for itself (the network only sees bytes). There is a real identification problem on what is paid. Indeed, the added value of new services remains in the information treatment peripherical to transmission. "*The commodity that should be priced becomes a decision variable. From a cost point of view, the unit of output becomes irrelevant. What matters most is its relationship to the customer's willingness to pay. Pricing by type of services may be difficult because it cannot be monitored by the network supplier. However, there appear to be opportunities to develop rate elements based on such factors as the occupancy of the data frame, the type of routing and switching, and additional processing provided by the network*" (Mitchell et al., 1991).

If the part of the network operator consists of transmitting information more than providing services, which is actually the case, it will progressively lose its authority in pricing, since its output is no more the heart of the service-added value. The network operator shall then become an element between others in a vertical structure characterized by broad competition in the downstream market of final services, by narrow competition in the supporting services, and by a monopoly or a small oligopoly in the upstream network infrastructures.

e. *Conclusion*

The obstacles to the spring of a universal network are numerous. Firstly, the technology of the universal network is very costly and leads to a financing problem as far as its introduction in the existing network is concerned.

Secondly, the universal network will be efficient only if it leads to a great number of various services. This efficient threshold constitutes what we could call the second critical mass.

A first critical mass (common to all universal networks) shall be reached, leading to a crossed auto development of services and demand on the network. It is also a necessity to create the conditions needed to reach the second critical mass "of efficiency" (to justify the switching). That is what will insure the universal network legitimacy or where each offer of the other networks becomes less attractive. The new network shall be the more attractive since alternative existing solutions may be considered by the companies (competing technologies or adaptations of existing technologies), motivated by the short term needs of the firms.

The atomization of "customized" solutions, efficient in regard of some groups of users, but inefficient considering a global point of view creates an inertia (linked with important sunk investments) slowing the spring of the universal network.

Viability of the universal network is also weakened by the difficulty to elaborate a pricing structure in the sense of service differentiation on the network. This pricing structure leads to an increased possibility of losing the constraint. For instance, the coexistence of both technologies and the pricing problems highlighted by this situation causes a loosening in the constraint on the consumer nondiscrimination.

Considering all those obstacles, is it possible to have an actor that guarantees the efficiency of the universal network, that is an entity which sets up some rules favouring the universalisation of networks (ease the technical consensus), and insures a satisfying service differentiation which encourages all attempts to create alternative networks prejudicial to efficiency?

It is worth thinking whether market mechanism set-ups are enough to insure the universal network efficiency, or conversely, must it be guaranteed by the regulator? This leads to the credibility of the regulator and his influence on the players to "play the game".

The broadband network emergence

a. *The diffusion process of services on multiservices networks*

The diffusion process of services on integrated networks is more tricky than on specialised networks. More precisely, we have shown the existence of a reciprocal network externality which hinders the development of services on integrated networks because the critical mass is bidimensionnal (Joram, 1993).

Some other papers (Henriet and Volle, 1987; Allen, 1988) show that the development of a diversified service supply can be eased, if the operator succeeds to create something close to a "market place". That is to say, an organisation around the network that facilitates the meeting of services suppliers and of the users, or, in other words, that bounds the transaction costs.

Beyond this idea of a "market place", there is a dynamic of coordination between agents focusing on the technical coherence of the network (equipment compatibility, terminal standardisation, etc.).

Last but not least, the problem we are confronted with is: who controls the market parameters, mainly the tariffs structure (network access charge for the service supplier, price of the service)?

b. *The transfer of users to the broadband network*

As the main part of the services supplied on the broadband network are already offered on specialized networks, the operator thus has to ensure the transfer of these users on to the broadband network. The incentive mechanisms of this transfer would be quite complex, since the switching costs from existing networks to the broadband network are high.

These problems brought up by the emergence of the broadband network have to be related with the pricing issues: can one subsidize both the start-up of the broadband network and the transfer of users from specialised networks?

A comparative analysis

A brief comparative synthesis of French videotex and ISDN experiences highlights the previous factors (organization, pricing, consumer switching). This synthesis gives some pieces of information on the success factors of multiservices networks.

In the Minitel case, the actors coordination is the basis of its success. Actually, the small services suppliers have accepted to discharge themselves from important firm's prerogatives (the prices are fixed by the network operator), but they obtained in return an easier access to the network, and the openness to a mass market (the Minitels were supplied free of charge). The large service suppliers which accepted to use this network, who were thought to be reluctant not to keep the control on prices, proposed some services with humble specific profitability but running for some other global company goals.

Finally, the Minitel is more than a multiservices network. It is a real product innovation for consumers, and a real process innovation for enterprises. These characteristics are as important as the organization of the market of services in the success of the Minitel.

Conversely, the difficult start-up of ISDN in France, shows the difficulties to make an integrated network viable when many substitutes exist. For instance, when the ISDN started, the advantages of integrated access did not appear obvious for firms. Firstly, the differentiation of services on ISDN was quite low in comparison with the differentiation on specialised networks (low option value). Secondly, at the beginning, ISDN did not propose really new services which would justify to switch to it, as was the case with Minitel.

To sum up, ISDN has two concurrent problems. A too low differentiation of services, and the lack of new services. Moreover, the penetration on existing services is very strong. This creates an inertia to switch to ISDN. However, the situation is changing with the price cuts last year tending to boost the ISDN market.

Conclusion

To summarize the previous points, we would like to fix the main viability conditions of the broadband network.

If the integrated network is efficient (on the welfare's side), coordination is necessary to its emergence, or, to its viability. Actually, it is while starting up that the viability of the broadband network can be easily contested. The difficulties in reaching the critical mass are sharpened if the transfer of consumers on the broadband network has to be ensured; setting up an incentive pricing structure suggests the use of cross-subsidies which would be efficient only if the operators are protected from skimming strategies.

The efficiency threshold reached, the broadband network might be thought less vulnerable. It is probably easier to let the competition play its part.

The goal of the broadband network possibly amounts not to oppose but to associate both strategies of coordination and competition between actors.

Notes

1. These costs can be monetary (differences in prices), can concern the change of using routines, or can appear if the switching involves equipment replacement.

References

Allen, D., 1988. New telecommunications services: networks externalitites and critical mass. *Telecommunications Policy*, September 12:257–271.

Antonelli, C. (ed), 1992. *The Economics on Information Networks*, (p. 477). Amsterdam: North-Holland.

Bancel-Charensol, L., 1990. Demande de raccordement au réseau et offre de service de télécommunication. *Annales des télécommunications*, 45(7–8):367–376.

Boiteux, M., 1956. Sur la gestion des monopoles publics astreints à l'équilibre budgétaire. *Econometrica*, 24:22–40.

Farrell, J., Saloner, G., 1986. Installed base and compatibility: innovation, product preannouncements, and predation. *American Economic Review*, 76:940–955.

Farrell J., Saloner G., 1985. Standardisation, compatibility and innovation. *Rand Journal of Economics*, 16:70–83.

Henriet, D., 1986. Externalité d'encombrement et tarification au trafic: approche théorique des justifications et de la vulnérabilité par la concurrence. *Annales des Télécommunications*, 42(1–2):13.

Henriet, D., Volle, M., 1987. Service de télécommunication: intégration technique et différenciation économique. *Revue Économique*, 2(3):459–474.

Joram, D., 1993. *Le passage des réseaux spécialisés aux réseaux multiservices: modèles de diffusion et restructuration du secteur des télécommunications*, Thèse de Doctorat de l'Université de Caen.

Katz, M.L., Shapiro C., 1985. Network externalities, competition, compatibility. *American Economic Review*, 75(3):424–440.

Katz, M.L., Shapiro C., 1986. Technology adoption in the presence of network externalities. *Journal of Political Economy*, (94):822–841.

Kridel, D.J., Lehman D.E., Weisman D.L., 1993. Option value, telecommunications demand, and policy. *Information Economics and Policy*, 5(2):125–144.

Mitchell, B.M., Vogelsang, I., 1991. *Telecommunications Pricing – Theory and Practice*. Cambridge University Press.

Phan, D., 1992. The diffusion process of value added services on a telecommunication network: the example of French videotex services on "kiosque Minitel". *ITS Conference*, 20 April 1992, Sophia Antopolis, France.

Tirole, J., 1988. *The Theory of Industrial Organisation*. Cambridge: MIT Press.

Beyond Competition: The Future of Telecommunications.
Edited by D. Lamberton.

CHAPTER 11

Telecommunications demand: a new approach to media choice problems

Hans Ouwersloot
Department of Spatial Economics, Faculty of Economics and Econometrics, Free University Amsterdam, Amsterdam, The Netherlands

1. Introduction

The concurrent developments of deregulation of the telecommunication markets and the rapid progress of telecommunication technology – often labelled with the term "telematics" – which encompasses more specifically the synthesis of telecommunication and computer technology, results in great policy and scientific interest in the field of telecommunication demand. Additionally, given the increasing importance of information – the object which is transmitted using telecommunication media – for modern postindustrial societies, it can be foreseen that the field will become of central importance to economics.

One of the interesting questions which arise in telecommunications demand is concerned with media choice. Such a choice is made both by the supplier and the consumer of information, and their choices are clearly interrelated. The supplier has to choose media (one or more simultaneously) to disseminate his information, while the consumer has to scan media to find the information he is looking for. As the supplier and consumer are, in many instances, independent, a matching problem may arise. This may be most clearly illustrated in the labor market: firms may advertise a vacancy in medium A while the best candidates can be found among the users of medium B. This example makes clear that both sides have an interest in the matching of information streams.

In this paper media choice is studied at the supplier's side on a micro level. Thus, given the need to communicate some information, the choice of the supplier for some medium is analyzed. In this context, there is no concern for matching problems as the receivers are all assumed to scan all media involved. Hence, this paper concentrates on the partial problem of media choice by the sender and analyzes the motives and reasons which influence this choice.

Section 2 starts by describing two conceptual models. The first puts media choice in the broader context of information exchange. The second presents and organizes

---

Address for correspondence: Hans Ouwersloot, Department of Regional Economics, Free University, de Boelelaan 1105, 1081 HV Amsterdam, The Netherlands. Tel.: +31-20-444-6093. Fax: +31-20-444-6005.

the aspects which influence the media choice, and their mutual relations. Media choice being a typical example of a discrete choice problem. Section 3 discusses these type of models. Both discomfort with the generally used logit model, and an impetus from the conceptual model of Section 2 lead to considering a less familiar model: the Elimination-by-Aspect (EBA) model, originally developed by Tversky (1972a, 1972b). This model is presented and discussed and its use in economics literature is reviewed.

In Section 4 the data is presented. Section 5 presents the analysis of the data. For reasons of familiarity and comparability, both the logit and the EBA model are used. The models are also tentatively compared. Section 6 concludes this paper.

2. A theoretical model of media choice

A basic model of information exchange is Fig. 1, taken from Fischer et al. (1992). The model describes the choices that have to be made given an intrinsic, exogenous need for communication, i.e., information exchange. This model is embedded in a network approach to communication. According to the model, the first decision is whether this need for communication can be satisfied within the existing network or not. If not, the sender needs a new node in his network; this branch of the model is treated elsewhere (Ouwersloot (1994), chapter 4, Ouwersloot and Rietveld (1993)).

When the communication need can be fulfilled within the existing network, the sender can choose between travelling and using (tele)communication media. The model shows that this choice depends on the supply of communication facilities

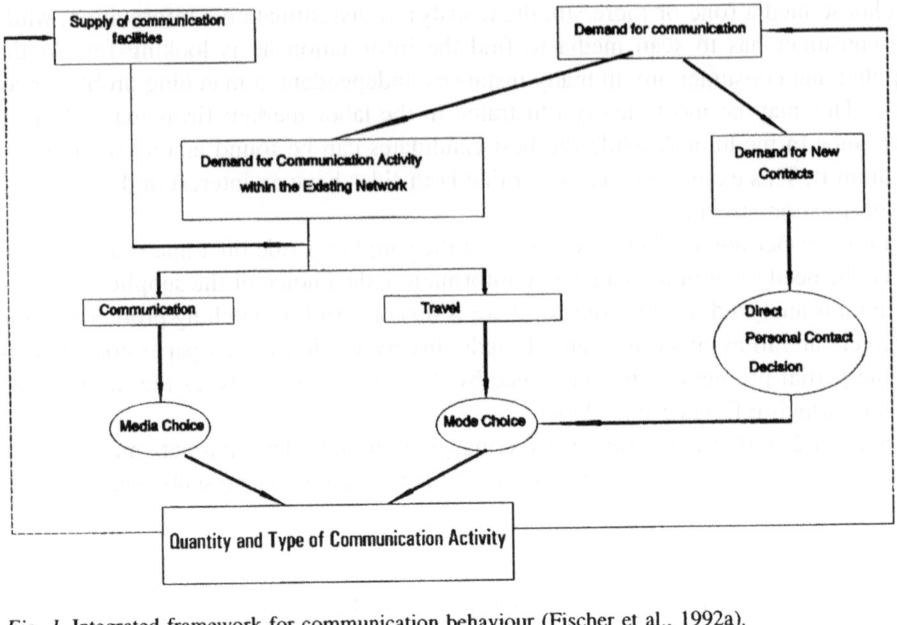

Fig. 1. Integrated framework for communication behaviour (Fischer et al., 1992a).

including travelling conditions. If the sender decides to travel he has to make a travel mode decision (car, train, plane etc.). When he decides to (tele)communicate he has to choose a medium (mail, phone, fax, video conferencing etc.). Finally, the model assumes that the result of the communication activity influences both the demand for and the supply of communication in the next period.

An interesting feature of this model is the integration of travel and communication decisions. It illustrates the well-known duality that travel can be both a substitute for and a complement to telecommunication. For communications within the network they operate as substitutes but, in order to add nodes to a network, travelling precedes telecommunication, and they are complementary.

In this paper we will concentrate on the media choice in a narrow sense, including travelling as an option, not for establishing a new contact, but merely for the information exchange. Fischer et al. (1993) present a model to analyze this media choice. Their model is inspired by Moore and Jovanis (1988), see Fig. 2. We use the Moore and Jovanis version as we consider it more appropriate for our purposes.

In this model a sharp distinction is made between the communication activity on the one hand, and media on the other. In the upper left part of the figure (boxes 1 to 4), the interplay of the indicated characteristics – communication activity characteristics, characteristics of the sender-receiver relationship, individual and environmental characteristics – leads to the determination of what is important in evaluating the media. On the other side of the figure (boxes 5 to 7), the media are evaluated: the attributes of the media are determined.

For example, the characteristics of the communication activity may be that the

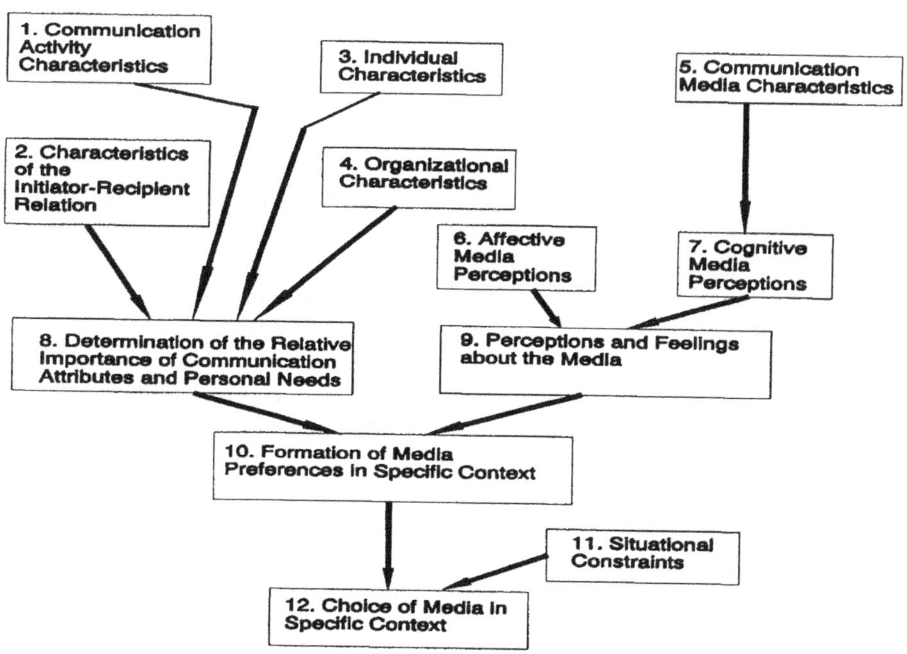

Fig. 2. Media choice model (Moore and Jovanis, 1988).

message is long, that transmission is urgent, that verification is important etc. A characteristic of the sender-receiver relation is, for example, whether the receiver belongs to the network of the sender. Attributes of media may be that transmission is fast, the medium is easy to operate, instantaneous interaction between sender and receiver is possible etc. Note that some characteristics of the communication activity correspond to attributes of the media, while others have no clear counterparts: urgency of the message corresponds to the speed of transmission, but no such direct correspondence exists for the length characteristic of the message, nor for the ease of operation attribute of the media.

The confrontation of characteristics of the message and attributes of the media leads to the formation of media preferences. The most preferred alternative is then determined. To arrive at a final choice, Moore and Jovanis – as well as Fischer et al. – include the possibility of situational constraints that may force the sender to choose a less preferred option. In our view, however, constraints are preferably incorporated in the evaluation of media attributes or characteristics of the communication activity, or even by restricting the choice set to available alternatives (see Section 5.1).

For an empirical application, the general model outlined above has to be operationalized. The most important point that needs to be specified is the mechanism that underlies the final formation of media preferences and the choice of media. As this is a typical example of discrete choice, in the next section we discuss some important features of discrete choice models, that, in the end, give support to our choice for the EBA model.

3. Operationalizing the model: discrete choice models

To use the model of Fig. 2 empirically, its three central concepts – perceptions of media (box 9), identification of characteristics of the communication activity (box 8), and formation of preferences (box 10) leading to a choice of a medium – have to be specified. Since the former two are closely related to the available data, their specification will be discussed later. Here we concentrate on the formation of preferences and choice on a microlevel. This is an application of multinomial discrete-choice models. In this section we will show that familiar models like logit and probit are less-well suited for the present problem, in particular when compared to the EBA model.

In his thorough discussion of the analysis of discrete choice, McFadden (1981) makes a distinction between Random Utility Maximization models (RUM) and Probabilistic Choice Systems (PCS).

The central feature of a RUM is that each alternative gives rise to a deterministic level of utility. For the correction of misconceptions of the model maker – specification and measurement errors – a random error term is added to the utility of each alternative. The overall levels of utility of each alternative are compared and the one with the highest utility is chosen.

A PCS, on the other hand, requires no utility-theoretic underpinning. It directly describes the probabilities with which the alternatives are chosen. Although a great deal of sophistication can be incorporated in the specification of the probabilities, leading to far from trivial models, these models are basically of a statistical nature.

Obviously, RUM provide a firmer theoretical basis for analysing multinomial choice problems, and are therefore principally preferred. Yet, for evaluation and estimation purposes they have to be transformed into PCS, which is achieved by choosing a probability distribution for the error term that leads to a tractable PCS formulation. On the other hand, it remains unanswered whether each PCS is, by definition, related to a RUM. It is conjectured that this is not the case. The discussion concerning PCS and RUM is carried much further in McFadden (1981), but the points above suffice for our purposes.

In practice, three (types of) PCS are considered: probit models, logit models and elimination models. Since logit and probit models belong to the standard package of researchers, we will only discuss those features of these models that we need for our purposes. On the other hand, we will pay some more attention to the less well-known elimination models to illustrate the contrast with the standard models.

Logit models are credited to Luce (1959) who derived them as a PCS from the Independence of Irrelevant Alternatives (IIA) axiom. Later, McFadden (1974) proved that the logit model is related to a RUM by specifying the mutually independent error terms to be additive to utility and having an Extreme Value (or Weibull Type I) Distribution.

In general, however, it is acknowledged that IIA is not a desirable property of the model, in particular where the application raises doubts about the independence of the alternatives. One way to solve the IIA problem is the use of Multinomial Probit Models (MNP) (for a general overview of models that solve IIA, see Timmermans (1984), or Timmermans et al. (1992)).

The probit model results from the assumption that the error terms in the RUM are additive to utility and have a multivariate normal distribution. This distribution is partly specified by the covariance matrix. When this matrix is assumed diagonal, the errors are independent and the model suffers from IIA again. However, assuming a nondiagonal covariance matrix implies correlation among the errors, which removes IIA.

The major drawback of the multinomial probit model is its computational infeasibility. In general, the estimation of n-alternative models involves the evaluation of (n–1)–multiple integrals. Multinomial probit is, therefore, only feasible for small problems, say with four or less alternatives.

The obvious way to overcome the IIA problem in a logit formulation is to drop the assumption of independence of the error terms. This leads to so-called Generalized Extreme Value (GEV) models (McFadden, 1981). These models allow for correlation among the errors of the alternatives, which is a way of modelling similarity of alternatives (cf. the MNP model).

The general structure is usually not developed as it is analytically complex. Instead, a hierarchical or tree structure is imposed on the choice set, in a way which represents the assumed similarity. In general, this tree can have several branches.

Each branch is then treated as a standard MNL model. The complete model, therefore, carries the name Nested MNL (NMNL).

The NMNL model overcomes the IIA problem but it requires two strong assumptions. In the first place, the tree structure must be defined beforehand. Although it is reasonable to suppose that decisions are made according to a tree structure, it is an unattractive feature of the model that the researcher has to define this tree. Secondly, the model assumes that the similarity between alternatives is independent of the circumstances. This latter drawback holds even for GEV models in general, as well as for MNP modelling, since they all assume a fixed covariance matrix.

Thus the underlying tree structure of decision making is supposed to be fixed. It is, however, not difficult to construct an example where this is not a reasonable assumption. Referring to the famous blue bus/red bus example,[1] suppose now that the difference between the buses is not their color, but whether or not they have a heating installation. Clearly this difference makes sense when a choice has to be made in wintertime, but it is irrelevant in summertime. Thus, heated bus and car will be considered as similar in the winter, while the two buses are the similar alternatives in the summer. A proper model should take such differences into account. This leads us to consider elimination models, which indeed have the flexibility to incorporate such subtleties.

Although variations exist, we will concentrate on, and use, the standard EBA model, as developed by Tversky (1972a, 1972b) exclusively. EBA is the purest form of the elimination models and, therefore, preliminary to its derivatives.

We assume here that the basic concept of EBA is known: alternatives are characterized by aspects and a choice is made by consecutively selecting aspects and eliminating those alternatives that do not possess the selected aspect until one alternative remains, which is then chosen. The random process by which aspects are selected is described by weights attached to these aspects.

Some basic but fundamental features of the model warrant attention. Firstly, in each round of evaluation, only those aspects are considered which are possessed by at least one of the alternatives. On the other hand, aspects that all alternatives possess are not considered. These rules guarantee that, upon consideration, at least one but not all alternatives are eliminated. Consequently, regardless of how many aspects are defined, at most n–1 of them need to be evaluated before a choice is made – n being the number of alternatives – because in each round at least one alternative is eliminated. Secondly, given n alternatives, there can be at most 2^n-2 aspects: when there are that many aspects, the model is called full. Finally, it makes no sense to incorporate alternatives that are characterized by identical subsets of aspects. It also makes no sense to incorporate an alternative whose collection of aspects is a proper subset of the set of aspects of another alternative since, according to the choice mechanism, the alternative with the smaller collection of aspects can never be chosen.

The description above of EBA is one of calculating the probabilities that a specific alternative is chosen. It is thus defined as a PCS, but is it also related to a RUM? McFadden (1981) shows that EBA is a special case of an even wider class of elimination models he calls Elimination-by-Strategy (EBS) models. Since these models can be shown to be related to random preference maximization models, where preferences are determined by a lexicographic ordering on the aspects, the EBA model is also related to random preference models, and consequently to RUM.

In the subsequent discussion of the EBA model we will use the following notation:

1. Let C_1, ..., C_n be the alternatives, and let A_1, ..., A_m be the aspects. Define the 0–1 matrix AA with n rows and m columns, the i-th row corresponding to C_i and the j-th column corresponding to A_j. Consequently, element a_{ij} of matrix AA equals 1 if alternative i possesses aspect j and 0 otherwise. The AA matrix completely describes the structure of a choice problem.
2. Next, an ordered sequence of aspects, the consideration of which leads to a unique choice, is called a *sequel*. For instance (A_1,A_3,A_7) is one sequel, but (A_7,A_1,A_3) is a different sequel, since the order is important. Also, sequels do not necessarily have the same length, although this length is bounded by n–1 as shown above. Notice that the alternative with which a certain sequel is connected at least possesses the aspects of this sequel, and possibly, others too.
3. Finally, define S_1, ..., S_n the sets of sequels that lead to the choice of alternatives 1, ..., n respectively. If a certain S_i is empty, we say that alternative i is dominated. Also, if two columns of AA are identical, we cannot discriminate between those two aspects and one is redundant. A *well-defined* problem implies no redundant aspects and no dominated alternatives. When a set of aspects leads to a well-defined problem, we also say that these aspects can discriminate between the alternatives, and this terminology is also used for subsets of aspects with respect to subsets of alternatives.

For further discussion of the EBA model, including modification for practical purposes, refer to Ouwersloot (1994, chapter 5).

The usual way of exploiting the EBA model is to define a full AA matrix and analyze which aspects are important. This can lead to severe data problems, as many data points are required. This problem has been recognized and modifications of the EBA model have been developed. For instance, Tversky and Sattath (1979) suggest imposing a hierarchical structure on the decision process. They call this model Elimination-by-Tree (EBT). Since it is mathematically equivalent to a so-called Hierarchical Elimination Model (HEM) which is associated with preference trees, they therefore call this type of model PRETREE. Maddala (1983) calls it the Hierarchical Elimination-by-Aspects model (HEBA). Thus HEBA = HEM = EBT = PRETREE.

We consider this model as less interesting as it has the strong disadvantage of imposing a fixed tree structure. Whenever this tree structure is relevant, the GEV model is more attractive in view of its computational advantage. Rather, the flexibility of the EBA model is its most attractive feature, giving it a comparative advantage over (Generalized) Extreme Value models (for a concise discussion of the HEBA model, see Maddala, 1983).

Another, more promising way to use the model is to hold on to the EBA specification, and to select on a priori grounds a restricted number of aspects for evaluation. In addition, the weights of the aspects are parameterized depending on exogenous variables, allowing the estimation of the parameters of these variables. The EBA model is used in this way in Section 5.

In logit and probit models the similarity of alternatives is captured in the correlation coefficients between the error terms that are associated with the alternatives. These correlations are a natural measure of the similarity of alternatives. Although EBA is a RUM, it is not explicitly defined what probability distributions are involved, and thus correlation parameters cannot be calculated. Ouwersloot (1994) suggests that the similarities between alternatives can be measured in a loose way by taking the weights they have in common divided by the sum of the weights, at least one of them possesses.

It should be clear that EBA circumvents the problems of MNL and MNP models. The IIA problem is solved as has been already shown by Tversky (1972a). In fact dissatisfaction with the IIA problem was one of the motives to develop the EBA model. Moreover, by having the weights depending on exogenous variables, this similarity is not fixed.

Also note that the EBA model fits very well to the conceptual model of Fig. 2. In concrete terms, Box 9 (Perceptions and feelings about the media) is reflected by the structure of the AA matrix, while the essential point of Box 8 (Determination of the relative importance of communication activities and personal needs) is captured in the parameterization of the weights of the aspects. Finally, Box 10 (Formation of media preferences in specific context) is modelled as the EBA choice mechanism. Thus, all three features are explicitly accounted for in the EBA model. In contrast, with MNL or MNP Boxes 8 and 9 are in effect taken together, leading to a "determination of relative attractivity of media".

We end this section by some additional remarks on the EBA model and a discussion of some of its applications.

A noteworthy feature of the EBA model is that the logit model can be defined as a special case of it. To see this, recall that the result of a logit model is the following formulation of the probability an alternative is chosen:

$$P_{\text{Logit}}(C_i \,\text{is chosen}) = \frac{e^{x_{c_i}\beta_{c_i}}}{\sum_j e^{x_{c_j}\beta_{c_j}}}$$

We compare this to the case that the AA matrix of the EBA problem is the identity matrix, and the weights are given by $WA_1, \ldots, WA_m$. In that case:

$$P_{\text{EBA}}(C_i \text{ is chosen}) = \frac{WA_i}{\sum_j WA_j}$$

Hence, when the weights in the EBA model are parameterized as:

$$WA_i = \exp(X_{C_i}\beta_{C_i})$$

the EBA model and the logit model coincide. So, the logit model is a special case of the EBA model, viz. when the EBA model is specified with only alternative specific aspects. In that case the AA matrix is the identity matrix.

A minor drawback of the EBA model is that the required parameterization of the weights involves these weights to be strictly positive, instead of nonnegative. It is, therefore, not straightforward to compare competing specifications of the EBA model by statistical tests, as they are not nested in the sense that one specification follows from another by setting the weights equal to zero. This holds, in particular, for testing the logit model as a special case of EBA.

As far as we know, the EBA model in its pure, general form has never been used. On the other hand, the PRETREE model, or hierarchical EBA, has occasionally been used. Tversky and Sattath (1979) give an application with experimental data. McFadden (1981) uses a PRETREE model for a comparison of this model with multinomial probit and logit. Gensch and Ghose (1992) give a number of references using one form or the other of PRETREE, particularly in the marketing literature.

Occasionally it is suggested that a disadvantage of elimination models is the use of attributes, rather than continuous variables. Manrai and Sinha (1989) addressed this problem by proposing an Elimination-by-Cutoffs (EBC) model. Instead of an aspect, a cut-off level of a continuous variable is selected: all alternatives that do not meet this level for the selected variable are eliminated. At a later stage in the elimination process, the same variable may be set at a higher level so that it eliminates at least one of the remaining alternatives. Clearly this can be interpreted as a reformulation of the pure EBA method.[2] The results of an empirical analysis gives slightly better statistics for EBC than for the multinomial logit, but recall that such comparisons have to be interpreted with care as the models are not nested.

Gensch and Goshe (1992) construct a preference tree in the Tversky and Sattath sense by factor-analysing attribute scores on the alternatives. They call their model Elimination-by-Dimensions (EBD) as the factor analysis results in the most important dimensions of variation among the alternatives. Similar to Manrai and Sinha they thus address the problem that EBA is only formulated in terms of attributes. The results of Gensch and Ghose are interesting. Their method, however, strongly depends on the supposed relevance of the tree structure, which is crucial for the PRETREE method. Moreover, we have already suggested that when a fixed tree structure may be assumed, the GEV model is to be preferred.

The model Fader and McAlister (1990) use comes closest to a true EBA model. Their approach is, however, limited as they identify only one aspect, viz. whether an alternative is being actively promoted or not. An interesting finding of Fader and McAlister is that the EBA model assigns very small probabilities to some aspects ("often in the order of 10^{-10}"), a result which is similar to our empirical findings (cf. subsection 5.2).

Summarizing the previous discussion, we have shown that the EBA model is a rich and flexible model, incorporating the familiar logit model as a special case. It does not suffer from the IIA syndrome, yet it is based on a random utility maximization model. In general we conjecture that the EBA model describes a decision structure which is highly relevant to real decision problems. At least to some extent, people make decisions by sequentially excluding alternatives from their choice set, by considering the aspects of the decision in decreasing order of importance.

A drawback of the EBA model, is that an aspects structure has to be superimposed in advance. On its own this is not too serious since analysts always have to make such decisions (e.g., which exogenous variables to include), but the serious point is that there is no possibility to test formally various versions of the model against each other, as competing models are not nested. This leaves the chosen specification subjective, which should preferably be avoided. In our view, however, this subjectivity is less restrictive than imposing a tree decision structure, as would be required for the analysis of GEV models.

Having discussed the models that may be used to analyze media choice, we now turn to the empirical analysis. In the next section we present our data set. Section five continues with the estimation results.

4. The data

The media choice decision is analyzed using data from a survey held among European academics in 1991/1992 at five locations – Vienna, Zurich, Amsterdam, Liverpool and Loughborough. This survey was administered by the NECTAR working group "Barriers to Communication", see Button et al. (1993).

In one of the questions, respondents were asked to make a media choice in a hypothetical situation in which they had to transmit a message. The following media were supposed to be available: traditional mail, courier service, telephone, fax, and e-mail. It was explicitly stated that the respondents could assume that the receiver had all the facilities associated with these media.

Different versions of the question sketched different situations, with respect to both the message and the circumstances. In this sense the message was long or short, complex or simple, while the circumstances were described as urgent or not-urgent, confidential or not-confidential. Altogether 16 (2^4) different versions were composed and randomly distributed among the respondents. In Loughborough and Amsterdam each respondent was asked to make a choice with respect to 4 different, pairwise opposite, versions. In the other samples respondents had to make a decision twice, also in opposite situations.

A typical example of a media choice situation is:
You have to let a colleague you have known for some time have a *confidential* message with the following content: "John Smith is applying for a job. Could you please provide a reference."

Your colleague is working at the University of Munich. The information has to reach him *tomorrow*. Which one of the communication media mentioned below would you choose? (It is assumed that the receiver of the message may use any of these media.)

In the opposite case the message would be nonconfidential, and would have to be received within two weeks. Similarly, the above being a short, simple message, a long message would be described as approximately 10 pages of text, and complexity is achieved by the presence of figures and tables. Since these characteristics can be represented in many ways, we give an overview of the cases in Appendix A.

Some potentially relevant characteristics were set at a constant level. "Your colleague" is always located in Munich,[3] and invariably "you have known him for some time". Thus, it is implicitly assumed that the semantic models of you and your colleague are sufficiently coordinated.[4]

Because of the repetition of observations on each respondent the 691 academics in the survey generated 1647 usable observations. Some econometric consideration of this repetition of observations on respondents are treated in Ouwersloot and Rietveld (1994). As these authors find no evidence that this peculiarity causes important effects, the point is neglected here.

Table 1 shows frequency distributions of the responses. From this table we observe quite some dispersion of media preferences between cities. Most notable in this respect is the large "market share" for fax in Loughborough and Liverpool, compared to Switzerland, and an opposite effect for telephone. A similar pattern, but less pronounced, can be observed for courier service and mail. Finally, Austrians seem to have more than average preference for e-mail.

5. Empirical analysis

In this section we will first use the familiar logit model to analyze the data on media choice. Besides straightforward analysis we will consider whether some extensions of the model improve its performance. Then we will turn to the application of the EBA model.

5.1 Logit analysis

As became clear in Section 4, our basic interest is with the effect of characteristics of the message and circumstances on media choice. Table 2 shows these effects. For

Table 1. Frequencies of media use.

Media	Absolute frequency	Relative frequency overall	Austria	Switzer land	Nether lands	Lough borough	Liver pool
Mail	626	38.0	36.9	43.5	40.2	35.3	30.5
Courier Service	175	10.6	7.5	7.5	12.4	11.9	12.8
Telephone	295	17.9	23.2	23.8	17.5	8.9	16.5
Fax	461	28.0	24.0	20.1	25.4	40.3	34.1
E-mail	90	5.5	8.4	5.1	4.5	3.6	6.1
N	1647		371	214	595	303	164

this table we only included the characteristics of the choice situations as described above. Note that for Tables 2–4 traditional mail is taken as the reference alternative.

In Table 2 superscripts are used to indicate which variables are significantly different from each other. The results generally confirm our expectations: urgency favors fast media; complexity and length disfavor telephone and e-mail; confidentiality disfavors fax; and courier service is in general not popular (measured by its constant term). There are also some effects that are harder to explain: Why does length favor courier over mail? Why doesn't confidentiality favor telephone and e-mail over the others?

Overall we find that mail is the most preferred alternative, and that only urgency really is a reason to use other media. This is an intriguing conclusion. At first sight,

Table 2. Logit estimates of media choice.

Variable	Estimate	(standard error)
Constant		
Courier service	–4.08	(0.34)$^{m, t, f, e}$
Telephone	–0.13	(0.17)$^{c, f, e}$
Fax	–1.40	(0.19)$^{m, c, t}$
E-mail	–1.70	(0.25)$^{m, c, t}$
Complex (1 if message is complex, 0 otherwise)		
Courier service	–0.26	(0.22)$^{t, e}$
Telephone	–2.76	(0.23)$^{m, c, f, e}$
Fax	0.09	(0.17)$^{t, e}$
E-mail	-1.19	(0.26)$^{m, c, t, f}$
Confidential (1 if message is confidential, 0 otherwise)		
Courier service	0.96	(0.24)$^{m, t, f, e}$
Telephone	0.15	(0.20)$^{c, f, e}$
Fax	–1.54	(0.18)$^{m, c, t, e}$
E-mail	-0.68	(0.25)$^{m, c, t, f}$
Long (1 if message is confidential, 0 otherwise)		
Courier service	1.02	(0.24)$^{m, t, f, e}$
Telephone	–3.66	(0.30)$^{m, c, f, e}$
Fax	0.07	(0.17)$^{c, t}$
E-mail	–0.35	(0.25)$^{c, t}$
Urgent (1 if message is urgent, 0 otherwise)		
Courier service		
Telephone	3.70	(0.25)$^{m, e}$
Fax	3.38	(0.23)$^{m, f, e}$
E-mail	3.91	(0.20)$^{m, t, e}$
	2.78	(0.26)$^{m, c, t, f}$

Number of observations:	1647
Log-likelihood:	–1548.44
Mean log-likelihood:	–0.94016

Note: For each variable it is indicated which of the estimates are significantly different (at the 5% level) from the other alternative dependent estimates for that variable. For instance, the constant term for courier service is significantly different from all other constants; the parameter of urgency for fax is significantly different from the urgency parameters of mail, telephone and e-mail etc.

courier, e-mail, and fax are media that each try to keep the characteristics of mail, but in addition offer rapid transmission. In EBA terminology: these media try to dominate mail. Obviously they do not succeed, but the logit analysis does not provide clues to the reason for this. One reason might be that we have not included a generalized cost measure, although it is the case that the newer media are more expensive or more time consuming. Unfortunately, however, we had no data on cost.

A number of additional analyzes were carried out with the logit model of which we will only mention the conclusions. (Ouwersloot (1994) discusses these analyses in more detail.)

When respondents report to have no access to certain media, this can be incorporated in the model by either including accessibility dummies or by restricting the choice sets of these respondents (cf. Section 2). The estimations showed the expected results (strong positive coefficients for the accessibility parameters, positive shift for the constant terms in the case of restricting the choice set) but the interesting result was that the latter way of treating was statistically superior to the former way, as measured by the increase in mean log-likelihood.

Inclusion of perceptions and feelings — in line with the Moore and Jovanis approach, see Section 2 — revealed that experience was important for fax and e-mail use, but not for courier and telephone. Other measures like appreciation were not successful in estimation.

The inclusion of personal characteristics mainly revealed that the probability of using modern media (in particular e-mail) decreases with age.

In studying the differences between the five locations at which the survey was held in the first place confirmed the tentative conclusions of the previous section. It further revealed that respondents from different locations show different perceptions about the confidentiality of telephone use: British academics find confidentiality a reason not to use the telephone, while Swiss and Austrians show the opposite pattern.

Most important, however, is the general impression that appears from these analyses, that the results are quite robust to varying specifications and that throughout the behavior of academics is quite homogenous over different locations.

5.2 Analysis with the EBA model

Use of the EBA model requires the determination of the alternative aspects matrix (AA) and the parameterization of the weights associated with the aspects. This last point was solved by assuming a log-linear form for the weights, which is in line with the logit model. Moreover, this form guarantees that the weights are positive which is required by the EBA model. Thus $W_i = \exp(X_i'\beta_i)$ for all aspects i = 1, ..., m. The weights are thus determined by the exogenous variables that can describe the characteristics of the message, or simply reflect individual or circumstantial points.

Our principal interest is explaining the choice of media, under the varying characteristics as described in Section 4, as we did in the preceding subsection. Therefore, we proceed by constructing an AA matrix by referring to the characteristics of the messages. In particular, we try to identify those aspects which the given characteristics appeal to and make statements about which alternatives possess these aspects.

Complexity
Complexity relates to the presence of tables and figures in the case of (the outline of) a paper or pictures etc. in a brochure. In one respect these are comparable forms of complexity, since they are not easily transposed on to another carrier. That means, assuming that the message is on paper, they are not easily described by mouth, nor are they easily put into a file for sending them by e-mail. It must, however, be recognized that it is very likely that the tables and figures are already present on file, contrary to the brochure. This implies that this complexity is not described unequivocally (the description of the case gives no clues about it). Nevertheless, in general complexity is associated with the necessity to use another carrier than the one used by the original message. We label this aspect *carrier*, meaning that the original carrier can be used. It is possessed by mail, courier, and fax.

Another interpretation of complexity refers to the format of the message. Fax and e-mail make use of precisely defined formats, whereas the others are far more flexible media. Again, messages not in the right format require a transformation which is by assumption more difficult for more complex messages. This aspect is labelled *flexibility* and mail, courier and telephone possess it.

Confidentiality
Confidentiality is associated with three aspects. In the first place it demands certainty of correct receipt. Secondly, it demands the protection of the contents of the message. Finally, it requires that no other persons are aware of the interaction of sender and receiver, or less strictly that there is control over who is aware of the interaction. Notice that the chosen forms of confidentiality all relate to the circumstances of the receiver. This is important since the sender has a great deal of control over his own circumstances but much less over the circumstances of the receiver.

We assume that the certainty of receipt is achieved by the addressability which is possessed by all media. It may, however, be desirable that this receipt is instantaneously confirmed which is most *directly* achieved by telephone only. Contents of the message are well protected by all media except fax. *Control* over awareness of interaction is best achieved by mail and courier.

Length
Length of the message refers to changing the carrier – like complexity – and so has to do with the earlier identified aspect *carrier*. A second point is whether the channels the media use put technical or practical constraints on the message. Mail and courier are almost unrestricted, whereas the others cannot cope with all volumes of messages. Hence, we define an aspect labelled *capacity*, which is possessed by mail and courier.

Urgency
Obviously transmission time is crucial. The urgent messages in the questionnaire required transmission in one day which is guaranteed by all media except mail. But there are still gradual differences between the remaining alternatives. Firstly, when something goes wrong using courier, there is no second opportunity. Secondly, when things fail with e-mail or fax the communication may still be completed in time but the procedures of transmission have to be repeated again. Telephone, on the other hand allows for direct control and confirmation of receipt. This gives rise to three levels of speed of transmission: the mere possibility of *timely* receipt (all except mail); the *certainty* of timely receipt (all except mail and courier); the *direct* confirmation of receipt, that does not require a repetition of procedures (telephone only). Compare this to the EBC model of Manrai and Sinha (1989), discussed in Section 3.

Apart from these aspects which are associated with the characteristic of the specific situation, we assume that two additional aspects are important. Firstly, because of their long history we assume that mail and telephone are far more *familiar* to the respondents than the other media. Secondly, for all media except telephone the result of transmission is something which is *physical*. The receiver can store the message, and read and reread it whenever he likes. This we consider as a desirable aspect.

So, we arrive at the following AA matrix (note that its transpose is given below):[5]

AA' =

MCTFE	Aspect	Due to
1 1 0 1 0	Carrier (1)	Complex, Length
1 1 1 0 0	Format (2)	Complex
0 0 1 0 0	Direct (3)	Confiden, Urgency
1 1 1 0 1	Contents (4)	Confiden
1 1 0 0 0	Control & Capacity (5)	Confiden, Length
0 1 1 1 1	Time (6)	Urgency
0 0 1 1 1	Certainty (7)	Urgency
1 1 0 1 1	Physical (8)	
1 0 1 0 0	Familiarity (9)	

The aspects control and capacity are taken together since they lead to identical columns. Verify that this problem is well defined.

We modelled the weights of the aspects following the above given considerations. We started by using the characteristics of the message for aspects (1) to (7) as indicated above, and a constant term for all aspects. In addition, we included age as an explanatory variable for the aspect familiarity, as a reflection of the generally held belief that younger persons are less hesitant to the use of new media. Finally, the physical aspect was taken as the reference aspect.

We estimated the above-described model, using a self-written computer program in the programming language Gauss. The estimation method is maximum likelihood.

For computational reasons we started the analysis on the Amsterdam sample only. After having identified the best fitting model (see below) we estimated this model for the entire sample. On that occasion the results for the complete sample turned out to be surprisingly similar to those of the Amsterdam sample.

We started the estimations by including all nine aspects given above. The weights of the aspects were parameterized as indicated, including a constant term for each aspect. The weight of the physical aspect was fixed at 1. By construction of the model all parameter estimates should be positive. When an estimate is negative this is interpreted as a misspecification of the model and the aspect has to be deleted from the model. For example, the results of the first estimation showed an estimate of –22.5 for the complexity variable of the aspect flexibility. Consequently this aspect was removed at the second round.

The same happened to the aspect Direct. The second round estimations showed parameter values of –5.0 for the confidentiality variable, and of -5.4 for urgency. Therefore, Direct was removed in the third round. Notice that the model remains well defined without the two removed aspects.

The third round no longer showed negative parameter values. It was, however, not possible to calculate the covariance matrix of the estimates due to singularity problems, which raised doubts about this specification. Upon inspection of the estimation results, it appeared that all point estimates of the variables remained practically the same from round two to round three, with the exception of the estimates for the control & capacity aspect. This suggested it to be desirable to perform a fourth round with the C&C aspect removed.

In addition, the constant term of the carrier aspect was fixed at the value 10 in this fourth round, for reasons which will be made clear below. With these adaptations, the estimations resulted in a well-defined and interpretable model, which we subsequently estimated for the complete sample. As indicated above, the results were only marginally sensitive to this increase in the number of observations. The results of the estimation on the entire sample are reported in Table 3.

From Table 3 it appears that the constant terms for the aspects carrier, contents, and certainty, are quite large. Recall that the weights are defined in a log-linear way. In fact, irrespective of the values of the characteristics, the weights for these three aspects are so large that one of them will always be selected first in the EBA choice process, and another one of them second. As an illustration the weights of the aspects are given in Table 4 for the various combinations of characteristics.

Selecting two of these three aspects always makes the third one redundant. Thus, when carrier and contents are selected as the first two aspects, the choice set is restricted to mail and courier, and as neither of them possess the aspect certainty, this aspect has become redundant. Similarly, when contents and certainty are selected, only telephone and e-mail remain in the choice set, and as neither of them possesses carrier, this aspect has become redundant. Finally, when carrier and contents are selected, fax is the only remaining alternative and consequently is chosen. Thus, after the first two aspects are selected, the choice set is restricted to either {fax}, or {telephone, e-mail}, or {mail, courier}. The other aspects – time, familiarity, and physical

Table 3. Estimation results of the Elimination-by-Aspects model.

Aspect	Media that possess aspect	Variables exogenous to aspect	Estimate	Standard error
Carrier	M/C/F[a]	(Constant	10.00	----)[b]
		Complexity	1.83	0.13*
		Length	1.97	0.13*
Contents	M/C/T/E	Constant	10.80	0.11*
		Confidence	1.38	0.11*
Time	C/T/F/E	Constant	–1.68	0.22*
		Urgency	3.88	0.24*
Certainty	T/F/E	Constant	9.81	0.11*
		Urgency	2.15	0.11*
Familiarity	M/T	Constant	1.57	0.24*
		Age[c]		
		<30	–1.03	0.30*
		30–40	–0.29	0.27
		40–50	–0.26	0.27
Number of observations:	1647			
Meal log-likelihood:	–1.009			

Note: The reference aspect is Physical, possessed by Mail, Courier, Fax, and E-mail.
*Significant at the 5% level.
[a]M = mail, C = courier service, T = telephone, F = fax, E = e-mail.
[b]The constant term for this aspect is fixed at the value 10; see text.
[c]Reference group for Age variable is older than 50.

— are then used to make a choice when two alternatives are left. Thus, time and familiarity are used to discriminate between mail and courier, and familiarity and physical are used to choose between telephone and e-mail. The complete choice process is illustrated in Fig. 3.

The above analysis can be summarized as a two-layer process. In the first layer the aspects carrier, contents and certainty are exploited, in the second layer the other aspects. However, this means that one of the constant terms of the aspects of layer one has to be set at a fixed value, since otherwise any value, large enough to make this two layer structure effective could result, meaning that the model is in practice not identified. It was exactly this problem that forced us to fix the value of the constant term of carrier at 10 in the fourth round. This value was suggested by the results of an estimation in the fourth round, where a value of 10.8 was found, but which did not lead to the identification of a covariance matrix. The other parameter estimates were not influenced by the arbitrary choice of ten — with the exception, of course, of the constant terms for the other two aspects of the first layer, contents and certainty. We also experimented with the value 5 for the constant term of carrier, but

Table 4. Weights of the aspects for various combinations of characteristics.

Characteristics				Absolute weights EBA						Relative weights primary aspects			Relative weights secondary aspects		
Com	Cof	Len	Urg	Carr	Cont	Time	Cert	Fam	Phys	Carr	Cont	Cert	Time	Fam	Phys
0	0	0	0	2.2×10^4	4.9×10^4	.186	1.8×10^4	3.60	1.0	.247	.549	.204	.039	.752	.209
1	0	0	0	1.4×10^5	4.9×10^4	.186	1.8×10^4	3.60	1.0	.671	.240	.089	.039	.752	.209
0	1	0	0	2.2×10^4	1.9×10^5	.186	1.8×10^4	3.60	1.0	.094	.829	.077	.039	.752	.209
0	0	1	0	1.6×10^5	4.9×10^4	.186	1.8×10^4	3.60	1.0	.701	.218	.081	.039	.752	.209
0	0	0	1	2.2×10^4	4.9×10^4	9.03	1.6×10^5	3.60	1.0	.097	.216	.688	.663	.264	.073
1	1	0	0	1.4×10^5	1.9×10^5	.186	1.8×10^4	3.60	1.0	.392	.556	.052	.039	.752	.209
1	0	1	0	9.8×10^5	4.9×10^4	.186	1.8×10^4	3.60	1.0	.936	.047	.017	.039	.752	.209
1	0	0	1	1.4×10^5	4.9×10^4	9.03	1.6×10^5	3.60	1.0	.401	.143	.456	.663	.264	.073
0	1	1	0	1.6×10^5	1.9×10^5	.186	1.8×10^4	3.60	1.0	.426	.525	.049	.039	.752	.209
0	1	0	1	2.2×10^4	1.9×10^5	9.03	1.6×10^5	3.60	1.0	.059	.522	.419	.663	.264	.073
0	0	1	1	1.6×10^5	4.9×10^4	9.03	1.6×10^5	3.60	1.0	.435	.135	.430	.663	.264	.073
1	1	1	0	9.8×10^5	1.9×10^5	.186	1.8×10^4	3.60	1.0	.822	.163	.015	.039	.752	.209
1	1	0	1	1.4×10^5	1.9×10^5	9.03	1.6×10^5	3.60	1.0	.281	.399	.320	.663	.264	.073
1	0	1	1	9.8×10^5	4.9×10^4	9.03	1.6×10^5	3.60	1.0	.827	.041	.131	.663	.264	.073
0	1	1	1	1.6×10^5	1.9×10^5	9.03	1.6×10^5	3.60	1.0	.310	.383	.307	.663	.264	.073
1	1	1	1	9.8×10^5	1.9×10^5	9.03	1.6×10^5	3.60	1.0	.737	.146	.117	.663	.264	.073

Note: Com = Complexity; Cof = Confidentiality; Len = Length; Urg = Urgency; Carr = Carrier; Cont = Contents; Cert = Certainty; Fam = Familiarity; Phys = Physical.

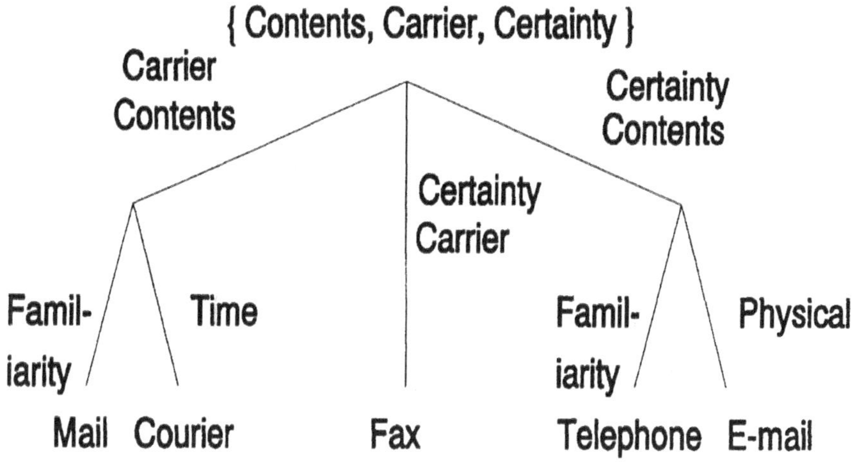

Fig. 3. Decision process according to the EBA model results.

this resulted in significant deviations from the estimations with the constant fixed at 10, and a lower value for the log-likelihood.

Continuing the discussion of Table 3, we find that the parameter estimates for the characteristics all have the expected (positive) sign and are significant. Also the age variable shows the expected result: older academics attach more weight to the familiarity of the media, as possessed by telephone and mail.

For a further interpretation of the outcomes of Table 3, and for comparison with logit outcomes, we constructed Tables 5 and 6. The computations for the logit model underlying these tables are based on the results of an estimation, where the probability of each alternative to be chosen is explained by the characteristics of the choice situation, and the age variable in 3 categories. Thus the same variables as in the EBA model were used, yet the logit model demands that each variable is used for each alternative, leading to 32 parameters, whereas the EBA model uses only 12 parameters. This parsimonious use of exogenous variables in EBA partly explains why the logit model performs better according to the value of the log-likelihood at convergence. For the EBA model this value was –1661.8, while for the appropriate logit model it was –1516.9. Notice that the Likelihood Ratio Test is not applicable here as the models are not nested. Aikake has suggested to that for nonformal testing the log-likelihoods for nonnested models should be corrected by the difference in number of used parameters: Akaike's Information Criterion, see Amemiya (1981). Clearly, the logit model outperforms the EBA model in that case too.

Table 5 shows the probabilities of each of the alternatives to be chosen for the various combinations of characteristics for both the EBA and the logit model. These probabilities can be compared by the observed frequencies which are also given. We do this by computing the squared errors of the prediction for each of the alternatives for each of the combinations of characteristics. The sum of these squared errors are

Table 5. Probabilities of choice from EBA model, logit model and observations.

Characteristics				Probabilities based on EBA					SSE
Com	Cof	Len	Urg	Mail	Courier	Teleph	Fax	e-mail	
0	0	0	0	.457	.024	.305	.013	.085	.019
1	0	0	0	.666	.035	.040	.247	.011	.009
0	1	0	0	.513	.027	.348	.016	.097	.022
0	0	1	0	.672	.035	.033	.252	.009	.023
0	0	0	1	.014	.036	.519	.287	.144	.057
1	1	0	0	.808	.042	.075	.055	.021	.002
1	0	1	0	.690	.036	.001	.270	.000	.012
1	0	0	1	.046	.116	.154	.641	.043	.042
0	1	1	0	.818	.042	.064	.058	.018	.006
0	1	0	1	.028	.069	.653	.069	.181	.029
0	0	1	1	.049	.123	.132	.659	.037	.049
1	1	1	0	.870	.045	.004	.083	.001	.017
1	1	0	1	.098	.245	.313	.257	.087	.037
1	0	1	1	.066	.167	.009	.755	.003	.007
0	1	1	1	.104	.261	.282	.276	.078	.131
1	1	1	1	.152	.382	.031	.426	.009	.012
									Sum .472
				Probabilities based on logit					
0	0	0	0	.448	.004	.453	.053	.042	.029
1	0	0	0	.830	.005	.017	.128	.021	.020
0	1	0	0	.469	.010	.478	.007	.036	.002
0	0	1	0	.787	.016	.006	.121	.071	.002
0	0	0	1	.022	.017	.690	.225	.046	.002
1	1	0	0	.928	.015	.019	.019	.019	.015
1	0	1	0	.806	.012	.000	.163	.019	.003
1	0	0	1	.063	.035	.039	.829	.034	.005
0	1	1	0	.865	.049	.006	.018	.063	.004
0	1	0	1	.027	.058	.835	.035	.045	.035
0	0	1	1	.055	.107	.012	.720	.107	.010
1	1	1	0	.920	.038	.000	.024	.018	.032
1	1	0	1	.186	.292	.118	.323	.081	.063
1	0	1	1	.049	.070	.000	.855	.025	.006
0	1	1	1	.100	.547	.022	.173	.158	.008
1	1	1	1	.130	.519	.001	.296	.054	.037
									Sum .272

(continued)

given in the last column for each situation. We find that on average the logit model performs better: the sum of all squared errors for EBA is .47, and .27 for logit.

In Table 6 we further explore these data by considering the effects of the characteristics. We study these effects by comparing the change in probabilities according to the observations, with those generated by the two models. The table

Table 5. (continued).

Characteristics				Probabilities based on sample				
Com	Cof	Len	Urg	Mail	Courier	Teleph	Fax	e-mail
0	0	0	0	.350	.010	.379	.136	.126
1	0	0	0	.724	.010	.000	.210	.057
0	1	0	0	.443	.021	.464	.031	.041
0	0	1	0	.776	.019	.019	.150	.037
0	0	0	1	.000	.018	.723	.205	.054
1	1	0	0	.837	.019	.087	.058	.000
1	0	1	0	.760	.019	.000	.192	.029
1	0	0	1	.030	.089	.030	.802	.050
0	1	1	0	.832	.089	.020	.020	.040
0	1	0	1	.050	.040	.700	.160	.050
0	0	1	1	.038	.048	.000	.798	.115
1	1	1	0	.767	.117	.010	.068	.039
1	1	0	1	.243	.140	.299	.290	.028
1	0	1	1	.067	.100	.011	.789	.033
0	1	1	1	.058	.548	.087	.192	.115
1	1	1	1	.068	.417	.010	.447	.058

shows the average change in probability for each alternative, and for each characteristic. Overall the logit model predicts better, with a minor exception for the effect of urgency.

We can draw three important conclusions from this section. In the first place, it has proven possible to operationalize the EBA model, and estimate a parameterization of it. Secondly, there is no doubt that on statistical grounds the logit model outperforms the EBA model. However, and this we consider the most important conclusion, the EBA model provides additional and useful insight into how a decision process like media choice takes place. This last conclusion offers new perspectives for the analysis of discrete choice problems.

6. Summary and concluding remarks

In this paper we studied the media choice problem. We started by describing a conceptual model. The first part of the model puts the media choice in a broader context of information exchange. Media choice may be different for an identical demand for communication within or outside the existing network.

Concentrating on media choice in a strict sense, we discussed a framework developed by Moore and Jovanis (1988), which we consider highly relevant for media choice. This model confronts requirements of the media, due to the characteristics of the communication activity, with the "cognitive and affective aspects" of the media. We noticed that characteristics may correspond directly to aspects ("urgency" and "transmission time"), but that the opposite can also be the case (the characteristic

Table 6. Comparing observed effects of characteristics with effects resulting from EBA and logit models

	Mean marginal effects				Absolute deviations	
	EBA	Logit	Sample		EBA	Logit
Complexity						
Mail	.093	.143	.119		.026	.024
Courier	.056	.022	.015		.041	.007
Telephone	-.213	-.289	-.243		.030	.046
Fax	.124	.161	.145		.021	.016
E-mail	-.059	-.037	-.036		.023	.001
				Sum	**.144**	**.094**
Confidentiality						
Mail	.090	.071	.069		.021	.002
Courier	.068	.158	.135		.067	.023
Telephone	.072	.033	.064		.008	.031
Fax	-.250	-.275	-.252		.002	.023
E-mail	.020	.014	-.016		.036	.030
				Sum	**.134**	**.109**
Length						
Mail	.099	.092	.086		.013	.006
Courier	.062	.115	.126		.064	.011
Telephone	-.231	-.325	-.316		.085	.009
Fax	.135	.094	.096		.039	.002
E-mail	-.064	.024	.008		.072	.016
				Sum	**.273**	**.044**
Urgency						
Mail	-.617	-.678	-.618		.001	.060
Courier	.139	.187	.137		.002	.050
Telephone	.153	.092	.110		.043	.018
Fax	.282	.366	.352		.070	.014
E-mail	.043	.033	.017		.026	.016
				Sum	**.146**	**.158**

Note: This tables shows the mean changes in probablities that alternatives are chosen due to changes in the indicated characteristic. The mean values of these changes are taken over all combinations of the other characteristics.

"complexity" has no obvious corresponding aspect; neither is the aspect "ease of operation" closely related to some characteristic).

The next section discussed discrete choice models. In particular, it concentrated on the logit model, as the standard for multinomial discrete choice models, and an alternative model, the EBA model. This latter model neatly fits the Moore and Jovanis framework as it defines aspects which are possessed by alternatives. Moreover, the relative importance of these aspects, i.e., the probability that an aspect is used in the selection of an alternative, may be explained by exogenous variables, in particular the characteristics of the communication activity. As a consequence it forces the researcher to decide beforehand how the media are defined in terms of aspects, which is a minor drawback of the model.

After a presentation of the data set, we reported on estimation results for both models. This empirical analysis covers two themes: an analysis of media choice and an investigation of the EBA model.

For the analysis of media choice, the logit model, is used first despite the conceptual preference for the EBA model. The results showed the effects as expected by intuition: urgency of the communication favors fast media, confidentiality disfavors fax, long and complex messages reduce the probability of using the telephone etc. In general, we found that traditional mail is used unless a specific reason urges the use of another medium. We also mentioned some additional analyzes with extensions of the basic model. Firstly, restricting the choice set to those media which are accessible for respondents gave better results than including an accessibility dummy. Secondly, including some background data to describe perceptions about media, inspired by the Moore and Jovanis model, gave modest improvements of the model. Thirdly, including personal characteristics and estimating media choice for each subsample separately, only showed that the results are consistent over all specifications.

Finally we investigated the EBA model. First, on theoretical considerations, the EBA model was specified. Thus, it was indicated what requirements could be derived from the characteristics of the communication activity, and which media possessed these aspects. We identified 9 aspects, 6 of which were used in the final estimations. The results allowed an interesting interpretation. They proved that choice followed a two layer process. In the first round choice is restricted to one of the three sets {mail, courier service}, {fax}, {telephone, e-mail} on the basis of three aspects (contents, carrier, confidentiality — for definitions see text). In the second round the choice is ultimately made by considering the other three aspects (familiarity, time, physical).

Comparing the EBA and the logit model we conclude that in a statistical sense, in the present case, logit performs better. Yet, the theoretical considerations, together with the much richer possibilities of interpreting the results make the EBA model an interesting and serious alternative to the logit model. Therefore EBA warrants more attention than it has received in research so far.

Appendix A

In this appendix we present the various versions of the messages the respondents had to consider for sending. The messages varied in the dimensions complexity, confidentiality, length and urgency. The words in **bold print** are literal quotations from the English version of the questionnaire.

Urgency is the most straightforward dimension. For each message it was indicated that the message had to reach the receiver either **tomorrow** or **within a time period of one to two weeks**.

The dimensions confidentiality, length and complexity are usually not stated explicitly in the questions, but are implicitly defined in the formulation of the message. Only occasionally is a message called confidential explicitly. A consequence is that the characteristics are to some extent subject to a personal interpretation. This should be kept in mind in reading the following.

The combination of the dimensions *complexity*, *confidentiality*, and *length* leads to the following messages.

1. a)The message: **John Smith is applying for a job. Could you provide a reference** is short, simple and confidential.
   b) A message meant to let your colleague know that you **intend to go to the airport to collect him/her by car** is short and simple but not confidential.
2. A long, simple message concerned a **10 page document** of only text. In the confidential case **it contains material concerning a promotion.** In the nonconfidential case it was **a written version of your lecture.**
3. a) A short, complex message is an abstract of a lecture – **two pages, containing a short text, one illustration and one table,** or even only **an important illustration** which is missing from a document already sent. This is characterized as nonconfidential.
   b) In the confidential case it is supposed that your colleague visits your place of work. He also plans to combine this with some tourist activities, yet the trip is paid for by his university. Therefore, **a short brochure,** the contents of which you want to transmit to your colleague, is to be considered confidential, short and complex.
4. A **paper** which is **12 pages long and contains several charts, illustrations and tables** is of the long and complex type. As such the message is nonconfidential. Confidentiality is achieved by requiring that **he/she should get this paper personally, and other members of his/her institution should not catch sight of it.** Curiously this specification does not mention the number of pages as in the nonconfidential case.

A few remarks are in order. In the first place, the difference between the messages described in 3(a) and 3(b) is greater than just the difference in confidentiality. The problems are mainly with the brochure: Is it ubiquitously considered short? Do respondents realize that only the contents of the brochure have to be transmitted, in which case they still may regard summarizing it too big an effort? Do respondents have similar ideas about the possibilities of faxing it? Are the contents really complex, e.g., when it concerns factual information etc? There is also a principal difference with the other messages. Whereas the brochure is material originally produced by a third party, all other messages originate with the presumed sender himself.

This leads to our second remark. It is not specified in the case of papers, figures, etc. whether the respondent may assume that the message is stored in a computer file or not. This makes a significant difference with respect to the option of using e-mail since files can be simply attached to an accompanying message. In any case, it is clear that the brochure is not on file.

Apart from the points raised with respect to the brochure, we conclude that the questionnaire is well set-up, since differences in messages really reflect the characteristics they intend to show.

Notes

1. This example can be found in any standard text on discrete choice models.
2. To see that this is so, suppose that there are three alternatives with values 5, 10, and 15 for the continuous variable. In the AA matrix this variable can be represented by 2 columns, the first (0,1,1)', the second (0,0,1)'. Setting the cut-off level at a level below 10, means selecting the aspect given by the first vector; setting this level above 10, leads to selecting the second vector. Indeed, selecting the second vector, makes the first one redundant.
3. There is a minor inconsistency in the two British versions. The Swiss and Austrian questionnaire used Munich, and the Dutch situated the receiver in Antwerp. The British version also used Munich whereas a choice like Dublin would have been more in line with the other versions (language similarity, short physical distance).
4. Semantic models are coordinated if the sender formulates the message within the same context that the receiver will use to interpret the content of the message. Ouwersloot (1994), discusses the importance of the concept of semantic models, following Nielen (1988).
5. M = Mail, C = Courier, T = Telephone, F = Fax, and E = E-mail.

References

Amemiya, T., 1981. Qualitative Response Models: a survey, *Journal of Economic Literature*, 19:1483–1536.

Button, K., Brown, P., Fischer, M., Maggi, R., Ouwersloot, H., Rammer, C., Rietveld, P., Salomon, I., 1993. *Academic Links and Communications*. Aldershot: Avebury.

Fader, P.S., McAlister L., 1990. An elimination by aspects model of consumer response to promotion calibrated on UPC scanner data. *Journal of Marketing Research*, 27:322–332.

Fischer, M.M., Maggi R., Rammer, C., 1992. Stated preference models of contact decision behavior in academia. *Papers in Regional Science*, 71:359–371.

Fischer, M.M., Maggi, R., Rammer, C., 1993. Scholarly communication in Europe: stated communication media choice and contact decision models based on laboratory choice experiments in universities. In: Nijkamp, P. (ed) *Europe on the Move*, (pp. 311–329). Aldershot: Avebury.

Gensch, D.H., Goshe S., 1992. Elimination by dimensions. *Journal of Marketing Research*, 29:417–429.

Luce, R.D., 1959. *Individual choice behavior – a theoretical analysis*. New York: John Wiley & Sons.

Maddala, G.S., 1983. *Limited-dependent and qualitative variables in econometrics*. Cambridge: Cambridge University Press.

Manrai, A.K., Sinha, P., 1989. Elimination-by-cutoffs. *Marketing Science*, 8:133–152.

McFadden, D., 1974. Conditional logit analysis of qualitative choice behavior. In: Zarembka, P. (ed) *Frontiers in Econometrics*, (pp. 105–142). New York: Academic Press.

McFadden, D., 1981. Econometric models of probabilistic choice. In: Manski, C.F., McFadden, D. (eds) *Structural Analysis of Discrete Data with Econometric Applications*, (pp. 198–272). Cambridge, MA: MIT Press.

Moore, A., Jovanis, P.P., 1988. Modelling media choices in business organizations: implications for analyzing telcommunication-transportation interactions. *Transportation Research-A*, 22:257–273.

Nielen, G.C., 1988. *Gegevensleer*, Samson, Alphen a/d Rijn (*Data Theory*).

Ouwersloot, H., 1994. Information and communication from an economic perspective. *PhD. Thesis*, Free University Amsterdam.

Ouwersloot, H., Rietveld, P., 1993. Formation and maintenance of knowledge-based networks – the case of university contact patterns. In: Bakes, H., Abler R., Roche, E.M. (eds) *Corporate Networks, International Telecommunications and Interdependence*. London: Belhaven Press.

Ouwersloot, H., Rietveld, P., 1994. *A note on estimation of discrete choice models with repeated observations, mimeo*, Free University Amsterdam.

Timmermans, H., 1984. Decision models for predicting preferences among multiattribute choice alternatives. In: Bahrenberg, P., Fischer, M.M., Nijkamp, P. (eds) *Recent Developments in Spatial Data Analysis*, (pp. 337–354). Aldershot: Gower.

Timmermans, H., Borgers, A., Van der Waerden, P., 1992. Mother logit analysis of substitution effects in consumer shopping destination choice. *Journal of Business Research*, 24:177–189.

Tversky, A., 1972a. Choice by elimination. *Journal of Mathematical Psychology*, 9:341–367.

Tversky, A., 1972b. Elimination by aspects: a theory of choice. *Psychological Review*, 79:281–299.

Tversky, A., Sattath S., 1979. Preference trees. *Psychological Review*, 86:542–573.

Beyond Competition: The Future of Telecommunications.
Edited by D. Lamberton.

CHAPTER 12

An activity-based model for communication-intensive companies as an approach to appraising information technology outsourcing

J.P. Briffaut[1] and F. Spitz[2]
[1]*Institut National des Telécommunications, Evry, France; and* [2]*AT&T – ISTEL*

Issues raised by the fast development of telecommunication capabilities in the management of information technology

Telecommunications have converted information from a business asset into a business environment

The various fields of information technology have not expanded at the same pace. The wide spectrum of capabilities offered by telecommunication technologies has overwhelmed contributions coming from other information cotechnologies in the past few years.Their impacts on business organisations are clearly perceived and expressed (Charam, 1991).

Information technology managers have to cope with fast changes in computing and data-base systems alongside with telecommunications which induce new patterns of interlinking and interworking processes. In addition they have to learn how to deliver user-oriented services which have to be dynamically stable and cost-effective, while building on long-term stable servicing competences.

In business nowadays it is not unusual to find information technology managers with inadequate experience and qualifications addressing all the technical fields involved. A solution to this situation is outsourcing information technology, in an attempt to bring technical and financial problems under full control. A survey carried out by the American consultancy firm INPUT shows how this service market is growing in Europe.

In this article we analyse how information technology can be rationally outsourced on the basis of an activity-based business model. After explaining why cost accountants made use of the activity-based business model, we describe its features in terms of the issues under consideration and introduce a fundamental distinction between information technology, information system and information management. In the following part, on the basis of data collected from technical literature and

---

Address for correspondence: J.P. Briffaut, Department of Management Sciences, Institut National des Telécommunications, 9 rue Charles Fourier, 91011 Evry Cedex, France. Tel.: +33-1-60764684. Fax: +33-1-60784018.

interviews of company representatives, we work out an analytical method to take coherent decisions on the best ways and means to outsource information technology. The same methodology can be used to investigate the market from the standpoint of service suppliers.

Computers, networks and organisational company structure

Malone and Rockart (1991) think that after a first-order effect of substitution (a pure mimic of human activities by computer programs) information technology will bring about a change in coordinating business activities through the effect of networked computers (called a second-order effect) which will most likely overcome the first-order effect. The result will be the emergence of communication-intensive economic structures.

It becomes clear that a new style of management and another approach to business structure will be needed to tap the productivity gains offered by the second-order effect of information technology. Many of the productivity gains of the early twentieth century were achieved with the help of new ways of defining and costing work, notably F.J. Taylor's "time and motion" studies.

The recognised inadequacy of present costing methods can be ascribed to the thrust of information technology into traditional business activities and their administrative procedures. Johnson and Kaplan in their well-known book "Relevance Lost" (1987) explain in their damning analysis of accounting that companies get misleading information from their cost accountants. The problem is "overheads", costs incurred by the business as a whole and having to be apportioned to products/services in order to yield cost prices. Up to now this step has been performed through an apportion rate, the relevance of which is highly questionable and questioned.

Why is information technology responsible for the increase of "common" costs, without being the only reason for it? At least two arguments can be put forward to explain the linkage between the elements under consideration:

- Many multipurpose equipments have been made instrumental by computer-controlled systems in order to achieve flexibility. Flexibility has become for companies a key factor in gaining and sustaining competitive advantages. It allows firms to offer a wider range of customised products and services to clients.
- Telecommunication services have facilitated the integration of business functions by sharing data and their processing. They appear to be essential to complete "time to market" operations.

This has resulted in an increase of common costs to such an extent that in some cases they exceed direct costs.

Cooper (1987) pioneered a solution called "activity-based costing". He designed a model to convert "common" costs into direct costs. Business resources are "consumed" to support activities which are configured to reach a goal. These linked activities are used to put out a product or a service. Costs incurred by each activity are apportioned to products/services via "cost drivers" or "cost vectors". This business model can be extended and applied to organise, manage and appraise the whole of company operations with special attention to the coordinating role of information.

Activity-based business model

Setting broad goals is easier than reaching them. When information technology keeps moving the goal posts, it is essential to develop courses of action on the basis of an adequate, sound architecture derived from a relevant business model.

Within the framework of the activity-based model company operations are viewed as networked activities engineered with and supported by resources. In order to gain control of complex activity patterns activities are configured into processes, each serving well-defined purposes. The whole or parts of traditional business functions (marketing, manufacturing, sales, etc.) are instrumental in the activities of various primary processes. These primary processes are portions of a central global process from suppliers to customers along the product/service axis.

Another dimension is required to complete the model, i.e., the managerial aspects. A major distinction is affirmed in the model between process management and resource management. Process management defines the environment of the execution process, specifies the desired outputs and the required inputs, organises and schedules the execution process, including its resource usage. Resource management takes care of the resources required for the execution processes. Five main resources are managed i.e., money, people, equipments, buildings, and data. The model can be conceptualized as a house with resources and processes on the ground floor and the management divisions on the roof. A special coordinating role is assigned to information. This is described in Fig. 1.

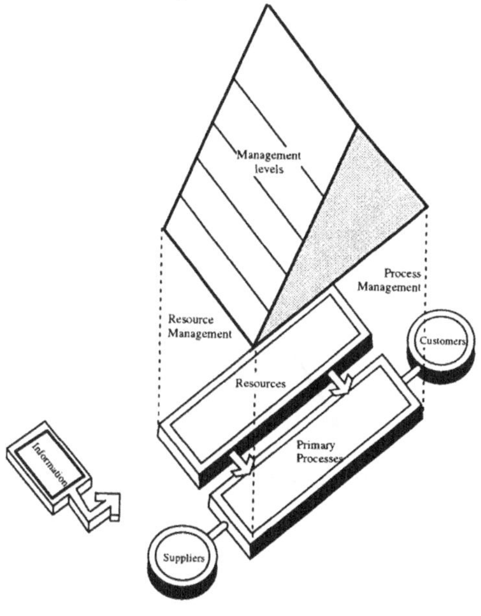

Fig. 1.

This model integrates well with Porter's (1985) model of value chain and illustrates on what basis an activity-based management (Morrow, 1992) can be implemented.

Information technology management versus information management

Firstly, a common confusion between information technology and information systems has to be eliminated. Information technology is the means and information systems are the ends. The delineation between ends and means or between applications and delivery can be made more explicit by introducing the concept of information management (Earl, 1989) as shown in Fig. 2.

Information management is concerned with the role and uses of information technology in the organisation. It focuses on relationships between users and specialists. It is concerned with the way users' requirements are economically met by data processing and data communications while avoiding constraints on the evolution of organisational structures. It ensures that information technology decisions are made within a policy framework. Studies have concluded that there is no universal best way for effectively and efficiently structuring the information technology function in companies (Feeny et al., 1987).

The activity-based model previously described can provide an adequate framework to structure and delineate the information technology (IT) function and the information management function: the information technology function supplies resources to business activities configured to reach a goal whereas their information

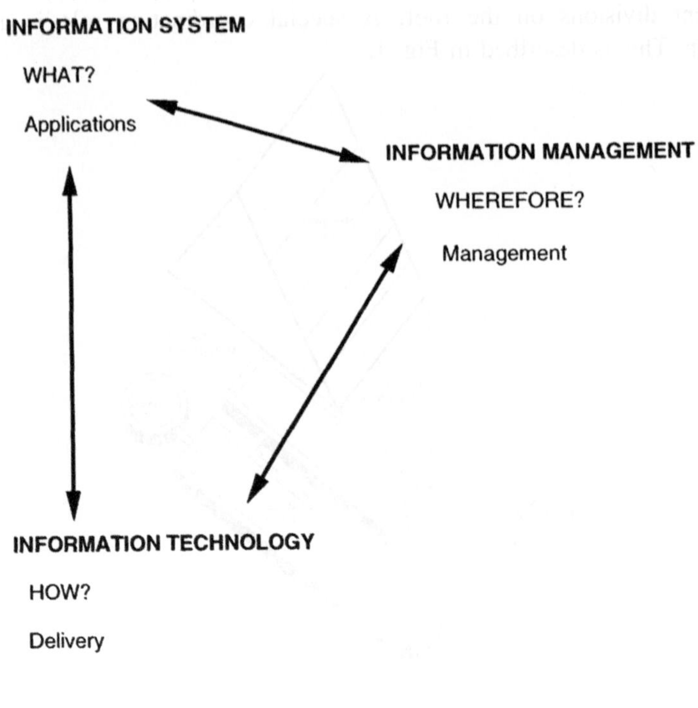

Fig. 2.

systems (IS) are monitored by the information management function. IS is the way IT is exploited to support business activities in terms of data. In other words IS intermediates between IT and end-users to secure a proper alignment of data needs and resources. IT and IS are closely linked and will always be mentioned together later on in this article. The four interdependent resources (coresources) of information technology and systems are:

- computing (processing hardware and its associated operating system software);
- communications (telecommunication networks for interlinking and interworking);
- data (data assets and their procedures for use, access, control and storage);
- applications (application systems of the organisation, their development methods and their relationships with business activities).

Some words clarified

In continental Europe outsourcing of information technology (IT) and information systems (IS), and facilities management are two expressions used without any clear-cut distinction between them. "Facilities management" was coined in the USA to encompass the management of all useful services (utilities) i.e., building, energy, maintenance, EDP, etc., offered to office blocks. In an early stage EDS made use of this expression to address outsourcing of IT-IS. It later changed the wording to "systems management" (SM) in which information could be logically added to the word systems.

Later on we shall use the term "outsourcing" to describe IT-IS services performed by a third party, rather than by the firm's "in-house" IT-IS employees. This wording not only avoids some misleading interpretation, but allows for an extended form of outsourcing i.e., cosourcing. Industrial partnership as a substitute for subcontracting was engineered by Japanese firms hitherto within closed circles. The idea was to weave strong links of common interest in a confident framework. European firms embraced the concept with an "openness" approach. That has resulted in the establishment of joint-ventures between competitors in order to share the manufacturing of common parts. The surge in alliances between firms that we are witnessing nowadays can be ascribed to this school of thought. So it is not surprising that firms in Europe are looking for service partners (not just vendors) to outsource their information technology and systems.

Two types of outsourcing, i.e., internal or external, can be distinguished without an operational difference between them. By internal it is meant that the purveyor of services is integrated into the firm as a self-governing department or a subsidiary. But in both cases the relationships between suppliers and users should be handled in principle in the same way. The concept of "IT-IS outsourcing" embodies several segments which can be referred to one or more activities of the value chain (design-implementation-operations-maintenance) associated with each coresource of IT-IS. These segments are:

- consulting for planning, designing, auditing IT and IS;

– migration to a new technical system;
– development and maintenance of software applications and data bases;
– network management;
– electronic data processing (production operations by hosts and microcomputers).

Data gathered on the state of IT and IS management in companies

Data gathered and meant for analysis originate from two sources, i.e., technical literature and interviews of company representatives. They are organised by main items in order to facilitate understanding:

1. In spite of the development of software engineering there are a large number of software projects that fail to satisfy users. Many stories can be told about this problem and the costs incurred. Examples will be restricted to two cases.
   – Figure 3 (source: US General Accounting Office) shows a breakdown of the fate of US federal software projects. According to this breakdown less than two per cent of software products were used as delivered.
   – The French national railway company introduced a multi-million FFr computer-based booking system, called SOCRATES, in 1992. The use of the system by booking clerks and travellers was so difficult that the traffic fell by seven per cent in 1993. Reworking of the system had to be carried out.
2. A market survey of IT-IS outsourcing in Europe was carried out by INPUT consultancy. Its main results were published in a technical journal (1993), and can be summarized as follows:
   – The market is decomposed into four segments, i.e., main frame computing, applications, network management and microcomputer systems. This market segmentation fits into our proposed splitting of IT-IS resources, with the proviso that data-base management be itemised with applications and that main frame computing and microcomputer systems be brought together under the heading "electronic data processing".

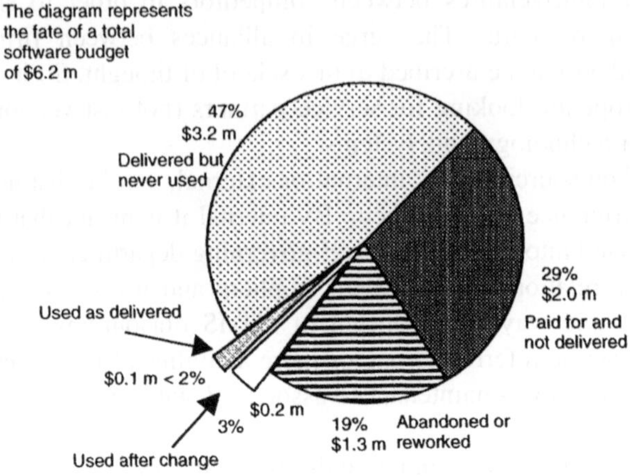

Fig. 3.

— The global turn-overs of companies in this market were, in 1992, for each segment respectively US$1260 million, US$880 million, US$410 million, US$260 million.
— The benefits expected by companies and derived from IT-IS outsourcing can be classified into five categories (Fig. 4).

3. The main results of a survey on outsourcing carried out by Peat Marwick Consultants (France) have been just published in a technical journal (1994). Ten EDP directors were involved in the investigation. The purpose of the study was to define a structured step-by-step approach to outsourcing by companies. The main conclusions can be set out as follows:
   — Companies turn to IT-IS outsourcing to get a grip on their IT-IS costs and to benefit from up-to-date expertise and experience in new technologies. These results confirm those obtained by INPUT.
   — Before taking any course of action business management should commission an audit on how IT-IS services are organised and what they cost. This audit has to form a necessary basis to decide upon outsourcing by allowing for benchmarking the existing situation against vendors'offers and to delineate the IT-IS services to be possibly contracted out.
   — In order to define relevant service requirements to seek formal bids from potential vendors, and to keep their evolution under the firm's full control, a master plan should be worked out. A steering committee is highly recommended for a mastered follow-up of outsourcing arrangements.
   — An outsourcing supplier should not be chosen on cost consideration alone. The supplier must share the firm's culture and become a partner committed and dedicated to shared goals.
   — "Good accounting makes good friends": service-level agreements and billing procedures are bound to be written down.
4. It is useful to point out what types of vendors are present on the supply market. They can be split into three categories on the ground of their core business field:
   — Subsidiaries of large companies having externalized their IT-IS services and wanting to turn them into a marketable profitable activity: Polydata of Philips, Debis Systemhaus of Daimler-Benz, Telesystème of France Telecom, EDS of

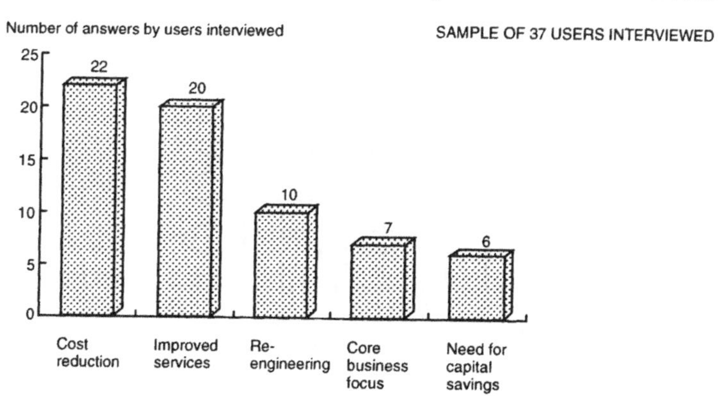

Fig. 4.

GM, etc.

- Software houses looking for new market opportunities and diversification of their range of products.
- Computer companies such as IBM, Bull and Digital.

5. Thomson-CSF is a subsidiary of Thomson with its main industrial operations in France. It produces defence-related equipment (radars, telecommunication systems, missiles) within forty divisions, organised into five business units. It now has a three-dimension matrix organisational form. A two-dimension matrix portion is addressed to functions (specialised skills and equipment) and process channels corresponding to product families. The process channels control the contributions of the various functions in order to secure the appropriate integration of capabilities and the proper allocation of resources. The operating logic of this portion of the matrix form is similar to that of project-oriented management. The third dimension groups a series of teams devoted to contracts which are always awarded as a combination of more or less customised systems. These teams maintain necessary linkages between clients and the process channels. A picture of the situation is sketched on Fig. 5.

No centralised coordinating structure for information technology and systems was in operation till about eighteen months ago when an IT-IS director was appointed at corporate level. The assignment he was given by the CEO was to bring expenditures (capital outlay and running expenses) under full control. The only existing metric to measure IT-IS costs at the corporate level was a yearly global budget granted to divisions, each having its own policy and practice.

The terms of reference of the IT-IS director's mission cover organisational leadership, control of expenditure (capital and costs) and technical leadership. Decisions being implemented can be summarised as follows:

- Internal outsourcing of IT and IS (implementation in progress)
- Technical orientations: client/server architecture-networking of all microcomputers systems-putting an end to software developments for scientific applications as well as management applications and use of standard packages or kernels-downsizing and convergence to UNIX operating system.

The competitive advantages expected by the CEO from the mission given to the

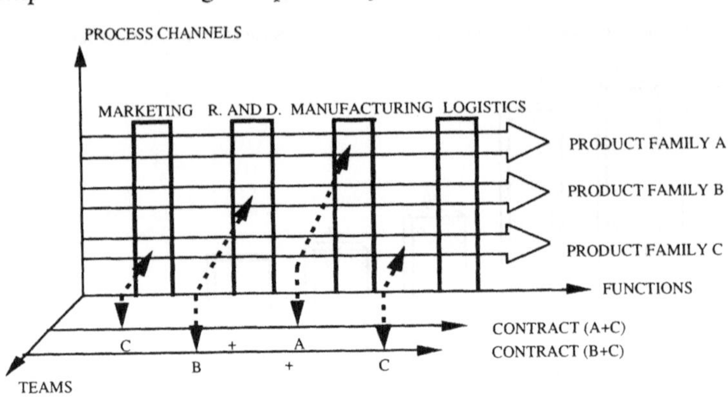

Fig. 5.

IT-IS director are 2-fold, i.e., the mastering of IT-IS costs (FFr200 million cut in spending is targeted in 1994) and better workflow resulting from compatible communication between systems.

An interesting feature is the enforced recommendation to use standard packages or kernels: in fact it results in a "de facto" outsourcing of software developments. The IT-IS director concentrates on the fact that packages must be chosen to fit existing or evolving operational procedures of the company. Many managers still fail to perceive the importance of this factor and the need to deploy information systems in pursuit of the goals of the firm and in support of its organisational structure.

6. Europcar is a French service company renting cars. It operates in nine European countries through subsidiaries and associates. It expanded by take-overs so that it was faced with communication problems between the fifty five software applications distributed over the company network. In order to solve technical problems in a viable economic and economical way and to define a policy, business management decided to carry out a master plan for information technology and systems with the idea of cost control in the back of its mind, feeling that global in-house competence was insufficient.

   The outcome of this exercise evolved multiple decisions:
   - Setting up a subsidiary called "Europcar information services" the shareholders of which are Europcar and its foreign branches, and scheduled to maintain in-house intelligence in IT and IS.
   - Global outsourcing of IT and IS awarded to Perot Systems Europe against fees based on a percentage of Europcar's yearly turn-over.
   - Internal billing of end users by Europcar information services on the ground of functional transactions.

7. HP-France has opted for outsourcing internally its IT-IS services, supporting its own operations. The activity-based pattern was implicitly used for designing and implementing the supply center of IT-IS resources. This center is run as a cost center.

   Tasks allotted to the service center are:
   - Management of infrastructure (data processing machines, networks, applications, databases).
   - Engineering of IT-IS activities into processes to trace the embedded contribution of each activity to achieving specific output goals.
   - Activity accounting establishing costs and performance, and allowing for charging costs of services to users.

About 30 services are operational today and are explained along with charges entailed in booklets made available to HP employees. There are four types, i.e., spoken telephone services, office computer facilities, datacommunications, and EDP operations.

The IT-IS policy, if expressed in terms of costs, can be best understood by data displayed on Fig. 6.

A feature worth drawing attention is the scheduled drastic reduction in "in-

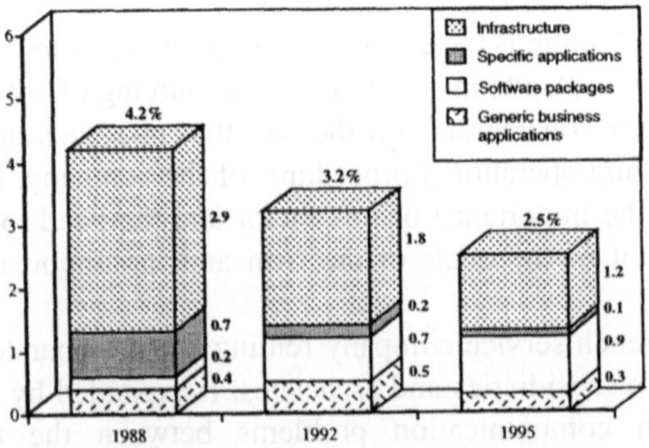

Fig. 6. Change in IT-IS costs with respect to HP's turnover in Europe.

house" software development and the concomitant increase in using software packages.

Concepts of management accounting for control of IT and IS

The central issue raised by responsibility accounting for IT-IS is to choose the relationships between IT-IS and the rest of the organisation. The basic question to be answered is: "what sort of responsibility center should the IT-IS function become?" Three traditional approaches are generally adopted to conduct management accounting for control of IT-IS resources. They are service center, cost center and profit center. Their characteristics will be briefly described.

The service center concept implies that:
- users are not charged for IT-IS resources consumed;
- IT-IS are funded by a budget granted on forecast requirements;
- use of IT-IS services is stimulated.

The cost center concept implies that:
- costs incurred by IT-IS services are charged to customers who are made responsible for IT-IS resources used;
- IT-IS are funded on cost-benefit analysis.

The profit center concept implies that:
- users are charged at cost plus prices (market prices);
- capital outlay is targeted to yield a return through cost recovery from users;
- business users can go external if offers by third parties are more attractive.

Each of these approaches has advantages and disadvantages, the appraisal of which is beyond the scope of this article. But within the framework of the activity-based business model IT-IS outsourcing is understood as a provision of resources by one or several suppliers. As a consequence costs thereof are logically charged to users and are perceived by them as managed with a cost-center philosophy. This procedure is applied by HP which operates activity-based costing. This chargeout system can be sophisticated so that users can be charged for some services at a cost price different from actual payments made to suppliers. This procedure allows through cost consciousness for a forced change in the behaviour of users to encourage adoption of a new technology or to deter from an old technology.

The activity-based business model as an analytical implement of data gathered

Global considerations

Within the activity-based business model, information technology and information systems are viewed as resources supporting activities. As such they are managed with the aim of fulfilling the requirements of end users which are embodied in applications and data bases. Normally the linkage between these resources and users should be secured by a support function we previously labelled "information management". Quality management, planning, logistics and management accounting can be quoted as other support functions. Figures 1 and 7 complement each other to illustrate a business organisation.

When outsourcing of IT-IS resources is brought under study, three types of crucial elements have to be dealt with, i.e., the sort of activities involved, resources to be externalized and information management.

1. The pattern of primary business processes has to be established so that some activities can be estimated critical with respect to IT-IS resources, insofar as embedded know-how is a core expertise to be maintained and enhanced and is at risk if the supply of some IT-IS resources are contracted out. It must be clearly apprehended that external outsourcing brings suppliers in the value network of the firm's activities. Network members generally participate in the market outside their relationship with the firm under consideration. It may, for this firm, be an easy path to win access to technological expertise and flexibility and to be kept abreast with innovation in products and services. On the other hand the firm must be aware of the fact that their external relationships can be instrumental in

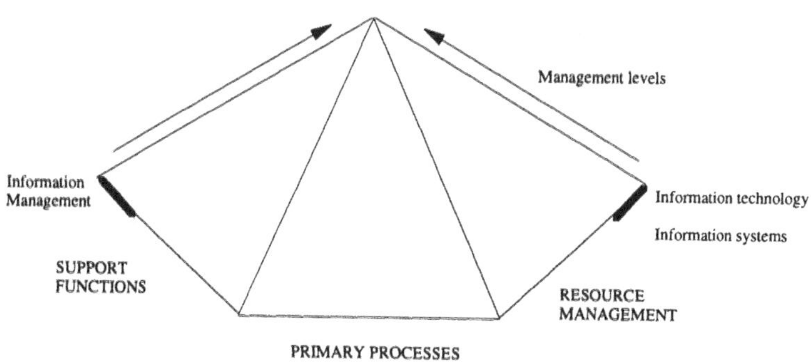

Fig. 7.

disseminating some core expertise or experience outside. These relationships have to be carefully and dynamically handled.

From gathered data it appears that potential medium-term and long-term effects of knowledge leakage and know-how losses, induced by outsourcing, are not clearly perceived, if at all, by managers. In addition it is not sure that they are well aware of experiencing a new organisational business form made up of a cluster of specialist units in order to arrange and coordinate resources. If its logic as well as its strengths and weaknesses are not well understood, the potential for its misuse and, as a consequence, for its failure increases (Miles & Snow, 1992).

2. In theory the supply of IT-IS resources can be separately contracted out for each process or any activity thereof. The situation can be best appraised with the help of the analytical matrix given in Fig. 8.

   HP made use of these criteria to decide upon outsourcing to third parties: support to users was appraised as strategic whereas hardware maintenance was contracted out. On the same decision pattern Elf Aquitaine has decided to contract out hardware and software maintenance, users' training and support of its microcomputers.

   Other decisional and decisive factors such as cost vectors and performance indicators should be made available and used for benchmarking. In a general way these data are not available as explained later on.

3. The information management function is best described as a multilingual translator between user experience/requirements and technical specifications. When IT-IS outsourcing takes place, the existence of this support function is all the more important. The information manager has to have his assignment extended to responsibilities of steering "in-house" "watchdog" competence in IT-IS, relationships with supplier partners and strategic planning of IT-IS applications. Attention must be drawn on the fact that outsourcing should not be an excuse for not establishing strategic plans for IT-IS; IT-IS are enabling tools to serve the firm's success factors. This key function empowers the top management to be provided with discerning, prospective pieces of advice and strategic plans. In other words it can be said that it is in charge of keeping internal managerial intelligence in IT-IS.

   It seems from collected data that outsourcing does not automatically lead to change existing IT-IS functions in companies. In fact, they should be logically transformed but it is too early to assess this evolution which is likely to be slow.

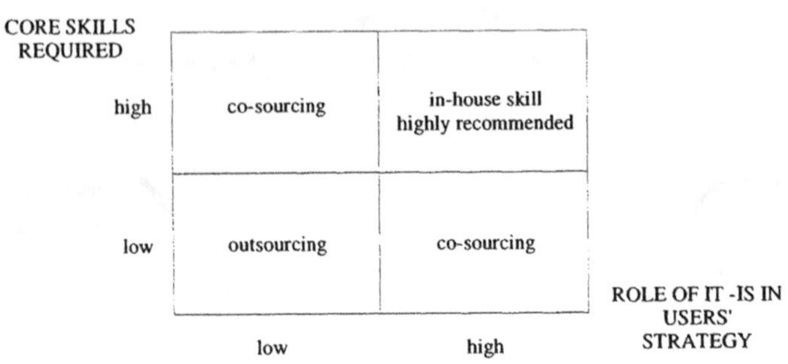

Fig. 8.

Focused interpretations of some data gathered

Cost-driven move and hidden factors

The economic challenge faced by many companies in the present crisis has pushed executives to limit overhead expenditures. Information technology and systems fall into this category. Their related costs are hard to value, which means difficult to be derived from usage of resources and, as a result, to be brought under full control. From collected data it appears that this context has triggered the outsourcing of IT-IS. The move taken converts overheads into direct costs. This is precisely the thought behind processes associated with the activity-based costing developed by Cooper (1987).

The practical paths followed to achieve controlled supply of information services are meant either to go external or to set up a business in the business. Steps taken by executives on the grounds of sought-after cost control reveal in fact deep uncertainty with respect to on-going technical and organisational evolution. Signals issued by miscellaneous sources have reached boardrooms. On one hand management gurus are preaching both the chasing of "hidden" costs and the implementation of business "re-engineering" which are supposed to be a route to build a new profit era. On the other hand IT-IS specialists advocate the benefits of the new IT Gospel, the basic features of which are client/server architecture and UNIX operating systems, and strongly rely on telecommunications. Impressive figures are put forward: according to Aydabirian (1994), managing director of HP France, it can be proven that the lead time for developing a software application can be reduced from three years to five or six months when UNIX platforms are used. Within this context IT-IS outsourcing is not only a way to achieve cost control, but also a hope to reap the benefits of new technologies: this is indirectly admitted through the search for improved services (see Fig. 2). Any decision taken in a non-stable environment entails greater risks. These risks are secured by contracts and the pertinent choice of suppliers. When terms of contracts signed with IT-IS suppliers are scrutinised, a "reversibility" clause is always present and is supposed to put management's mind at ease. Outsourcing should be viewed as a two-way relationship having to be based on trust, competence and the sharing of common "cultural" values. It is advised by many consultants to consider suppliers as partners if stable resource flows are to be established.

A telling insight of the combination organisational re-engineering and IT-IS is "workflow". In essence the workflow concept mimics the way people communicate, as they alternate between being first the speaker and then the listener. Workflow methods have been around for nearly a decade, but they are only now becoming practical, thanks to the arrival of new programming tools offered by a dozen software firms (Action technologies, Lotus, Delrina and Reach, etc.). Over the next three years, as more companies pension off their old mainframe computers in favour of PC-based "client-server" networks, the workflow market is expected to explode into a $2 billion-a-year world-wide business.

Change in the market structure of the computer-services industry

If IT-IS outsourcing gains momentum and becomes a wide-spread managerial practice,it is likely to be influential in shaping a new pattern of demand for most IT-IS products and services. Upon this assumption suppliers of hardware and software commodities would have to deal with a reduced number of clients, i.e., companies providing outsourced IT-IS facilities to end users.

New factors explaining buyers' decisions would have to be aggregated for an understanding of the market. In spite of the fact that these new factors embrace many determinants, in our opinion two fundamental factors induced by the behaviour of end users should have a major effect. These factors are cost and innovation. The influence of the anticipation of end users for lower prices and wider spectra of services should drive the consistency in, and transitivity between the preferences of outsourcing suppliers. These factors address mixed technicalities such as the interoperability of operating systems, the portability of software programs and easily-customised software packages. As far as this last commodity is concerned, companies show strong desire for an extended use of software packages as a substitute for specific software developments; development costs have a natural propensity for "running away".

It is worthwhile remarking upon the presence of the major hardware manufacturers in the market place of outsourced IT-IS commodities. One may wonder whether they are trying to occupy a new opportunity space to counterbalance their declining traditional activities or to play a leading role in restructuring the computer-related industry. Time will tell.

The structural form of the whole computer-services economic sector could experience a change similar to that undergone by the microcomputer industry. The "new" pattern of the microcomputer sector, proposed by Grove of INTEL (1993) is split into a series of horizontal layers (microprocessors, platforms, operating and communication systems software, application programs and distribution), each layer containing several companies instead of vertically integrated single-company towers of proprietary packages.

The delineation of market segments we propose for the computer-services market is consistent with present market offers and management thinking in terms of the value chain of operations associated with IT-IS resources. Figure 9 features the arrangement of market layers and the content of the value chain.

Consulting and migration are crosswise segments. Consulting refers mainly to the functional design activity of the value chain. By "functional design" is meant the expression of users' requirements in terms of IT-IS resources. Migration covers all the activities of the value chain in order to switch from a technical system to another.

How to approach outsourcing of IT-IS?

Outsourcing of IT-IS resources can be described and analysed with a three-dimension

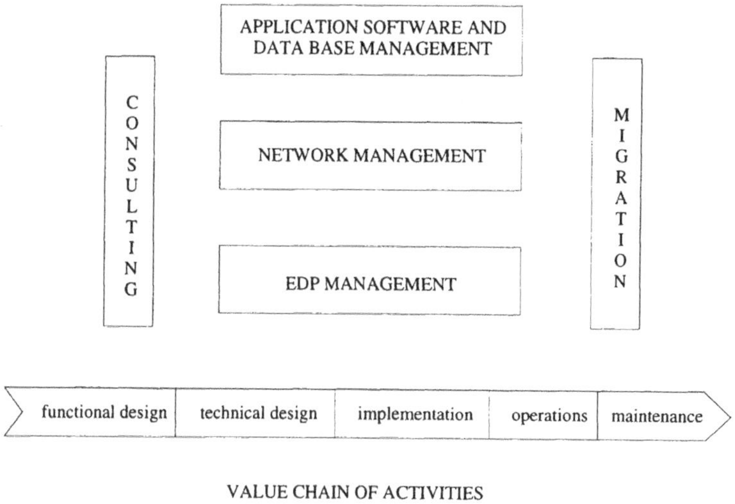

Fig. 9.

matrix allowing for a comprehensive logic of the situation.

1. The *first dimension* is addressed by the domain of activities. Activities supported by IT-IS resources which may be outsourced must be traced back and localised in the value chain of business activities. Their nature with respect to the firm's goals, embedded strategic know-how, confidentiality must be characterised. They have to be established as core or non-core competence with the help of the analytical matrix of Fig. 8 before the analysis proceeds to the next steps.
2. The *second and third dimensions* are closely linked up. The second dimension refers to the sort of information technology and systems resources to be contracted out. They are the three horizontal upper layers of Fig. 9. The third dimension addresses outsourced services having respect to IT-IS resources. They can be categorised as activities or group of activities along the value chain. A common-sense clustering of activities into services is:
   - Consulting (planning and functional design).
   - Technical partnering (technical design and implementation).
   - Production operations.
   - Maintenance operations.
   - Migration.

Executives want scoreboards of performance indicators translating goals and courses of action taken into numbers and ratios. Performance indicators must reflect the appraisal of resources as well as that of activities consuming resources. It has to be clear that good quality level of resources does not imply automatically good quality level of services making use of considered resources. Three interlocked frames of management thinking are required to assess out- or cosourced IT-IS services: information, assignment and control.

Each frame is addressed by questions used as blue prints to derive managerial indicators.

Information frame

Questions pending are:
– Are IT-IS services provided with the right infrastructure and the proper skills?
– Do we benefit from the right services?
Proposed performance indicators are:
– List of services provided versus viable existing services.
– Global reporting of services availability and reliability (response time/failures/ maintenance, etc.) versus service-level agreements.
– Security (access control/environmental control/contingency planning).
– Support to users.

Assignment frame

Questions pending are:
– How do IT-IS technologies service the firm's organisational structure?
– How are IT-IS technologies enabled to meet business changes?
Proposed performance indicator is the firm's technological model showing how IT-IS resources underpin its organisation, and broken down into the following items:
– individual → description of tasks supported;
– workgroup → explanation of workflows to coordinate decisions and actions;
– business unit → automatised procedures of manufacturing and delivery mechanisms for its products and services;
– firm → competitive advantage(s) derived from IT-IS;

Control frame

Questions pending are:
– How do cost structures of services compare with alternatives?
– How are service costs charged to end users?
Proposed performance indicators are:
Per month: – expenditures versus budgets;
– itemised billing per service;
– itemised billing per user.
Per annum: – cost pattern and cost vector per service provided, established with an activity-based approach;
– evolution of costs per resource consumed;
– global IT-IS cost per product/ order whatever unit is representative of the firm's business activities.

Conclusion

The sluggish economic context experienced by Western firms in recent years has yielded a triggering effect to generate a new range of "products" offered by computer-related companies and a reaction of business management to soaring costs of information technology and systems. This macro change will most likely have a leverage effect by enabling and multiplying "micro" changes at work in firms.

Outsourcing of information technology and systems may provide new opportunities for firms which are not yet mature in harnessing information technology and systems, to do it with a creative vision within a collaborative pattern. It may bring in its wake more competitive exploitation of information technology combined with appropriate behavioural involvement of all stakeholders. In order to reap the full benefits of information technology it is essential that the firm be perceived by its decision-makers through a business model reflecting reality in the way management information is compiled and reported: activity-based management hinging on a clear distinction between resources and activities fulfills this requirement. Information system applications cannot generate competitive advantages by themselves unless organisational opportunities for the firm's operations are identified and implemented. Skills are always fostered by incremental learning. But top management must be aware that when decision is taken to build on out- or cosourcing, the firm's organisational structure is implicitly changed by the rules of game.

There is no longer a role model in information technology and systems. All that can be done is to promote co-working between end users and specialists, within the framework of an information-effective, activity-based business model under the conduct of the information manager: (s)he holds a key support function proper to ensure alignment of business data needs and IT-IS resources.

Acknowledgements

We want to thank Mrs Benech (INPUT), Mrs Steidel-Florence (AT&T-GIS), Mr Briquet (Thomson-CSF), Mr Mainguy (HP-France) and Mr Terrade (Europcar) for fruitful discussions.

References

Aydabirian, R., 1994. *IX Magazine,* (interview) 15 January.

Charam, R., 1991. How networks reshape organisations for results. *Harvard Business Review*, Sept/Oct.

Cooper, R., 1987. The two-stage procedure in cost accounting. (a) *Journal of Cost Management*, I:43–51. (b) *Journal of Cost Management*, I:39–45.

Earl, M., 1989. *Management strategies for information technology*. Prentice Hall.

Feeny, D., Edwards, B., Earl, M., 1987. *Complex organisations and the information system function in companies*. Oxford: Templeton College (Research paper).

Grove, A., 1993. The computer industry. *The Economist*, 27 February (quoted in the survey).

INPUT, 1993. *Le Monde Informatique,* April.

Johnson, H.T, Kaplan, R.S., 1987. Relevance lost. *Harvard Business School Press*.

Malone, Th., Rockart, J.F., 1991. Computers, networks and the corporation. *Scientific American*, Sept.

Miles, R.E, Snow, Ch.C., 1992. Causes of failure in network organisations. *California Management*

Review, Summer.
Morrow, M., 1992. *Activity-based-management*. Woodhead Faulkner.
Peat Marwick Consultants Report, 1994. *01 Informatique* (report by Blanc, S.), 11 February.
Porter, M., 1985. Competitive advantage. *The Free Press*.

Part III: Regulation

Beyond Competition: The Future of Telecommunications.
Edited by D. Lamberton.
1995 Elsevier Science B.V.

CHAPTER 13

The AT&T divestiture: a 10-year retrospective

Marcellus S. Snow*
Department of Economics, University of Hawaii at Manoa, Honolulu, Hawaii, USA

1. Introduction

More than 13 years have passed since AT&T, under pressure from USA antitrust authorities, the federal judiciary, numerous present and potential competitors, and public sentiment, agreed to the divestiture of most of its assets and to the reorganization of many of those it was allowed to keep. From the perspective of the 11 years since its entry into force, it is clear that this massive restructuring of private wealth is the signal act of American antitrust policy since the Standard Oil divestiture during the Progressive era nearly 90 years ago; arguably, it will be judged as the major global event of competition policy during the century about to end.

The sheer order of magnitude of the productive forces that were redeployed on 1 January 1984 is so staggering as to invite comparison with entire medium-sized national economies rather than with other public or private enterprises. Indeed, such a contrast is instructive. If we use 1990 as a base year and project forward and backward assuming equal rates of growth, the annual revenue of AT&T before its divestiture approximated the gross domestic product of Denmark. After the divestiture, AT&T's revenue was close to the GDP of Singapore, while the revenue figures of the seven Bell regional holding companies carved out of AT&T's assets can notionally be represented by the GDPs of Cameroon, Costa Rica, Ecuador, Kenya, Luxembourg, Oman, and Syria.

The purpose of this essay is to evaluate the costs and benefits of the AT&T divestiture from the perspective of the 11 years since it entered into force. Section 2 examines the early history of AT&T and seeks to account for how the firm came to have so few strong supporters by the early 1980s. Section 3 presents in considerable detail the events immediately leading up to the divestiture and outlines the mechanics of the agreement. As a sample case of the many kinds of potentially useful evidence that did not inform the antitrust action leading to the divestiture, Section 4 discusses the theory of natural monopoly as it relates to telecommunications, and the empirical

---

Address for correspondence: Prof Marcellus S. Snow, Department of Economics, University of Hawaii at Manoa, Porteus Hall 542, 2424 Maile Way, Honolulu, HI 96822, USA. Tel.: +1-808-956-7521. Fax: +1-808-956-4347. e-mail: snow@uhunix.uhcc.hawaii.edu.

*I am indebted to David Allen, Visiting Fellow in Telecommunications, East-West Center, Honolulu, for comments that materially improved this paper. I am responsible for all remaining errors.

work carried out after the divestiture. In Section 5, evaluations of the divestiture, as well as the broad lines of its evolution from the mid-1980s to the present, will be summarized. Section 6 provides a framework for evaluation; samples expert views about natural monopoly and the divestiture; discusses, as another sample case, the Internet's possible evolution in the context of the divestiture; provides a somewhat random updating of recent business and judicial events under the divestiture; and offers some final reflections.

2. AT&T's century of operation before the breakup[1]

AT&T was founded in 1885 as a subsidiary of the American Bell Telephone Company for the carriage of telephone traffic between the local networks of the Bell Companies. As its fundamental telephony patents expired in the late 1800s, AT&T began acquiring independent telephone companies to expand its network. During that same period the Interstate Commerce Commission (ICC) was created (1888), which was briefly AT&T's first regulator. Passage of the Sherman Act (1890) brought AT&T under the antitrust scrutiny of the Department of Justice (DOJ), where it remains today. At the turn of the century, AT&T became the holding company parent of the Bell System. With the acquisition of Western Electric in 1881 as the Bell System's manufacturing arm and the establishment of Bell Telephone Laboratories (owned jointly by AT&T and Western Electric) in 1924, the Bell System had assumed the familiar structure it would retain until 1984.

From these early beginnings to the present day, Congress did not tell the federal government whether to regulate or to prosecute monopolies. Thus, such decisions were left to be made through bargains and implicit understandings between federal and state regulatory agencies and the Department of Justice (Temin, 1987, p. 9). As AT&T acquired more and more independents early in this century, Theodore Vail, its president, envisaged a universal, interdependent and intercommunicating telephone system. Rather than favoring market competition or a government-owned monopoly, he called for a private monopoly subject to regulation. By 1910, independents were challenging the Bell System through state and federal antitrust legislation. A federal antitrust investigation conducted by the ICC and the DOJ brought AT&T to the bargaining table in 1913 for its first landmark compromise. Under the resulting Kingsbury Commitment, AT&T promised to dispose of its stock in Western Union; to allow all independents to interconnect with its toll facilities; and to refrain from further acquisitions of directly competing independent telephone companies.

During the 1920s both Vail and the independents withdrew from certain markets in which direct competition had resulted. This policy of detente allowed several large independents to emerge. President Woodrow Wilson briefly took control of the telephone network in 1918 as a wartime measure, which further advanced Vail's goal of a single unified system. The Bell System's purposes were also served in 1921, when Congress authorized the ICC to approve the consolidation of telephone company holdings by acquisition, thereby weakening the effect of the Kingsbury

Commitment; and the following year, when AT&T pledged to work cooperatively with the independents. State and federal regulation became more pervasive during the 1930s, and by 1934, when the Federal Communications Commission (FCC) was established and replaced the ICC as Bell's federal regulator, AT&T was once again dominant in the telephone market.

The 1935 formation of the Walker Commission at the FCC, broadly mandated by President Franklin D. Roosevelt to investigate AT&T and its competitors, produced a draft report in 1938 quite critical of the Bell System. It alleged that Western Electric had overcharged for its equipment, criticized state regulatory authority as inadequate, and suggested that government ownership be considered if regulation failed. Broadly negative criticism of the report (including that by other FCC commissioners) and a voluminous, point-by-point rebuttal by AT&T led to a final report in 1939 which merely urged close and careful regulation of Bell. Nonetheless, after the passage of the Depression and World War II, the smoldering embers of the Walker Report ignited in 1949 into the government's first full-dress antitrust suit against AT&T and Western Electric. It repeated the allegations of Western Electric's overcharging; in addition it asserted that AT&T had conspired to monopolize the telephone equipment market, thereby excluding other manufacturers and earning monopoly profits. As a remedy, the DOJ asked the courts to separate Western Electric from the Bell System; to divide it into three competing units; and to require that the Bell system engage in competitive bidding. The inauguration of the more economically conservative Eisenhower administration in 1953 produced a landmark 1956 Consent Decree, which, while sparing AT&T many of the rigors that had been called for and forestalling court action, can be regarded as a watershed of antitrust activism that ushered the Bell System into the modern era. The Decree permitted AT&T to retain Western Electric, which at the time still manufactured 80% of all USA telephone equipment. It also required AT&T and Bell Labs to grant nonexclusive licenses for their present and future patents to any applicants, including competitors. Most importantly, the Bell System's business activity was limited to providing regulated public communication facilities and services, a constraint that was progressively onerous during the 1960s and 1970s as unregulated computer and information services became increasingly widespread and profitable.

Over the next quarter century, patterns of cost, technology, demand, and innovation in telecommunications and information conspired gradually to erode AT&T's markets and market share and to sour America's love affair with Ma Bell. Microwave technology, which matured during World War II, was the basis of the FCC's "Above 890" decision in 1959 allowing companies other than AT&T to provide services at microwave frequencies. Within 3 years, a small firm named Microwave Communications Inc. (MCI) had applied for, and was subsequently awarded, permission to provide private microwave service between Chicago and St. Louis. In 1956, a court allowed the Hush-A-Phone apparatus, a nonelectronic cup-like device, to be attached to telephone handsets despite Bell's far-fetched protests that its system integrity would thereby be compromised. The FCC's Carterfone decision in 1968 in effect permitted mobile radio/telephone messages to be patched into the

Bell System. The Specialized Common Carriers decision of 1971 allowed more competitive entry. The FCC's First Computer Inquiry, completed in 1971, ruled that new data communications should not be regulated, thereby foreclosing AT&T from that market under the 1956 Consent Decree. The Second Computer Inquiry, terminating in 1979, determined that all customer premises equipment (CPE) should be unregulated; it also adopted certain accounting and structural separation requirements and required AT&T to form a wholly separate subsidiary to offer any unregulated services or equipment. The report of the Task Force on Communications Policy, issued in the final days of the Johnson administration, was generally pro-competitive and provided little solace for AT&T. Commercial satellite systems, originating with INTELSAT in 1965 internationally and during the early 1970s domestically under the FCC's Open Skies decision, were likewise not foreclosed to AT&T's competitors, although AT&T was for years a minority shareholder of the Communications Satellite Corporation (Comsat). Most ominously for the future, the FCC's Execunet Decision in 1977 allowed MCI to offer Execunet Service on a message toll basis, the deepest incursion to date into AT&T's monopoly.

AT&T's response to these challenges was often erratic and inconsistent. Its Hi-Low tariff in the mid-1970s, departing from nationwide average cost pricing to underprice its competition, was declared void by the FCC, which found it without cost justification. Its highly discounted Telpak tariff, offered in the early 1960s to counter microwave competitors, was in effect for 15 years before AT&T discontinued it due to the actions of arbitragers (resellers and sharers). During that period the FCC was unable to come to grips with many fundamental pricing issues, including most notably cost allocation procedures. WATS tariffs, AT&T's most successful, continue today. AT&T's interface with the political arena, which had been brilliantly managed for nearly a century, began, under the chairmanship of John deButts, to give way to a more strident, circle-the-wagons mentality. The euphemistically titled "Consumer Communications Reform Act of 1976" was in fact introduced by Bell in an effort to reaffirm the need for a single national network; to restore the status quo ante in equipment competition, entry barriers, and pricing; and to limit the FCC's authority over the Bell System. It was one of 11 major bills introduced but not acted upon by Congress between 1976 and 1981.

Meanwhile, late in 1974, after the paralysis of the Watergate era had passed, the DOJ again sued AT&T for violation of antitrust laws, including (1) exploitation of its "bottleneck" monopoly in local services to give it an advantage in long distance, equipment, and information markets; (2) overcharging other firms for interconnection to the Bell System; and (3) cross-subsidizing its competitive services by shifting costs to noncompetitive regulated markets, thereby allowing it to set prices below costs in competitive markets and to force rivals (including those with low costs and efficient services) to withdraw. This time severe structural remedies were asked for, including the divestiture of Western Electric and some or all local Bell Operating Companies (BOCs) from AT&T Long Lines. The accession of the more conciliatory Charles L. Brown as AT&T's president in 1977 and chairman and CEO in 1979 set the stage for the momentous events of the early 1980s.

3. Dynamics of the divestiture

Brown quickly shifted course, calling for "a new realism" and acknowledging that competition "is here and it is growing." AT&T, he declared, was ready without preconceptions to explore alternative futures. For its part, the federal judiciary, in the person of Judge Harold H. Greene, denied AT&T's motion for dismissal of the antitrust case with enough vehemence to "send shock waves through the defense". William Baxter, assistant attorney general for antitrust at the DOJ, declared his readiness to "litigate to the eyeballs".

Circumstances conspired to allow a federal judge and a law professor on leave at the DOJ—neither of whom had any special expertise in telecommunications matters—to wield extraordinary power in shaping the outcome of the government's antitrust suit. Judge Greene insisted on oversight authority over any agreement between the DOJ and AT&T that might avert further legal action. Assistant Attorney General Baxter, a law professor at Stanford University, had come to the DOJ in the early weeks of the Reagan administration. Both of his superiors (including the Attorney General), who ordinarily would have maintained a deep personal involvement in so massive a federal lawsuit, recused themselves from participation because of potential conflicts of interest arising from past private involvement as lawyers in telecommunications matters. Thus a third-tier Justice official, less than a year into his tenure of office, functioned as the de facto Attorney General in crafting an agreement that would lead to the demise of the world's largest private firm. Furthermore, two cabinet members, the Secretary of Defense and the Secretary of Commerce, engaged in highly public but ultimately unsuccessful efforts to have the suit against AT&T dropped in the interest of natural security and international competitiveness. Ordinarily, such high-level political pressure might have been more effective. At the time, however, the influence of Congress on telecommunications matters had been dissipated by yet another failure to rewrite the out-of-date legislation that had established the FCC in 1934; and executive branch officials generally were wary of exercising even the tiniest pressure on the outcome of the DOJ's suit owing to provisions of the recently passed Tunney Act requiring meticulous disclosure of actions leading to a consent decree.

Faced with strong prosecutorial evidence and the determination of both judge and prosecutor to litigate the suit to a conclusion if necessary, AT&T's chairman decided to execute the previously unthinkable act of negotiating an agreement that would spell the end of AT&T as he knew it. On the basis of delicate negotiations with the DOJ[2] an agreement was announced on 8 January 1982, averting an impending and quite probably hostile decision by the court. The major provisions of the agreement – which was technically an alteration of the 1956 Consent Decree and thereby called the Modification of Final Judgment (MFJ) – were the following:

1. AT&T would divest itself of its twenty-two local Bell Operating Companies (BOCs), which would be regrouped under seven regional holding companies (RBOCs). Within 6 months, AT&T would file a plan indicating how this divestiture would be carried out by the deadline of 1 January 1984.

2. The BOCs would be allowed only to provide monopoly local services within 161 newly formed local access and transport areas (LATAs). Inter-LATA service would be provided by AT&T and its competitors, but AT&T would continue to be much more heavily regulated than its long-distance rivals.
3. AT&T would provide all other interexchange carriers and information providers "equal access" to its network in terms of type, quality, and price.
4. The twenty-two divested BOCs and their respective RBOCs would market but would not manufacture equipment. In addition, they would not discriminate in equipment procurement.
5. AT&T would retain Bell Labs and Western Electric.
6. The RBOCs would retain control of the Yellow Pages.
7. The RBOCs would retain the right to use the Bell trademark and logo, while AT&T could do so only in connection with Bell Labs.
8. The RBOCs were allowed to file for waivers allowing them to enter unregulated, competitive "lines of business" in which the court determined that they could not exercise monopoly power.
9. AT&T was freed from the strictures of the 1956 Consent Decree and allowed to enter information services and other markets outside of regulated telecommunications.

4. Case study I: natural monopoly

As seen in the previous section, the MFJ was negotiated under unusual circumstances. There was no protracted judicial proceedings, although both sides presented voluminous evidence to the court. For its part, AT&T used the excellent microeconomists it employed at Bell Labs and retained from academia. Ironically, as it turned out, it was primarily this group of economists which developed, during the duration of the antitrust suit from 1974 through the MFJ in 1982, the theory of multiproduct natural monopoly. Yet the earliest published empirical investigation of multiproduct natural monopoly, conducted by Evans and Heckman (1983a, 1984), appeared after the MFJ had been agreed to. Evans and Heckman, using a second-order Taylor series approximation of a translog cost function, concluded that AT&T did not enjoy natural monopoly during a period of several decades before 1982–in other words, its cost function as estimated was not subadditive, meaning that some combinations of outputs could have been produced more cheaply by several firms than by AT&T alone. Such a finding might arguably have exerted an even more procompetitive influence on Bell's antitrust case and might also have disposed AT&T to make even more concessions in the MFJ to avert completion of the government's lawsuit.

Snow (1987a, 1987b), in a study partially motivated by a desire to examine the robustness of Evans and Heckman's techniques to new data, applied their econometric technology to INTELSAT and found that the global commercial satellite cooperative did exhibit natural monopoly during the period 1967–84. In 1988, however, Charnes, Cooper, and Sueyoshi (1988) used goal programming techniques on what they claimed were the same data and functional forms as Evans and Heckman but arrived at the opposite conclusion regarding AT&T and natural monopoly. In a rejoinder, Evans and Heckman (1988) argued that differences in data and the model used rendered the comparison impractical, noting that better data rather than alternative econometric techniques were indicated. Röller (1990a, 1990b), however, further narrowed the scope of the Evans and Heckman natural monopoly test by constraining the cost function to a region in which it was "proper," namely exhibiting nonnegative marginal costs, nonnegativity, homogeneity, monotonicity, and concavity in input prices. In doing so, Röller—like Charnes, Cooper, and Sueyoshi—reversed the findings of Evans and Heckman and concluded that AT&T had been a natural monopoly before its divestiture.

Waverman (1989), in extensive work on natural monopoly in AT&T, found that "the [natural monopoly] test for aggregate AT&T data is so sensitive to data and to econometric technique that it cannot be relied on for making policy." His view was that "neither scale nor scope was significant in the 1947—77 period at the level of a firm such as AT&T before divestiture....It is unlikely that this mammoth organization experienced economies of scale and scope" or that "significant economies of scale existed in interexchange service between 1950 and 1980" (p. 94).

The most recent contribution to this debate is that of Shin and Ying (1992), who noted that past empirical studies of natural monopoly had problems with both the choice and the aggregation level of data. To examine a richer data set they used a pooled cross-sectional sample of 58 local exchange carriers (LECs) from 1976 to 1983. They found strong evidence against the presence of natural monopoly in the local exchange carriers and argued that if the LECs had no natural monopoly, the pre-divestiture Bell system, of which they constituted a large proper subset, could hardly have been a natural monopoly either. As Shin and Ying noted, their results "suggest that the benefits to breaking up the monopoly outputs of existing local exchange carriers substantially outweigh the potential losses in efficiency....Local telephone companies are unnatural monopolies in undeservedly protected monopoly markets" (pp. 181—182).

One is struck by the complexity of the econometrics needed to test the relatively simple hypothesis of subadditivity. Of five studies of AT&T's cost function subadditivity, three (Evans/Heckman, Waverman, and Shin/Ying) refute subadditivity, while the other two (Cooper/Charnes/Sueyoshi and Röller) confirm it. Would these conflicting results have been of any use to either side in the antitrust proceeding leading to the MFJ? Even if an unambiguous confirmation of natural monopoly had emerged from this line of research, one must acknowledge some of the difficulties the concept entails. For example, as a purely static phenomenon the notion of natural monopoly cannot measure or incorporate the dynamic benefits of novel and qualitatively different services and equipment that emerge more readily under a regime of competition. In addition, since the cost function depends on underlying technologies and input prices, it could conceivably alternate between sub- and superadditivity within very brief periods of time, making the possibly temporary presence of natural monopoly an uncertain policy guide as to whether costly and irreversible structural remedies on the scale of the AT&T divestiture should be undertaken in the name of static efficiencies.

5. Evaluations of the divestiture

a) Competition and prices in telecommunications markets

As one would expect, researchers in the late 1980s found evidence of significantly increased competition. Waverman (1989, p. 111) concludes that the inter-LATA interexchange market is competitive, although imperfectly so. Judge Greene himself (Baxter et al. 1993, p. 49) regarded increases in competition in various markets (manufacturing, long distance, information services) as among the greatest successes of the MFJ. Egan and Waverman (1991, p. 133) found "no indication of serious competition among [local exchange carriers] for the basic local market and little evidence for intra-LATA toll services." In most other markets, however, they detected the presence or emergence of workably competitive conditions. Egan and Waverman find access markets "uniquely competitive" for large users, since the alternative of bypass exists. The CPE market, competitive before the divestiture, today exhibits "numerous buyers and sellers, none of which exercises market power" (Egan and Waverman, 1991, p. 138). The market for switching and transmission systems, while dominated by Western Electric and Northern Electric, is competitive because of a vigorous fringe and the close substitutability of the two major incumbents' products (Egan and Waverman, 1991, p. 143). Again, price competition for cellular telephone equipment is "fierce by any measure" (Egan and Waverman, 1991, p. 145). Finally, information services broadly conceived are seen as quite competitive with price competition and numerous suppliers (Egan and Waverman, 1991, p. 148).

Levin (1991) agrees for the most part with these conclusions. Too much regulation and too little demand made the enhanced services market uncompetitive at the time he was writing. In general, he saw the extent of competition as depending crucially on the pace at which Judge Greene continued to relax various MFJ restrictions on the RBOCs (Levin, 1991, pp. 153–155). Since Levin's comments in the late 1980s, it is fair to conjecture that greater competition has emerged, due both to Judge Greene's rulings and other factors.

In the 4-year period following the AT&T divestiture, local service prices increased by 41.0%, while interstate toll prices fell by 34.1% and access charges by 52.6% (Egan and Waverman, 1991, p. 121). These often painful realignments, while approximating the "correct" cost-based prices, laid the resulting political repercussions squarely at the doorstep of state regulators. Noll and Smart (1991, p. 187), writing with particular reference to federal/state regulatory relations, argue that "the primary effect of divestiture and federal deregulation was reduced prices for customer equipment and for services that were becoming competitive."

Taylor and Taylor (1993) argue econometrically that reduction in interstate long-distance prices and expanding interstate demand are more than explained by reductions in access charges paid by the long-distance carriers to local telephone companies. In other words, competition in the interstate long-distance market had no role in this expansion of demand. In addition, they found that net of access charges, AT&T's nominal tariffed prices increased at an annual rate of 1.2% between 1984 and 1992, a period in which it was losing market share. Net of access charges, real interstate toll rates declined about twice as rapidly during the 7 years before the divestiture than in the 7 years following. These data, note Taylor and Taylor (1993, p. 187), are "more consistent with the presence of a regulated price umbrella than with a competitive market." Instead, one finds "regulated competition," a term anathema to Crandall, Brown, and others, and "asymmetric regulation" of AT&T, which "have yet to bring the benefits of lower prices and expanded demand to all interstate telephone customers" (ibid., p. 189). In support of Taylor and Taylor, A. Michael Noll (1993) argues that technical innovation and increasing productivity, rather than increased competition due to the divestiture, account for the continuing decline in long-distance rates. He concludes that "today's stimulated and forced competition is a sham benefitting AT&T's competitors and not the public."

Price-cap regulation, popular in the United Kingdom since the early 1980s, was adopted by the FCC in 1989 for AT&T as an alternative to the well-known theoretical and practical difficulties of rate-of-return regulation. In 1992, the FCC reported a basically positive experience with price-cap regulation. Nevertheless, Braeutigam and Panzar (1993, p. 197) conclude that "(price-cap) regulation is a classic case in which practice is far out ahead of theory....From the public-policy perspective, the choice between PC-style incentive regulation and rate-of-return regulation remains very much an empirical question." At the state level, price caps have been shown to produce lower costs than does rate-of-return regulation, but state regulators are reluctant to adopt incentive schemes that may generate profit windfalls for the firms they regulate (ibid., pp. 194—195).

b) Federal-state regulatory dynamics

Intrastate telecommunications services are the province of state rather than federal regulation. Historically, state regulatory authorities have often lacked the necessary resources to correctly perceive and act on many of the issues facing them. They have been overwhelmed by the expertise and resources of large carriers, primary AT&T and its local companies, operating within their borders. Since the divestiture increased the number of telecommunications carriers operating intra-LATA and intrastate, the challenges facing state regulators have if anything been compounded. These include the following. (1) The BOCs were prevented by the MFJ from participating in a number of competitive markets, thus reducing the power of the states to regulate those markets and reducing the present and future profits of local telephone companies from newly emerging competitive services. (2) The FCC's implementation of the MFJ shifted most of the burden of service cross-subsidization problems to state regulators and at the same time reduced the flow of service subsidies from the federal to the state level. (3) The "equal access" requirement imposed on local telephone companies forced massive and unexpected investment burdens on local telephone companies by requiring them prematurely to retire switches that could not be cheaply converted to equal access configurations (Noll and Smart, 1991, p. 192). These difficulties, along with the "sluggish" growth of local service in a very mature and largely saturated market (Egan and Waverman, 1991, p. 119), have eroded the once cooperative federal-state relationship in telecommunications while increasing the relative influence of state regulation and forcing state regulatory boards to develop comprehensive pricing and competition policies for the first time in more than 50 years (Noll and Smart, 1991, pp. 185–186).

One may point to deregulatory efforts at the state level in recent years. In particular, about one-fourth of the states have deregulated inter-LATA but intrastate traffic, under the (still unconfirmed) assumption that such markets are competitive while intra-LATA interexchange links are not. This has effectively deregulated a number of small local telephone companies outside of RBOC ownership. For example, a 1985 Illinois law allowed local and interexchange carriers to "self-certify their services as competitive and to de-tariff them." A minimally regulated "social contract" approach to local service now prevails in Idaho, Vermont, and Nebraska. In 1985 the state of Washington reduced regulation where effective competition can be shown to exist (Waverman 1989, pp. 100–101; McGowan et al., 1991, pp. 68–69).

Despite these developments, Noll and Smart (1991, pp. 209–210), among other researchers, are not sanguine about the results of increased state regulation since divestiture. They find that most business customers are still charged above cost, while residential and business customers in small exchanges still receive a "very large" subsidy. Furthermore, state regulation since divestiture has still not moved prices toward costs; indeed, urban prices are deviating increasingly from costs. Such trends argue strongly against the contention that "pricing efficiency is a major force in state regulatory policy." Likewise, Hausman, Tardiff, and Belinfante (1993) find that state-regulated prices of intra-LATA calls are well in excess of cost and that long-distance calls are charged at well above cost. Inefficiencies in federal and state telephone regulation amount to "billions of dollars" and retard the advancement of the "Information Age" (ibid., p. 183).

c) Consumer and worker interests

The "political economy of deregulation" school[3] sees telecommunications deregulation in the United States as a rare instance of a victory by general interests—in essence, consumers and small firms—over "special" interest—large firms, labor unions, and politically entrepreneurial regulators.

As Hendricks and Sassalos (1991, pp. 392–394) note, deregulation resulting from the MFJ and other actions confronts telecommunications unions with two major problems. First, newly competitive circumstances oblige them to organize workers in new rather than existing firms, where many of the former are foreign-based or historically nonunion. Second, new divisions of existing firms, created specifically to administer newly deregulated markets, must be organized. Despite such challenges, these researchers find little evidence that deregulation itself will cause competition that would reduce wage rates in the industry. Equipment manufacturing would present the greatest possibilities for such downward wage pressure, but the more likely scenario would be RBOC competition with AT&T in the long-distance market. Since the RBOCs are large and unionized, union workers stand to lose much less than they did when the USA trucking and airline industries were deregulated in the 1970s. Loss of jobs due to import penetration has been the greatest threat to date. The Communications Workers of America, the industry's largest union, estimates that 120,000 jobs were lost in the 1982–1984 period alone. In addition, service sector jobs have been lost due largely to technological change. The tradeoff between the interests of workers and consumers in such cases is particularly clear and poignant. Crandall (1991b, pp. 408–415), to his surprise, finds that relative wage rates in the telephone service industry did not decline in the postdivestiture period, although in the telephone equipment industry they fell slightly relative to all manufacturing wages. Although capital equipment prices may have fallen since divestiture, wage rates in the industry have been only slightly affected.

The major negative fallout for telecommunications industry workers since the divestiture, then, has been job loss. Those who have maintained their employment have not been more badly affected than workers in other industries and have fared considerably better than workers in other deregulated industries.

The major consumer issue anticipated as a result of the divestiture was lagging telephone penetration as a result of rate rebalancing. Nevertheless, Hausman, Tardiff, and Belinfante (1993) conclude, in an econometric study, that long-distance prices exert a sufficiently strong effect on demand for basic access that a revenue-neutral rebalancing of telephone prices "would lead to both large gains in economic efficiency and increased telephone penetration" (ibid, p. 179; emphasis supplied). Logically, then, the authors see no tradeoff between economic efficiency and telephone penetration given their data and methods of analysis. It is conceivable, however, that their results are a local reversal of the more generally assumed Pareto-type trade-off between equity and efficiency in telecommunications markets, to the extent that difficulties in aggregating the social welfare function permit such an analysis.

d) Issues facing the RBOCs

The role of the seven RBOCs has remained particularly complex and controversial. Judge Greene, whose authority in this area continues, has been subject to pressures from AT&T to minimize the business activities allowed the RBOCs under the MFJ, and from the RBOCs themselves to be allowed maximum leeway. Academics, policy analysts, and other presumably neutral observers disagree on the merits of what has been done and what should be done. One school of thought sees the RBOCs as reincarnations or clones of the old AT&T. The late William G. McGowan of MCI, who since the late 1960s was AT&T's foremost competitive nemesis, wondered whether in a "reproductive miracle", a "highly public" but "hardly immaculate" conception has passed a "monopoly gene" from "Ma to Baby Bell" (McGowan et al., 1991, p. 58). Gerald R. Faulhaber, a University of Pennsylvania faculty member who was previously a planner and researcher for AT&T and Bell Labs, argues that AT&T has been unduly disadvantaged by overly generous regulatory favors to the RBOCs since divestiture (Faulhaber 1987; see Snow 1988a for a review). Examples of favoritism to the RBOCs include progressively more lavish line-of-business waivers and AT&T's foreclosure from the lucrative Yellow Pages markets. Faulhaber sees Congressional action as the best counterpoise to Judge Greene's judicial activism as the "de facto czar of telecommunications", who is "part of the problem even as he struggles to be its solution". Congress should replace rate-of-return regulation with price-cap rules; maintain equal access for all interexchange carriers under MFJ rules; and gradually eliminate all BOC carrier access charges except those specifically incurred to connect interexchange carriers to the local switch. Regulation "will not solve these problems"; rather, "regulation is the problem". Unless a miracle of Congressional cooperation provides meaningful legislation for the first time since 1934, the telecommunications industry "will probably continue to suffer from economic inefficiencies caused by the regulatory process" which will "both retard and distort trends toward new products and declining costs".

Waverman (1989) and Crandall (1989), by contrast, offer arguments on behalf of the RBOCs to be treated more liberally by Judge Greene and by any Congressional legislation. Crandall notes approvingly that Judge Greene has now removed all limits on BOC entry into unregulated, noncommunications markets. Comparing the costs and benefits of a welfare simulation, Crandall finds large welfare gains if the BOCs were allowed to enter the inter-LATA service market, even though this might risk "cross-subsidization induced by regulation". Further, Judge Greene ignores economic welfare by not allowing the RBOCs to expand their "line-of-business" activities (ibid., pp. 142–146). Waverman (1989, pp. 105–112) likewise argues for RBOC entry into inter-LATA interexchange services.

In retrospect, the gradual, phased relaxation of the original MFJ restrictions on line-of-business activities by the seven RBOCs' 22 local service subsidiaries seems as inevitable as it is desirable, so long as the usual antitrust safeguards against anticompetitive practices are adopted, preferably by Congressional action.

e) Technology issues

In the simpler days of the predivestiture Bell System, note Bolter and McConnaughey (1991, p. 290), a remarkable technological determinism lasted almost a century. Innovation was driven by the supply of technology, which in turn was determined largely by the engineers at Bell Labs. Since the divestiture there is ample evidence that demand by customers, particularly large and sophisticated business users, is driving the extent and pace of innovation. Such demand is by its nature less predictable. It is also clear that qualitatively new kinds of services and equipment have evolved primarily because of demand-side impetus. The time lag between technological capability and commercial innovation is declining markedly (ibid., p. 290).

During the 1960s and 1970s, indeed, major USA information-intensive firms found themselves obliged to develop private networks suited to their special needs because of the time lag required by the Bell System and other carriers to provide such services. Since divestiture, however, plant replacement and revitalization, within the elements of the old Bell System and elsewhere, have proceeded briskly with the realization that competition has become a fact of life. One facet of this technological renewal is the accelerating transition to digital technology and facilities in a network that from its inception was essentially engineered for analog transmission (ibid., p. 291).

Crandall (1989, pp. 138–139) presents data suggesting that the divestiture allowed the BOCs to reduce their dependence on Western Electric (now renamed AT&T Technologies). For example, AT&T's share of the USA switch market declined from 70 to 49% during the period 1982–86. Northern Telecom, a Canadian firm, now competes aggressively with AT&T for the central office market. In general, the BOCs can now purchase equipment from a variety of other suppliers, probably at lower costs.

To be sure, some aspects of the predivestiture technological determinism persist in various USA telecommunications markets. Noam (1989, p. 276) and Bolter and McConnaughey (1991, pp. 297–299) interpret emerging ISDN services in this light. Noam takes particular exception to the integration of services within ISDN and surmises that competition among diverse service suppliers could induce a dynamic, downward shift of the cost curve that would more than counterbalance any static economy gains from integrating services within a single ISDN network. Bolter and McConnaughey (1991) see ISDN as "technically feasible" but lacking demand, a quintessential "technology push" innovation. They argue that only with "public guidance" and "appropriate public policy infrastructure" can ISDN be introduced so as to ensure full conversion without a degeneration into piecemeal convergence of partial and incompatible ISDN systems. Japan, France, and Germany, by contrast, now have ISDN universally available as the result of conscious policy in that direction.

Temin (1987, p. 360) holds that the "public goods" nature of basic research in telecommunications has increased since divestiture. In particular, the breakup of the Bell System makes it more difficult to prevent the benefits of such research by Bell Labs from being appropriated by firms that do not pay for it, thereby reducing AT&T's incentive to engage in such research in the first place. The textbook solution is for the government to bear the burden of financing a public good such as basic research.

f) International telecommunications trade and services

Given the progressive globalization of telecommunications markets, the breakup of AT&T can be seen as the "beginning of the end" of the USA antitrust policy that flourished during the 1960s and 1970s (Temin, 1987, p. 345). The impersonal forces of global competition may now, in other words, replace conscious antitrust policy at the national level, particularly if such policy is seen to disadvantage the largest and most efficient USA firms in the international arena. A major impact of divestiture has been to provide an even larger and more open market for non-USA suppliers of telecommunications equipment and services. Inevitably the issue of trade reciprocity arises. Ironically, "some of the strongest advocates of protectionist policy in...procurement subsequently began to seek their fortune in the newly liberalized USA market". In the service industries, likewise, "the relative ease with which services can be offered in the United States contrasts strikingly with the barriers that prevent American carriers from even reaching international markets" (Noam, 1989, pp. 288–289). By the early 1990s, however, a gradual turnaround was evident. AT&T, through its international subsidiary, began to emerge as a successful competitor in many international markets, including China. In addition, various USA long distance carriers, notably MCI and USA Sprint, have competed for several years with AT&T to negotiate international long-distance agreements with dozens of countries.

On balance, however, international trade flows in telecommunications are negative for the United States and pose a significant irritant to trade relations generally. International trade in services is still incompletely understood from a theoretical point of view. Often, for example, it is confused with direct foreign investment. In any event, the topic, which was the subject of working group attention during the early years of GATT's Uruguay Round, was unceremoniously dumped from GATT's agenda in the headlong rush to reach an agreement by the December 1993 deadline.

6. A balance sheet

a) Regulation and competition

Two competing forces collided to produce the MFJ. One, the populist, antimonopoly, antibig bigness zeal that periodically dominates American policy and politics, was embodied in the DOJ's lawsuit against AT&T and obviously prevailed. The second, a desire for efficiency, is more of an engineer's and economist's ideal and resides in the FCC. Charles L. Brown, AT&T's chairman and CEO at the time of the divestiture, argued that the Bell System had opted to preserve its social contract with America – maintaining its political, if not its structural continuity – by "conforming the business to the desires of the public"[4] (Brown, 1984, p. 1). Noam (1993, pp. 439–440) has suggested a useful two-dimensional taxonomy by which to analyze the MFJ: proregulatory and deregulatory sentiment, as well as promonopoly and antimonopoly views. Among those with proregulatory, antimonopoly views he lists Judge Greene, as well as "those economists who believe that market power may require a curb by both structural and regulatory intervention". The specter of "regulated competition", however, is anathema to many, including Brown (1984, p. 2) and Crandall (1991a, pp. 162–163). Promonopoly, proregulatory advocates in the economic tradition of John Kenneth Galbraith, by contrast, "believe in the necessity of large firms in order to perform complex tasks..." and consider it "misguided to dismantle the efficient and socially redistributive AT&T system." They expected cost and price increases, distributional inequities, technical incompatibilities, and reduced R&D and were skeptical about the possibilities for competition following the divestiture. The antimonopoly, deregulatory mindset, according to Noam, is exemplified by William Baxter, the DOJ's chief advocate for AT&T's divestiture. Those in this camp are "procompetition centrists or free market advocates who believe that competition needs to be established by intervention" and that "market forces and competition (can) assert themselves" when monopoly and its attendant inefficiency are removed. Benefits include lower costs, cost-based pricing, innovation, and accelerating productivity and modernization. The final taxonomic cubbyhole is inhabited by promonopoly, deregulatory advocates, including "some members of the Chicago and Princeton schools of thought". In the Chicago view, "competition rather than bureaucrats should have reduced AT&T's power"; the Princeton school "opposed attacking an efficient natural monopoly".

Noam's view of events since the divestiture are quite sanguine (Noam 1993, pp. 446–448). Those with promonopoly views "were not supported" by ensuing events; "the divestiture accelerated competitive entry considerably and revealed the extent of accumulated inefficiencies". Along the regulatory axis, "deregulators were more often right than the proregulators". Regulation proved its residual value, however, in areas such as "lifeline" services for the poor, open network architecture, and interconnection of alternative local telephone companies. In summary, "the system proved itself capable of adjusting to major and rapid change... the overall impact (of the MFJ), as a whole, was positive". The key future task is to "balance the institutional diversification of the network system with technical integration".

b) Case study II: the Internet and the National Information Infrastructure (NII)

USA Vice President Albert Gore has become a visible spokesman for developing a nation-wide "information superhighway", although patterns of cost-sharing and other vital details have yet to be worked out. In December 1993, under Gore's encouragement, 28 companies, including AT&T, IBM and other computer firms, and several RBOCs and their subsidiaries, formed a coalition to ensure uniformity of equipment and technical standards in the NII (Honolulu Star-Bulletin, 1993, p. C4). It would be instructive to consider the role of the Internet, the worldwide network of computer networks, in evolving into part of the NII given the constraints and opportunities afforded by the postdivestiture telecommunications market in the United States. USA telecommunications companies, for example, are seen as well-suited to replace the present chaos by providing basic Internet services. Yet their ability to do so will depend on the vagaries of obtaining appropriately broad waivers for such a new "line of business" from Judge Greene (Klingenstein, 1993, p. 141). McGarty (1993, p. 227), discussing the integration of network elements under the evolving NII, notes that as early as the FCC's Above 890 decision, diseconomies rather than economies characterized the comprehensive integrated network. Natural monopoly, he contends, no longer exists in market interconnection. Faulhaber (1993, pp. 284–285) likewise rejects the natural monopoly argument for the Bell System and argues against subsidies for the evolving Internet: "any assertion that Internet...requires continued government intervention because it is a natural monopoly deserves very close scrutiny". Nevertheless, one sees a fundamentally cooperative approach to the Internet in the USA, while ISDN markets there have been fiercely laissez-faire with little government intervention.

c) The experts appraise the divestiture

Allen (1993, Appendices 1 and 2) provides brief informal responses from 30 key economists, industry leaders, lawyers, consultants, and government officials close to the divestiture about their views of (a) the extent of natural monopoly in predivestiture AT&T and (b) the divestiture in general. Their responses, given in 1989, are sufficiently compelling and diverse to merit further analysis. Eleven of the 30 were asked their views on the presence of natural monopoly in AT&T. Table 1 is my impressionistic evaluation of their responses.

All of the 30 were asked by Allen about their views on the divestiture itself. Table 2 shows their ranked responses on a scale of 0 to 10, where 10 is the highest evaluation.

The mean response among the seven noneconomists was 5.375, insignificantly different from the overall response mean of 5.4. It is arguable that a poll of the same individuals today would disclose a somewhat more favorable rating of the divestiture, due both to the failure of various horror scenarios to materialize and to the general robustness of the USA telecommunications sector a decade after the MFJ entered into force.

d) Update at random

A haphazard glance at recent media accounts gives ample testimony of the content and pace of developments involving the corporate dramatis personae in the 1984 divestiture. Judge Greene, still presiding over his handiwork, has ruled that the $12 billion acquisition of McCaw Cellular by AT&T would violate the MFJ, but is leaving the door open for waiver requests (Honolulu Advertiser, 6 April 1994, p. B1). The FCC's order in February 1994 that cable firms cut prices by 7% has frustrated a $4.9 billion partnership of Southwestern Bell with Cox Cable, as well as a $33

Table 1.

Is there natural monopoly in AT&T?	
Yes	(1)
Yes, locally only	(1)
Yes, integration economies only	(1)
Probably	(1)
Equivocal	(4)
No	(3)

Table 2.

Rating of divestiture (10 = highest)	
Rating	Number
10	0
9	1
8	4
7	7
6	3
5	6
4	3
3	2
2	2
1	2
0	0

Note: Responding = 30; mean response = 5.4; standard deviation = 2.19.

billion merger of Bell Atlantic with Tele-Communications Inc., the country's largest cable TV company (ibid.). To AT&T's consternation, the RBOCs, which paid $40 million for lobbying efforts in 1992 alone, seem close to legislative relief from remaining line-of-business restrictions under the MFJ (Wall Street Journal, 15 March 1994, p. A1). Intent on avoiding bypass of their local networks, the RBOCs now provide service to about half of the 14 million cellular telephone customers in the USA (ibid., 11 February 1994, p. R20). Ameritech, seeking to be the first RBOC to enter the $60 billion long-distance market presently dominated by AT&T, Sprint, and MCI, has offered to surrender its monopoly rights over local traffic to do so (ibid., 12 December 1993, p. A3). The USA Court of Appeals reversed the ruling of a lower court that had fined Nynex $1 million for violating the MFJ (New York Times, 17 November 1993, p. D1). Top computer and telecommunications executives met to discuss new alliances and deals, something inconceivable not long ago (ibid., 21 October 1993, p. D1). After years of fearing and fighting the RBOCs, newspapers are now forging partnerships with them under increasingly favorable decisions by the courts regarding RBOC entry into information service businesses (ibid., 5 July 1993, p. A11). The FCC opened up the $90 billion local telephone exchange market to new competition (Boston Globe, 18 September 1992, p. 76). The Wall Street Journal (23 June 1992, p. A20) complained editorially that telephone and cable TV firms use too many resources fighting antitrust battles; market competition is preferable.

e) Finale

The late Ithiel de Sola Pool (1984), who in his Technologies of Freedom showed that most forecasts of the course of technology turned out to be wrong, was himself a somewhat more successful prognosticator concerning the future of the AT&T divestiture. "Ten years from now", he wrote 10 years ago, "a substantial revisionist literature will undoubtedly have emerged lambasting the Consent Decree (MFJ) of 1982 as short-sighted and ill-considered....quick-freezing the realities of 1982 into a decree will not provide an appropriate arrangement for the technologies of 1992" (ibid., p. 112). Among the developments he presciently foresaw are the proliferation of electronic mail; mounting competition between phone and cable television companies; worldwide electronic access to library holdings; and the increasing urgency of the standards issue (ibid., pp. 124, 125, 127 and 119, respectively). Our glance at recent newspaper accounts of activities by AT&T and the RBOCs conveys the impression that even a perfectly crafted MFJ by 1982 standards would be fraying at the edges by the mid-1990s, given the dizzying pace of developments in USA telecommunications markets. Indeed, the very competitive forces unleashed by the divestiture gave it a built-in obsolescence rivaling the half-life of the latest generation of computer game software. Thus, the appropriate question does not appear to be "was the AT&T divestiture of 1984 a good idea in the light of subsequent events", but "what can be learned from the divestiture and subsequent events to affect an economically, politically, and socially optimal transition of USA telecommunications into the twenty-first century?"

Notes

1. For background on Sections 2 and 3 see Allen (1993); Cole (1991b); Crandall (1991a); Shooshan (1984a); Snow (1991); Temin (1987); and Teske (1990).
2. Baxter, who was vacationing in Utah during the first week of 1982, continued his input from a warming hut on the slopes of a ski resort and from the Salt Lake City airport.
3. See Noll and Owen (1983) for an elaboration of this concept.
4. There is an interesting analogy here with the "social economy" school still influential in continental Europe. According to its main proponent, public opinion is the primary legitimizing determinant of whether major sectors of a country's economic infrastructure should remain under public ownership or be privatized (Snow, 1988b).

References

Allen, D., 1994. *Competing Visions for World Telecommunications: the Global Evolution of Industry Structure*. Manuscript.

Allen, D., 1993. *Breakup of the American Telephone Company: The Intellectual Conflict, Before and Since*. Manuscript.

Baxter, W.F. et al., 1991. In: Cole, B.G. (ed) *Questions and Answers with the Three Major Figures of Divestiture*, (pp. 22–49). New York: Columbia University Press.

Bolter, W.G., McConnaughey, J.W, 1991. In: Cole, B.G. (ed) *Innovation and New Services*. (pp. 285–325). New York: Columbia University Press.

Braeutigam, R.R., Panzar, J.C., 1993. Effects of the Change from Rate-of-Return to Price-Cap Regulation. *American Economic Review*, 83(2):191–198 (May).

Branscomb, L.M., 1993. In: Kahin, B. (ed) *Information Infrastructure for the 1990s: A Public Policy Perspective*. (pp. 15–30). New York: McGraw Hill.

Brown, C.L., 1984. In: Shooshan, H.M. III (ed) *A Personal Introduction*. (pp. 1–7). New York: Pergamon.

Charnes, A., Cooper, W.W., Sueyoshi, T., 1988. A Goal Programming/Constrained Regression Review of the Bell System Breakup. *Management Science*, 34:1–26.

Cole, B.G., 1991a. *After the Breakup: Assessing the New Post–AT&T Divestiture Era*. New York: Columbia University Press.

Cole, B.G., 1991b. *Introduction*, (pp. 1–18). New York: Columbia University Press.

Coll, S., 1986. *The Deal of the Century: The Breakup of AT&T*. New York: Atheneum.

Crandall, R.W., 1991a. *After the Breakup: U.S. Telecommunications in a More Competitive Era*. Washington: Brookings.

Crandall, R.W., 1991b. In: Cole, B.G. (ed) *Efficiency and Productivity*, (pp. 408–427). New York: Columbia University Press.

Crandall, R.W., 1989. In: Crandall, R.W., Flamm, K. (eds) *The Role of the U.S. Local Operating Companies*, (pp. 114–146). Washington: Brookings.

Crandall, R.W., Flamm, K., 1989a. (eds) *Changing the Rules: Technological Change, International Competition, and Regulation in Communications*. Washington: Brookings.

Crandall, R.W., Flamm, K., 1989b. *Overview*, (pp. 1–10). Washington: Brookings.

Crandall, R.W., Owen, B.M., 1984. In: Shooshan, H.M. III (ed) *The Marketplace: Economic Implications of Divestiture*, (pp. 47–70). New York: Pergamon.

Egan, B.L., Waverman, L., 1991. In: Cole, B.G. (ed) *The State of Competition in Telecommunications*, (pp. 115–182). New York: Columbia University Press.

Evans, D.S., 1983. *Breaking Up Bell: Essays on Industrial Organization and Regulation*. New York: North-Holland.

Evans, D.S., Heckman, J.J., 1988. Natural monopoly and the Bell system: response to Charnes, Cooper and Sueyoshi. *Management Science*, 34:27–38.

Evans, D.S., Heckman, J.J., 1984. A test for subadditivity of the cost function with an application to the Bell system. *American Economic Review*, 74(4):615–623 (Sept).

Evans, D.S., Heckman, J.J., 1983a. In: Evans, D.S. (ed) *Multi-Product Cost Function Estimates and Natural Monopoly Tests for the Bell System*, (pp. 252–282). New York: North-Holland.

Evans, D.S., Heckman, J.J., 1983b. *Natural Monopoly*, (pp. 127–156). New York: North-Holland.

Faulhaber, G.L., 1993. In: Kahin, B. (ed) *Pricing Internet: The Efficient Subsidy*, (pp. 271–295). New York: McGraw Hill.

Faulhaber, G.R., 1987. *Telecommunications in Turmoil: Technology and Public Policy*. Cambridge, MA: Ballinger.

Geller, H., 1984. In: Shooshan, H.M. III (ed) *Regulatory and Public Policy after Divestiture: The FCC and Congress*, (pp. 83–99). New York: Pergamon.

Hausman, J., Tardiff, T., Belinfante, A., 1993. The Effects of the Breakup of AT&T on Telephone Penetration in the United States. *American Economic Review*, 83(2):178–184 (May).

Hendricks, W., Sassalos, S.C., 1991. In: Cole, B.G. (ed) *Labor, Employment, and Wages*, (pp. 389–407). New York: Columbia University Press.

Kahin, B., 1993. *Building Information Infrastructure: Issues in the Development of the National Research and Information Network*. New York: McGraw Hill.

Klingenstein, K., 1993. In: Kahin, B. (ed) *A Coming of Age: The Design of the Low-End Internet*, (pp. 119–143). New York: McGraw Hill.

Levin, S.L., 1991. In: Cole, B.G. (ed) *The State of Competition in Telecommunications*, (pp. 149–155). New York: Columbia University Press.

MacAvoy, P., Robinson, K., 1983. Winning by Losing: The AT&T Settlement and Its Impact on Telecommunications. *Yale Journal on Regulation*, 1(1):1–42.

McGowan, W.G., Sikes, A.C., Nelson, S.L., 1991. In: Cole, B.G. (ed) *Policy Directions for the Future*, (pp. 50–76). New York: Columbia University Press.

McGarty, T.P., 1993. In: Kahin, B. (ed) *Alternative Networking Architectures: Pricing, Policy, and Competition*, (pp. 218–270). New York: McGraw Hill.

Noam, E.M., 1993. Assessing divestiture and deregulation in telecommunications. *Southern Economic Journal*, 59(3):439–449 (January).

Noam, E.M., 1989. In: Crandall, R.W., Flamm, K. (eds) *International Telecommunications in Transition*, (pp. 257–297). Washington: Brookings.

Noll, A.M., 1993. *A Study of Long-Distance Rates: Divestiture Revisited*. Manuscript, Annenberg School of Communication, University of Southern California.

Noll, R.G., Owen, B.M., 1983. *The Political Economy of Deregulation: Interest Groups in the Regulatory Process*. Washington: American Enterprise Institute.

Noll, R.G., Smart, S.R., 1991. In: Cole, B.G. (ed) *Pricing of Telephone Services*, (pp. 185–234). New York: Columbia University Press.

Röller, L.-H, 1990a. Modelling Cost Structure: The Bell System Revisited. *Applied Economics*, 22:1661–1674.

Röller, L.-H, 1990b. Proper Quadratic Cost Functions with an Application to the Bell System. *Review of Economics and Statistics*, 72:202–210.

Shin, R.T., Ying, J.S., 1992. Unnatural Monopolies in Local Telephone. *RAND Journal of Economics*, 23(2):171–183 (Summer).

Shooshan, H.M. III, 1984a. *Disconnecting Bell: The Impact of the AT&T Divestiture*. New York: Pergamon.

Shooshan, H.M. III, 1984b. In: Shooshan, H.M. III (ed) *The Bell Breakup: Putting It in Perspective*, (pp. 8–22). New York: Pergamon.

Snow, M.S., 1991. Economic Effects of Deregulation of Telecommunications Services in the United States. Paper presented at *International Conference on Economic Effects of Deregulation in Telecommunications Services*, Sogang University, Seoul (October).

Snow, M.S., 1988a. Review of telecommunications in turmoil: technology and public policy by G.R. Faulhaber (Cambridge, MA: Ballinger, 1987) *Telecommunications Policy*, 12(1);93–94 (March).

Snow, M.S., 1988b. The State as Stopgap: Social Economy and Sustainability of Monopoly in the Telecommunications Sector. *Review of Social Economy*, 46(2):1–23 (April).

Snow, M.S., 1987. An economic issue in international telecommunications: natural monopoly and commercial satellite systems. In: Macauley, M. (ed) *Economics and Technology in U.S. Space Policy*, (pp. 219–241). Washington: Resources for the Future/National Academy of Engineering.

Snow, M.S., 1987b. Natural monopoly in INTELSAT: cost estimation and policy implications for the separate systems issue. *Telematics and Informatics*, 4(2):133–150.

de Sola Pool, I., 1984. In: Shooshan, H.M. III (ed) *Competition and Universal Service: Can We Get There From Here?*, (pp. 112–131).

Taylor, W.E., Taylor, L.D., 1993. Postdivestiture Long-Distance Competition in the United States, *American Economic Review*, 83(2):185–190 (May).

Temin, P., 1987. *The Fall of the Bell System: A Study in Prices and Politics*. New York: Cambridge University Press.

Teske, P.E., 1990. *After Divestiture: The Political Economy of State Telecommunications Deregulation*. New York: State University of New York Press.

Waverman, L., 1989. In: Crandall, R.W., Flamm, K. (eds) *U.S. Interexchange Competition*, (pp. 62–113). Washington: Brookings.

Wiley, R.E., 1984. In: Shooshan, H.M. III (ed) *The End of Monopoly: Regulatory Change and the Promotion of Competition*, (pp. 23–46). New York: Pergamon.

Beyond Competition: The Future of Telecommunications.
Edited by D. Lamberton.

CHAPTER 14

The economic consequence of the introduction and regulation of international resale of telecommunications services

Martin Cave
Department of Economics, Brunel, University of West London, Uxbridge, Middlesex, UK

1. Introduction

This paper is concerned with the impact which international resale, especially international simple resale, may have upon the tariffs charged to subscribers for international telecommunications traffic and on commercial relationships between operators in the originating and terminating countries. It also discusses how these effects are influenced by the regulatory approach to international simple resale adopted by the authorities in the two countries involved, or adopted by countries acting on a multilateral basis.

In the traditional model, international telecommunications and telephony services were provided through a collaborative arrangement between what were in most instances monopoly operators. A call by a customer in the originating country would be delivered by the operator to an international gateway. It would then be conveyed to the international gateway of the destination country, using cable – or sometimes satellite – links jointly owned by both operators. The call would then be delivered by the operator in the destination country. This is the simplest case. In other cases the call was delivered through a transit country, or countries, involving two international segments and the participation of a third operator.

In the interwar period, a framework was developed for the remuneration of the operator in the receiving country. The customer originating the call would pay its domestic operator a collection rate and that operator would then pass on some of the revenues to the terminating operator, in consideration of the services rendered. In the case of intercontinental traffic, the payment made was based upon an accounting rate agreed between the two operators, initially intended broadly to reflect the total cost of completing all stages of the call. The accounting rate was shared between the two operators, normally on a 50-50 basis. The payment received by the terminating operator was known as the settlement rate. In cases where traffic between the two countries is balanced, no financial transfers are required with a 50:50 split, as the settlements cancel out. Where traffic is out of balance, the operator with the larger

---

Address for correspondence: Prof Martin Cave, Department of Economics, Brunel, University of West London, Uxbridge, Middlesex UB8 3PH, UK. Tel.: +44-895-203-320. Fax: +44-895-274-697.

number of originating calls makes a payment to the partner operator. Within Europe and the Mediterranean basis, accounting rates are based on a formula which is intended to be cost based.

Although accounting rates were initially intended to reflect costs, when those costs fell sharply as a result of new technological developments, accounting rates fell with a lag and by a smaller amount. As a result, terminating international telephone calls became a highly profitable business, in some cases more profitable than originating them. Thus countries with a balance of incoming calls had little incentive to agree to reductions in the accounting rate. Operators with a surplus of outgoing calls chafed under the obligation to make substantial net settlement payments to their partners.

As well as achieving these transfers, settlement rates in excess of costs also introduce inefficiencies into telecommunications charges by forcing up collection rates. Since there is evidence that price elasticities of demand for international calls exceed those for other services, the effect of this may be to impose a considerable welfare loss on the economies concerned.

International simple resale (ISR) offers a potential method of bypassing the accounting rate system. Resale involves the purchase of transmission capacity in bulk and its resale at a discount relative to the price paid if the same service were purchased from the public telecommunications operator (PTO). Thus a domestic reseller could lease private circuits from an operator and sell services based upon them directly to its customers. International resale involves the leasing by the reseller of international private circuits. It can take a number of forms, depending upon the mechanism by which the call is conveyed on its national segments. Thus one variant involves the call being delivered to the reseller by the telecommunications operator of the originating subscriber. Having conveyed the call over its international segment on a private circuit, the reseller then passes it to the telecommunications operator in the receiving country for termination. The reseller thus "breaks out" into the public switched telecommunications network (PSTN) at both ends. This is called international simple resale. Alternatively, if conveyance on one of the two domestic legs is by a private circuit leased by the reseller, break out into the PSTN occurs at only one end. The third possibility is for a call to be conveyed in its entirety along private circuits leased by the reseller.

Resale is effectively an arbitrage operation: its profitability depends upon imbalances between the prices and costs of services provided by the telecommunications operator. But international simple resale has a particular characteristic which makes it potentially highly profitable. The reseller bypasses the accounting rate system and is not obliged to pay the terminating operator the settlement rate. Since the latter – for reasons noted above – in most cases significantly exceeds the price which resellers have to pay the operator for termination of calls (or the costs which they themselves incur in renting private circuits to deliver calls themselves) resellers potentially have a substantial cost advantage.

This paper investigates the likely effects of the introduction of international simple resale on the tariffs charged for international telecommunication services and upon the accounting rates agreed between PTOs. Whether ISR is permitted is essentially a decision taken by regulatory bodies in the originating and terminating countries. The policies adopted will influence the development of ISR and hence have consequences for collection and accounting rates. The paper thus investigates the likely effects of alternative regulatory policies.

Following this introduction, Section 2 gives an analysis of the current system for determining accounting and settlement rates, thus providing an explanation of why arbitrage opportunities exist for international resellers. This is followed in Section 3 by a discussion of possible forms of international resale, the costs associated with them, the demand likely to be present for such services and the impact of ISR upon collection rates and accounting rates. Finally, Section 4 discusses and evaluates possible regulatory options towards ISR. Some of the more technical economic analysis is relegated to Appendices 1 and 2.

2. The operation of the accounting rate system

This section sets out a framework for analyzing the derivation of accounting and settlement rates on the assumption that bypass of the system by ISR and other means is not available. The analysis of the accounting rate system is formulated in terms of an economic model of interactions between operators in various countries, subject to specified incentives. This framework is necessary to predict how the introduction of a new service such as international simple resale will affect operators' pricing policies in respect of both collection rates and accounting rates. It is, however, undoubtedly the case that the current system has been heavily conditioned by the history of its development, particularly in such matters as the division of the accounting rate into shares attributed to the originating and terminating operator.

In order to model the determination of accounting and settlement rates, we suppose that operators' objectives can be adequately approximated by maximisation of profit from international calls. It is recognised that many telecommunications companies, especially those in public ownership, pursue overall quite different objectives. Nonetheless, there has been a widely observed tendency for international calls to be highly profitable, the profits being used to subsidise domestic services. It is thus not unreasonable to hypothesise that in setting international telecommunication charges, operators are seeking to maximise profits.

The analysis begins with the assumption that two monopoly operators are bargaining over the accounting rate and the division of that rate into accounting rate shares or settlement rates (SRs) which need not be the same in both directions. Each side would ideally like to pay a zero settlement rate to its partner and receive a positive settlement rate on the call which it terminates. But the process of negotiation is likely to result in settlement rates which are broadly similar in each direction. Operators therefore have to decide upon the level of SRs on the basis that they will both receive and pay them.

It may seem at first sight that profits would be maximised by setting that settlement rate at a very high level, but this is not the case for several reasons. First, the originating operator should logically see the settlement rate as an element of its costs in producing an international call. The higher the settlement rate it faces, the higher its marginal cost and the higher the collection rate which it charges to maximise its own profits. The revenues which an operator receives for terminating calls are thus dependent on the settlement rate it imposes. On standard assumptions there is likely to be a unique settlement rate which maximises an operator's profits from this activity.

Second, there is some evidence that incoming calls into a country generate outgoing calls in response. An increase in call minutes in one direction will thus generate extra traffic in the reverse direction. Thus by keeping its settlement rate down – and thus reducing its partner's marginal costs and therefore, hopefully, also influencing its partner's collection rates – more incoming calls are generated leading to increased demand for profitable outgoing calls. This effect might lead to a reduction in the profit maximising collection rate. Additional outgoing calls could stimulate additional profitable incoming calls. This phenomenon is sometimes known as call reciprocation. Some studies have claimed to have detected its existence in the traffic data.

Third, there is increasing evidence for the proposition that differences in telephone charges between countries, depending upon the direction of the call, influence the relative magnitudes of incoming or outgoing traffic, with the call being made from the "cheaper" end (for an analysis based entirely on this assumption, see Hakim and Lu (1993)). This is sometimes known as call reversion or "call-back". The crudest method of achieving this objective is for the caller to make a very brief call, asking the called party to ring back. More recently, sophisticated systems have come into existence permitting callers to revert the call by direct contact with a number in the receiving country. This makes demand sensitive, not only to the level of the collection rate in one country, but to differences between collection rates in the country pair. A call-back service in country B can also be used to enable a caller in country A to make contact with country C. These arrangements have a major impact on traffic flows, but they are not the main topic of this paper.

The existence and relative strength of these factors will cause each operator in principle to have an ideal combination of settlement rates for incoming and outgoing calls. It is almost inconceivable, however, that these ideals will coincide. The process of agreeing an accounting rate share can thus be seen as a classic bargaining problem, which the parties have to resolve by agreeing a compromise. As Appendix 1 shows, they can either bargain non-cooperatively, each operator imposing a rate which maximises its profit individually, on the assumption that the other operator is behaving in the same way. Or they can arrive at a cooperative solution by mutual agreement, thus generally reducing settlement rates compared with the non-cooperative solution and increasing the profits of both operators.

As with most market structures, the effect of declining costs in delivering the call will be to reduce settlement and collection rates. A lower cost of termination reduces the receiving country's profit-maximising settlement rate, and that lower cost is then factored into the sending country's optimal collection rates. The analysis is thus consistent with the observation of declining rates as technical progress reduces costs.

There is no necessary reason that the standard solutions to the bargaining problem would yield a 50-50 split of the accounting rate. Rather, the 50-50 split appears to have emerged as a convention as result of special historical circumstances (see Eward, 1985, pp. 77–80). Alternatively, it may have been embraced generally and adopted as a CCITT recommendation (D.150) for reasons of administrative simplicity or because operators are anxious about possible unequal bargaining positions and therefore support the equal split as a means of minimising possible disadvantage to themselves in the future. If a 50-50 split is accepted as a constraint, then associated with each operator will be a settlement rate which maximises its profits. Given that both parties need to agree a rate, it is likely that the outcome will be the higher of the two preferred rates.

The bargaining model set out above has been formulated in terms of monopoly operators at either end, with no alternative routes between the countries. The introduction of competition at one end immediately gives the monopolist at the other a negotiating advantage, as it can then seek to "whipsaw" competitive rates down by threatening to send all its outgoing calls to the operator offering it the most advantageous rates. Faced with this prospect, regulators have traditionally responded by imposing a uniform settlement or parallel accounting regime under which the settlement arrangements made by competing operators with a foreign administration, must be the same. The effect of this arrangement is to remove the incentive on the part of a single competitor to accept a less advantageous settlement rate, as the regime will require the new rate to apply to all. Arrangements of this kind operate (with some exceptions) in the UK, the USA and, ex ante, in Australia. Parallel accounting is often accompanied by a system of proportionate return: under this arrangement the overseas operator sends traffic to competing operators in proportion to incoming traffic received from them.

Competition at one end alters the level of profitability for the monopolist associated with any (equal) settlement rate, as it changes incoming traffic flows. But there is no clear reason why the monopolist would choose to lower the settlement rate, especially as — with call reversion — the ratio of incoming to outgoing calls will increase.

The impact of competition upon collection rates depends upon the vigour with which the parties compete, the ease of switching from one competing carrier to another, and the regulatory system (see Appendix 2). The presence of a cost of switching carriers (discussed below) gives one or more operator a degree of market power to exploit, as does product differentiation. On the other hand, full price competition would bring collection rates down to, or even below, marginal cost.

Collection rates may in principle fall below cost as a result of the operation of the proportionate return rule. Under that arrangement, an outgoing call entitles an operator to receive a proportion of incoming calls, and to make profits from the excess of settlement rate over the cost of terminating of a call. As prices fall in the competitive country, not only does the number of outgoing calls increase, but also the price imbalance will encourage call reversion.

Thus intense competition at one end of a route may impose transfers on the competitive country which have to be set against, and may exceed, the benefits to its subscribers from lower collection rates. This potentially harmful effect arises as long as settlement rates exceed the associated costs. It is thus clear that settlement rates lie at the heart of the framework for setting international telecommunications charges. Any innovation which enables an operator to bypass the system is thus likely to change both settlement rates and collection rates. International simple resale is such a device, and the next section considers the forms it can take, the costs associated with it, and its likely effect.

In summary, this section has shown that, if accounting rates and accounting rate shares have to be agreed by the two sides, then some process of bargaining or compromise is required, with the outcome depending upon the degree of cooperation between the two sides. As costs fall, the settlement rate preferred by each side and the probable bargaining outcome will decline too. If a convention of a 50-50 split of the settlement rate applies, there may be a tendency for the negotiation to yield the higher of the two rates proposed on each side, as the operator concerned can prevent or delay reductions in the settlement rate associated with reductions in cost. (For evidence of this proposition see, for example, BTCE, 1993.)

3. The effect of international simple resale on accounting rates

International resale was defined above as an arrangement whereby a company leases international capacity from an operator, or a consortium of operators, and sells that capacity directly to customers, without the incorporation of a value-added element. In this section we analyze four types of resale, depending upon the extent to which the reseller makes use of the PSTN and upon whether the reselling company is owned in whole or in part by an operator offering switched service. On this classification there are four possible types of reseller:

1. An independently owned international simple reseller (using the PSTN at both ends).
2. A carrier-owned international simple reseller (using the PSTN at both ends).
3. A one-way independently owned reseller (using the PSTN at one end only).
4. A one-way carrier-owned reseller (using the PSTN at one end).

In cases 2 and 4 the reseller may be owned either by a carrier operating at either end, or by a third party. If the latter, the reseller can deliver traffic to its home base, thus exemplifying "hubbing". This is discussed below.

The reseller's costs

To simplify, resale involves the provision to the customer of three links: a domestic link at the originating end, an international link, and a domestic link at the terminating end. These segments can be provided in one of a number of ways, each with associated costs. Table 1 shows the options at each stage. A call will thus

Table 1.

Originating segment	International segment	Terminating segment
1. Private circuit tariff	5. International private circuit tariff	8. Private circuit tariff
2. Use of the PSTN at tariff rate	6. Use of private circuit at cost-based interconnection rate	9. Use of PSTN at tariff rate
3. Use of the PSTN at inter connect rate	7. Operators' cost of self-provision of half circuit	10. Use of the PSTN at inter connect rate
4. Operators' costs of self-provision		11. Operators' costs of self-provision

impose costs equal to one option from column one, one from column two and one from column three. (The existence of alternative routes involving third countries is considered in the next section.)

The form of resale, as classified above, will affect costs. A company operating full independent resale will incur the costs of the international private circuit and will remunerate the operators at the originating and terminating ends. In such cases it becomes critical how the reseller buys the originating and terminating ends from the relevant operators. If it has to buy at tariff rates, resellers may find their cost advantages reduced. If they can buy at interconnect rates, the cost advantage may be far greater. The basis of charging makes an important difference. Thus in Australia, interconnect charges are half tariff rates, whereas in Japan they exceed them (WIK, 1994, p. 240).

Thus domestic regulatory policy on access to "wholesale" interconnect rates has a major impact upon the profitability of ISR. This consideration is exemplified in the UK. In order to provide high quality services to large customers, it is necessary for a reseller to interconnect with the telecommunications operator in each country in such a way that network information is conveyed easily between networks and an appropriate signalling standard is used. In practice, this means that such operators require "high level" interconnect which makes available such features as CCITT switching system No. 7, Calling Line Identification and short dial tone. However, a reseller requiring high level interconnect becomes, under UK license conditions, eligible for wholesale interconnect rates but liable to pay access deficit contributions (ADCs) which, on international calls, are exceptionally high, reflecting their relative profitability (WIK, 1994, pp. 193–195). The obligation to pay such contributions (if they were not waived by the Director General of Telecommunications) would substantially reduce the profitability of the operation.

Other types of reselling activity, apart from ISR, will incur different costs. Thus one-way resale obliges the reseller to lease circuits at one end. A carrier-owned reseller will have the additional advantage of being able to self-provide its own domestic circuits and half the international circuit, at a cost.

The form of resale available will also influence the market to which the reseller has access. It is useful if slightly arbitrary to divide the sources of demand for international telecommunication services into three categories.

The first comprises large international organizations such as multinational service or manufacturing companies or governmental institutions requiring an advanced range of high quality voice and data services. Such organizations are prepared to pay to receive the quality they require but not prepared to accept a poor quality service. They are also able to exercise a substantial market power vis a vis the operators.

The second category consists of medium sized and small businesses (SMEs) or other organizations with significant demand for international services, but lacking the need either for exceptional quality levels or for an exceptional range of services. The third category comprises subscribers (typically household) with a relatively small demand for international telecommunications services.

Observation suggests that firms in the first category, with their substantial international buying power and capacity to negotiate with competing consortia, have the greatest opportunities for bargaining successfully with telecommunications operators. The nature of their businesses often involves a substantial concentration of either originating or terminating traffic (or both) from a relatively small number of sites. Such organizations are likely to be attracted by the notion of having a single operator responsible on a regional or global basis for the provision of their services. At the same time, their large expenditure makes them highly desirable customers.

Medium-sized enterprises with a more limited demand for services can often be directly connected to an international supplier via a private circuit. As a consequence, their needs can in many cases be satisfied on the basis of a leased line at the originating end and use of the PSTN to terminate calls to a wide range of destinations.

The third category of "small" users of services is most likely to require connection using the PSTN at both ends. The level of demand is unlikely to be sufficient to justify the cost of a private circuit. However, such customers may find it worthwhile to access an international reseller via the PSTN, possibly being billed by the latter on a prepayment basis.

If this classification of demand is accepted, then we can identify a tentative matching of customers with possible suppliers. The largest customers are most likely to have their needs met by global providers of telecommunications services satisfying a broad range of demands at both the origination and termination end. To do this they are likely to need high level interconnection and signalling systems to ensure the required quality of service. This may have regulatory consequences as we note below. The natural suppliers of such large scale services are the major international operators on each route, and carrier-owned resellers at the other end. Resellers operating at this scale would be likely to be operators or consortia including major international operators (such as Concert, Euretcom and Unisource) with the capacity and experience to supply the market either directly or through a reselling subsidiary.

The second category of customers – small and medium sized businesses – are capable of being supplied by resellers breaking out into the PSTN at one or both ends. Given the nature of the demand, there is no obvious reason why operators themselves have an advantage in supplying this market, except possibly in terms of cost. This will depend on the degree of transparency in and the level of interconnection charges.

Finally, the third source of demand – small users – cannot readily be satisfied except by resellers using the PSTN for both the origination and termination of calls. As well as needing to have the appropriate regulatory permissions, such resellers will also have to overcome the handicap of having to cover costs of access to the network, charged either on an interconnection or on a tariff basis, and of covering additional billing costs. Inertia is likely to be greater here too: the savings from switching supplier do not outweigh the effort required, if the total bill is small.

This account suggests that the international telecommunications market is characterised both by product differentiation and by costs of switching suppliers. Product differentiation arises because some customers require a range of services which can realistically only be satisfied by the largest operators, or consortia of such operators. Switching costs arise because – especially in the case of smaller customers – the act of transferring international business to a reseller imposes additional costs – transactions costs on the part of the customer, which has to transfer the business, and a further cost on the part of the supplier, which has to bill the customer separately for its services. In addition, a reseller will have to cover the costs of access to origination and termination facilities.

The effects of authorising ISR at both ends

Liberalisation allowing all types of resale, including ISR, at both ends of a route previously monopolised would, if it occurred, have significant effects. These arise largely from the capacity of ISR to bypass the settlement rate system. Resellers will suffer a cost disadvantage, compared with licensed operators, in supplying certain of the services necessary to complete international telephone calls. Generally they will have to pay more for the originating and international segment. The former will cost more because they will have to pay for access at domestic tariff, while the latter is charged at international private leased circuit (IPLC) tariff rates. On the other hand, their costs in terminating a call are likely to be substantially less than existing rates because they avoid the need to pay settlement rates, paying instead domestic tariff or an interconnect rate. The balance of these two factors will determine the opportunities for profitable ISR.

However, the scope for entry by resellers into the various markets is limited, as a result of the characteristics of those markets. Thus the largest customers are likely to continue to require service from major operators or consortia of operators, doing business either by selling discounted services or through reselling subsidiaries. SMEs will be in a position to have their demand satisfied by resellers relying either on the PSTN at both ends or on a combination of the PSTN at one end and leased circuits at the other. The size of their expenditure is likely to make the costs of switching to a reseller – including billing costs – relatively insignificant. Moreover, the contestability of reselling – arising from the limited investments required to set up an operation – makes it likely that competition amongst suppliers will drive prices down. The third category of demand, small users, is likely to be disadvantaged by the high level of switching costs associated with reselling in relation to the absolute savings to be made by customers with low expenditure on international calls. This is particularly so if resale authorisations are restricted to particular routes. A customer would need to make considerable expenditures on the particular route to make the switch to a reseller worthwhile.

The impact of authorising resale is thus likely to lead to the emergence of a range of companies satisfying different segments of the market. In some cases, resellers associated with existing operators will have an advantage through being able to maintain higher service quality. In the case of smaller customers, switching costs will enable existing operators to maintain some degree of monopoly power.

The key point, however, is that the introduction of resale in such circumstances undermines the existing accounting rate system, because the opportunity for bypass which it creates radically alters incentives for administrations to maintain settlement rates substantially in excess of costs. If a country's settlement rates are too high, all incoming traffic will bypass them through a combination of resale for the international segment and access to the PSTN or use of domestic private circuits for termination. The cost advantage of the operator in the destination country, combined with the costs of switching to a reseller, will preserve some of the monopoly profits currently available, but their extent will be limited. In the extreme case, if there were no product differentiation and switching costs were low then the operator would be forced to reduce its termination charges to a level determined by the reseller's costs. But even if such extreme conditions are not satisfied, resale undermines the system of settlement rates, where these are set substantially in excess of termination costs, just as would unregulated negotiation between competitive carriers at either end of the route. Not surprisingly, this explains the strong correlation between a government's willingness to permit resale and its enthusiasm for facilities-based competition.

The implications of permitting resale in one direction only

The above analysis has assumed that resale has been liberalised in both directions. We now consider the implications of liberalisation in one direction only. We thus suppose that subscribers in country A have access to independent or carrier-owned resellers for the delivery of traffic to country B, while the traffic from B to A goes through the conventional accounting rate system.

It is immediately apparent that this outcome is highly skewed in favor of operators and customers in A. Traffic from A can bypass the settlement rate system and subscribers in A will benefit from reduced costs, provided country A allows a number of competing resellers to offer service. If the only operator allowed to offer service were the carrier, then the benefit would accrue to it rather than country A's customers. This same advantage will not, however, be available to operators and subscribers in B. They will continue to have to pay settlement rates to the operator in A substantially in excess of costs. Assuming collection rates fall in A, this effect will be to some extent mitigated by a change in the balance of traffic as the result of call reversion. Nonetheless, the new asymmetrical regime will involve a substantial transfer of resources from operators and (indirectly) from subscribers in B to the equivalent bodies in A.

Of course, the outcome described might not be politically sustainable, as the liberalising country would be experiencing an outflow of resources and would therefore put more pressure on its partner for a reduction in accounting rates. Such pressure may be successful, but it would require a willingness on the side of the partner to accept a lower level of profits than it could acquire by charging a high settlement rate on incoming traffic and by-passing the rate in the other direction.

Carrier resale

Carriers have opportunities and incentives to set up reselling operations which differ from those of independent operators. They have the capacity to offer high levels of service for the largest customers – not often available to independent resellers. Moreover, they are already to some degree beneficiaries of the accounting rate system, as they share with overseas operators the excess profits associated with high accounting and collection rates.

Despite carriers' dislike of ISR, the pattern of behavior now increasingly visible is for carriers to set up reselling subsidiaries in other countries. These are then used to target the largest customers in the other country, which are offered a high quality end to end or global service. The carrier can send traffic directly to its reseller without incurring the settlement rate. Equally, it can offer clients in the other country access to international service which bypasses its own settlement rates. Even though it loses the associated revenue, it may still be able to undercut the overseas customer's own operator.

This combination of circumstances means that carrier-owned resellers impose weaker pressures on the accounting rate system as the carriers seek to use reselling to practice price differentiation, bypassing the accounting rate for large customers while paying it on a reciprocal basis for small customers. The effect of carrier-based reselling alone on the accounting rate system would thus be muted.

4. Regulatory responses and extensions

The previous section has sought to analyze the effects of introducing ISR in a telecommunications link, both in one direction only, and in both directions simultaneously. We now examine the implications of alternative regulatory responses to ISR, with a view to predicting their effects on accounting and collection rates. We also examine the implications of other related regulatory measures such as restrictions on “refile”.

Based upon observation of current practice, the following regulatory policies towards international resale can be identified:

1. International resale can be banned in both countries.
2. International resale may be permitted, but only where it involves break out into the PSTN at one end (either the originating or the terminating end); access to the PSTN may be charged either at tariff or at interconnect rates.
3. ISR can be limited to relationships with countries operating “equivalent” regimes.
4. ISR is allowed to all countries, whether reciprocating or otherwise.

Where ISR is permitted, the issue also arises of how many operators to license. Given the ease of entry into the activity (and of exit from it), it seems likely that if entry is permitted to one operator, it will be permitted to all. Accordingly, resellers’ tariffs can be expected to be in line with resellers’ costs. If restrictions on entry do apply, resellers are likely to charge more and make positive profits. However, the analysis is only changed in an inessential way.

Banning ISR represents effectively the policy adopted by most regulatory authorities. Its implications have been analyzed in Section 2 above, which showed that, on the basis of current negotiation procedures between operators and the convention of a 50-50 split, the settlement rate was likely to be the higher of the rates preferred by the two operators but to fall as costs fall. In most cases, the country enjoying a net inflow of calls at the given collection rates prevailing in both countries will prefer a higher settlement rate than its partner. Thus the effect of banning ISR is to maintain for most country pairs the status quo – i.e., relatively high settlement rates in all cases, and high collection rates – at least in countries where competition does not operate. But both rates tend to fall as costs fall.

We also investigated in Section 3 above the implications of permitting international simple resale in one direction only – a possible outcome of policy number 4 above. As noted there, the effect of this arrangement is to allow one country to bypass the accounting rate system and to avoid making settlement payments to the other operator in respect of outgoing traffic. Incoming traffic, on the other hand, continues to be remunerated at the settlement rate. It is clear that this asymmetry works strongly to the advantage of the operator and country originating the traffic delivered by the reseller.

The implication of restricting ISR to cases where use is made of the PSTN at one end only are rather more complex. The effect of the policy is to limit the potential subscriber base available to the reseller. The restriction to the use of the PSTN at one end only obliges the reseller to make alternative arrangements using leased lines for either the originating or the terminating national leg of the course. Such arrangements will only be commercially profitable when the amount of traffic generated by each customer is sufficiently great to justify the leasing of a private circuit. Thus the restriction to one-way break out is likely to have the effect of denying many domestic subscribers access to the reseller. The restriction raises the reseller’s costs by an amount equal to the difference between the cost of leasing a line and the cost of use of the PSTN. This will vary from subscriber to subscriber depending upon the tariff, but there will always be a cut-off point below which the reseller cannot undercut the PTT.

A third possibility is for the regulator to restrict international simple resale to countries which operate what are regarded as equivalent regimes, where equivalence is essentially defined as the availability of bypass in both directions. The objective of this policy is clearly to prevent the harmful consequence to a country noted above, which arises when incoming traffic, serviced by a reseller, is able to bypass the accounting rate system, while outgoing traffic has to incur the full cost of the settlement rate system. The results of the argument to date are summarised in Table 2.

Transit arrangements, hubbing, refile, etc.

It was assumed above that there is a single international route between the two countries. In practice, that is clearly not the case as traffic can be conveyed by a number of alternative routes, involving third countries. Indeed, in some cases there may be no method of conveying a call between two countries without a transit arrangement of this kind. It is therefore essential to examine the effects of resale when the route taken by a call, by choice or necessity, involves a third country.

The existence of alternative indirect routes from country A to country B, involving transmission through country C inevitably imposes some limits on the settlement rate levied by country B on incoming traffic from country A. Thus if it were cheaper for the operator in A to route traffic through B, incurring a settlement rate imposed by B, and a second settlement rate imposed by C or traffic from B, then the operator would be tempted to do so.

In principle, this unilateral cost saving is prohibited by an article in Appendix 1 of the International Telecommunication Regulations, which states that: ".... where one or more routes have been established by agreement between administrations and where traffic is diverted unilaterally by the administration of origin to a route which has not been agreed with the administration of destination, the terminal shares payable to the administration of destination shall be the same as would have been due

Table 2.

Regulatory policy	Market segment by the reseller	Served settlement rates
a) International resale banned	None	Unchanged
b) International Resale Permitted with One-way break out	Larger subscribers	Reduced[a]
c) Full ISR to "equivalent countries"	Potentially all	Substantial reduction[a]
d) Full ISR without equivalence	Potentially all incoming traffic	Unchanged outward; substantially reduced inwards

Note: Although there are a small number of countries where policy towards international resale exemplifies policies (b) to (d), policy (a) remains the norm. [a]Reduction is greater if access to PSTN is at interconnect rather than tariff rates.

to it had the traffic been routed over the agreed primary route and the transit costs are borne by the administration of origin...."

Such activities are "refiling", and the regulation discourages competition for traffic among PTOs by requiring that refiled traffic is paid for at the normal accounting rate. However, an operator might wish to ignore this provision, or the operator terminating the call may be prevailed upon to accept the diversion; with the growth of market-based least-cost routing systems, the restriction is likely to become harder and harder to retain as there is no obvious method of enforcement. In such cases, the existence of an alternative route imposes some restraint on C's settlement rate with A. The apparently increasing competition amongst operators seeking to provide hubbing services suggests that third-country participation is likely to be of increasing importance.

The introduction of alternative routes available to resellers also introduces complexities. Suppose that country A permits resale between itself and country B on the basis of equivalence. Country B, however, is prepared to admit resale with country C on a nonequivalent basis – i.e., to accept incoming traffic by-passing the settlement system without having an equivalent arrangement in respect of traffic directed to C.

The effect of this combination of circumstances is to place A in a nonequivalent relationship in respect of its traffic to and from C via B. Traffic from C to A wholly bypasses the accounting rate system while traffic in the opposite direction has to pay the settlement rate. It is thus not surprising that a country, in taking a view as to whether a second country offers equivalent arrangements, takes into account that country's relationships with third parties.

In the two cases mentioned above, the introduction of a third country was optional. In many cases, by contrast, the route followed by a call necessarily involves transmission across another or other countries. In order to deal with this situation, separate regional arrangements have often come into operation under the auspices of the ITU providing for the allocation of payments among various countries involved: the originating country, the terminating country, and other countries en route. The best known of these arrangements is TEUREM, covering Europe and the Mediterranean basin. When calls within this region are switched through the PSTN of intermediate countries, the arrangement is known as transit. In such circumstances, the originating country pays a settlement rate to countries providing transit facilities and to the terminating operator. However, most PTOs now have direct connections with each other, achieved by permanently renting capacity from operators in intermediate countries.

Arrangements within TEUREM differ from those for intercontinental telecommunications services and many intracontinental services outside Europe, in that the division of the accounting rate between the originating operator, the terminating operator and any transit operators is intended to be cost based, rather than the outcome of bilateral or multilateral bargaining, or based on a simple principle such as equal shares between the originating and terminating operator. The division should, therefore, reflect the average transmission distance at either end. As a consequence, the settlement rate for terminating a call in a larger country should, other things being equal, be larger than that for termination in a small country. The precise relationship between the settlement rate shares is to some extent the outcome of negotiation. At the same time, there is considerable evidence that international telecommunications charges within TEUREM exceed costs by a substantial margin, creating a profit opportunity for resellers who are bypassing the settlement system. This remains true despite the substantial recent falls in termination charges, agreed within TEUREM. Thus, as in the case of intercontinental telecommunications services, international resale threatens the accounting rate system within TEUREM. The danger is, however, diminished by reduction in intraregional collection rates and rebalancing of tariffs by raising local call charges, which will increase resellers' costs.

5. Summary, conclusions and implications

The analysis above has demonstrated that monopoly operators collaborating to offer international telecommunications services within the framework of an accounting rate system have an incentive to establish termination charges or settlement rates in excess of the costs of providing the relevant termination services. The price they would wish to charge to maximise the profit from their monopoly control of call termination depends upon a variety of factors, including the impact of the settlement rate on the other country's collection rate – and thus the volume of incoming traffic, and the extent of call reversion – the degree to which the caller and callee are able to arrange the call to originate in the country with the lower collection rate.

While each operator would ideally like to impose an optimal settlement rate upon incoming calls, yet make no payment to its partner for outgoing calls, this is clearly impracticable. Accordingly, the two operators have to agree a compromise. For historical reasons, the convention has arisen of a 50-50 split of the accounting rate for most intercontinental calls. This arrangement also relieves the parties of some bargaining costs and frees them from the risks of a severely disadvantageous arrangement. In the case of certain regional arrangements, including those prevailing within TEUREM, accounting rate shares are loosely based upon costs, although there is evidence that they exceed those costs in absolute terms.

The introduction of competing operators at one end within the accounting rate framework places the remaining monopolist in an advantageous position to whipsaw downwards the settlement rate which it pays or raise the rate it receives. If this policy worked, its effect would be to transfer resources from the competitive to the monopolistic country. Accordingly, the regulators typically intervene by imposing a uniform settlement rate policy, often combined with a proportionate return principle. This reduces the incentive for a competitor to agree a less advantageous settlement rate, but simultaneously eliminates the downward pressure on those rates. In cases where there is competition at both ends (notably on the UK-USA route), the regulatory authorities have demonstrated a willingness to depart from the uniform settlement policy.

International resale is an arrangement whereby an operator leases international circuits and – combining these with the national ends – offers service in competition within the operator. The national links can be provided at one or both ends, either by fixed circuits or by access to the PSTN, for which the reseller pays either at tariff or at interconnection rates. Resale is thus a technique for arbitrating the telecommunications services provided by the operators. In particular, international resale offers an opportunity for bypassing the accounting rate system, as a reseller avoids payment of the settlement rate to the terminating operator.

Because a reseller bypasses the accounting rate system, its costs are likely to be less than those of the operator, which has to pay a settlement rate. This is counterbalanced to some degree by the fact that the reseller's costs of leasing lines or of access to the PSTN at tariff or at interconnect rates are likely to exceed the incremental costs of the operator in providing international service.

Different types of customers are likely to view international resale in different ways. The largest multinational companies will both demand an exceptionally high level of service and be in a position to bargain with telecommunications operators or consortia for the provision of global services. This market may well be served by carrier-owned resellers, leaving little room for small independents. Middle sized companies with substantial international traffic will be able to direct it over leased circuits to the reseller's international gateway, and it will in many cases be terminated on the PSTN of the receiving country. In the case of domestic subscribers, it will usually be essential for the reseller to have access to PSTN at both ends (ISR).

Because of this consideration, the regulatory approach adopted to international resale has a major impact upon its prospects. If resellers are allowed to break out into the PSTN at both ends, they are well placed to offer service to both medium sized and small customers. If, however, only one national link can be achieved on the PSTN, then the smaller customers are likely to remain a monopoly of the telecommunications operator.

As well as deciding what access international resellers should have to the PSTN, regulators also determine the countries to which ISR may be provided. If a country permits incoming calls to be supplied by a reseller, – bypassing its settlement rate system, while there is no equivalent opportunity for resellers to offer service in the opposite direction – then international resale can have an adverse effect upon the liberalising country. Accordingly, many countries, including the UK and the USA, only permit international simple resale arrangements in relations with other countries offering equivalent opportunities.

Thus we have shown above that the introduction of ISR places downward pressure upon settlement rates, the extent of that pressure depending upon the cost advantage of the operator compared with a reseller, the costs of switching to a reseller and the reseller's access to the PSTN.

Available evidence suggests that the settlement rates in many cases are substantially in excess of the costs incurred by the terminating country. There is thus reason to suppose that ISR is likely to strengthen existing downward pressure on settlement rates. A full estimate of its likely effects would require the costing of alternative services, which goes beyond the scope of the present paper.

The paper also discusses the link between international resale and other policies having the effect of putting pressure on settlement rates. The possibility of "hubbing", involving delivery of a call from A to B via a third country, C, introduces the possibility of competition in settlement rates along alternative routes. This competition, is, however, muted by ITU restrictions on the avoidance of settlement rates.

An alternative approach to reducing settlement rates is to permit facilities – based on competition at both ends. While facilities-based competition, such as that shown at both ends of the UK-USA route, has the potential to put downward pressure on settlement rates provided that the regulators permit the necessary departures from a uniform settlement policy, it is unlikely that the number of competitors will be large enough to bring rates into line with costs at once. The resale market is, by contrast, fully contestable if entry is liberalised because of the relatively low level of investment required, and the absence of major sunk costs. The pressure it is likely to exert on settlement rates thus has the potential to be greater than that provided by facilities-based competition, if international leased circuits are priced at cost. However, international call revenues are often used to cross-subsidise domestic services. In such cases, the price of international leased circuits may be deliberately kept high, to discourage international resale.

A final question which arises concerns the role of international resale once settlement rates have been brought into line with costs. The paper has shown that international resale is essentially an arbitrage activity. Typically, resellers will be able to incur the heavier costs than operators in leasing circuits or buying access to the PSTN, because of the substantial benefit they can derive from bypassing the settlement rate system. In the longer term, settlement rates are likely to fall in response to ISR and other pressures and, with a cost based settlement rate system, resellers would continue to face the cost disadvantage they presently enjoy without the compensation benefit of bypassing the settlement rate. They may continue to play a role in the market through an ability to bundle services in a way demanded by certain customers or through particular marketing skills, but – depending upon regulatory arrangements for setting tariffs for leased lines or access to the PSTN – they may face difficulties in the long term.

Acknowledgements

This chapter was written with assistance from a Research Grant under the UK Economic and Social Research Council's Contracts and Competition Programme.

Appendix 1

Bargaining over settlement rates

The aim of this Appendix is to set out in diagrammatic form the mechanism whereby profit-maximising monopoly telecommunications operators interact with one another in agreeing an accounting rate and accounting rate shares. The analysis is taken from Cave and Donnelly, 1994a.

Given demand relationships in both countries, we assume that Operator 1 would maximise its profits by paying a zero settlement rate to Operator 2, and by imposing upon Operator 2 the settlement rate which maximises its (Operator 1's) profits from the termination of calls, bearing in mind that the settlement rate will be regarded by Operator 2 as a component of its marginal costs, thus influencing Operator 2's collection rate and the inflow of calls. This point is represented by A in Fig. 1. Point B represents a similar profit maximising combination for Operator 2. The concentric curves centred on A show combinations of the two settlement rates yielding Operator 1 constant levels of profit; as the distance from A increases, Operator 1's profits decline. The concentric curves centred on B show constant levels of profit for Operator 2; profit declines as distance from B increases.

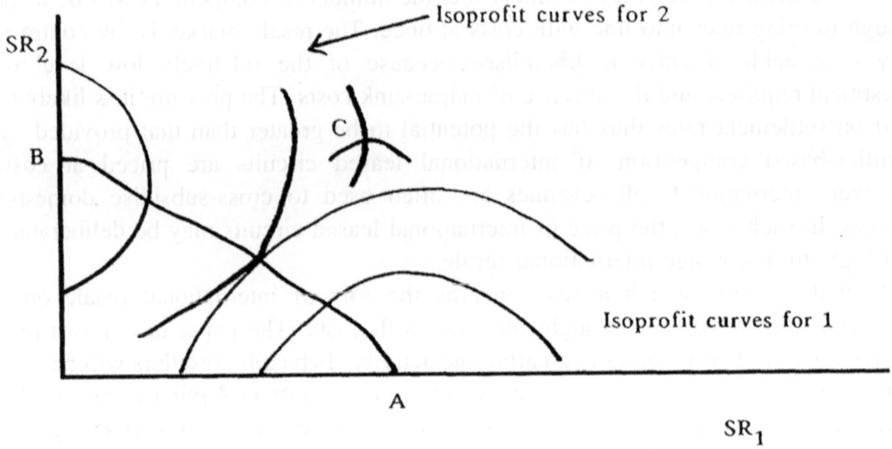

Fig. 1.

SR_1 Settlement rate payable to 1
SR_2 Settlement rate payable to 2
A Operator 1's optimal combination of settlement rates
B Operator 2's optimal combination of settlement rates
AB The "contract curve"
C The non-cooperative equilibrium

Given the operators' conflicting objectives the accounting rate and accounting rate shares have to be determined through a process of negotiation or agreed choice among possible cooperative solutions. These lie on the "contract curve" connecting points A and B, which is a line joining all points at which it is impossible for one operator to increase its profits without imposing lower profits on the other. These points represent points of tangency between constant profit curves of each operator. If, on the other hand, the operators behave non-cooperatively – each maximising its profits, taking the other's settlement rate as given – the outcome is represented by Point C in Fig. 1.

It is possible to identify a particular cooperative solution by invoking some bargaining equilibrium. Such an equilibrium could lie anywhere on the "contract curve". However, as we have seen, the conventions frequently involve a 50-50 split of the accounting rate into equal settlement rates; this is equivalent to requiring the solution to lie on the 45 degree line. This is illustrated in Fig. 2. Given the need for both parties to agree the accounting rate, it can be hypothesised that the outcome will be that point on the 45 degree line which maximises the profits of the operator seeking the higher settlement rate. In this instance, this is Operator 2 and common settlement rate adopted is represented by Point E.

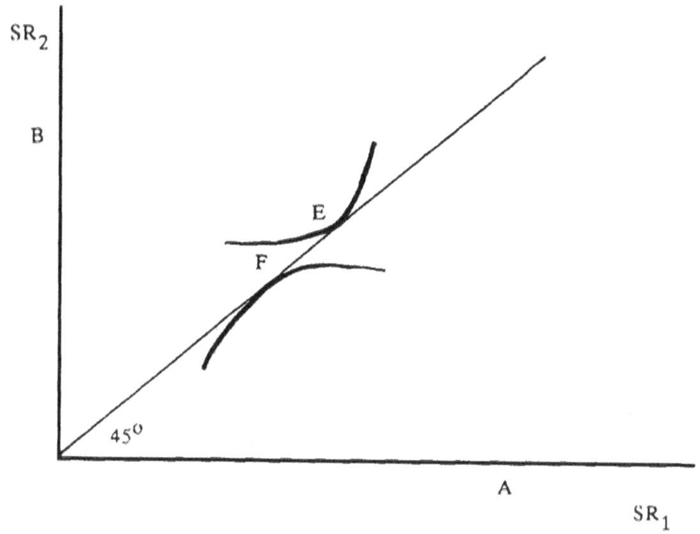

Fig. 2. E Settlement rate, based on 50-50 split, desired by Operator 2
F Settlement rate, based on 50-50 split, desired by Operator 1

Appendix 2

International operators' incentives for competition

This Appendix analyses competition between an operator and a reseller. For reasons noted in the text, the reseller's costs differ from those of the operator: the reseller faces higher costs on the originating segment and the international segment, but lower costs on the terminating segment, as a result of its capacity to bypass the settlement system. However, subscribers transferring to the reseller also experience a switching cost, arising from the need to establish billing arrangements, etc. For a full analysis see Cave and Donnelly 1994b.

Competition between the operator and reseller is modelled on the basis that each firm maximises its profits on the assumption that the other firm's prices remain constant. This is sometimes referred to as the Bertrand process. In the absence of switching costs, the equilibrium arising from competition among firms with identical costs is characterised by a common price equal to marginal cost and a random division of the market among the competitors. When one firm has a cost advantage, the equilibrium is fractionally below the marginal cost of the second most efficient firm, and the most efficient firm has a 100% market share.

The introduction of switching costs softens price competition between firms. Suppose that switching to the reseller imposes a cost upon the subscriber, which varies from subscriber to subscriber. A subscriber with a zero switching cost will always use the cheaper supplier, but subscribers with positive switching costs will be willing to pay more for the operator's service, provided that the price differential is less than their switching cost.

This can be illustrated in Fig. 3. The demand curve for international telecommunications services is represented by DD. The operator's price is P_o. The reseller then offers service at price P_r, reflecting its lower costs. By undercutting the reseller slightly, the operator could recapture the whole market. However,

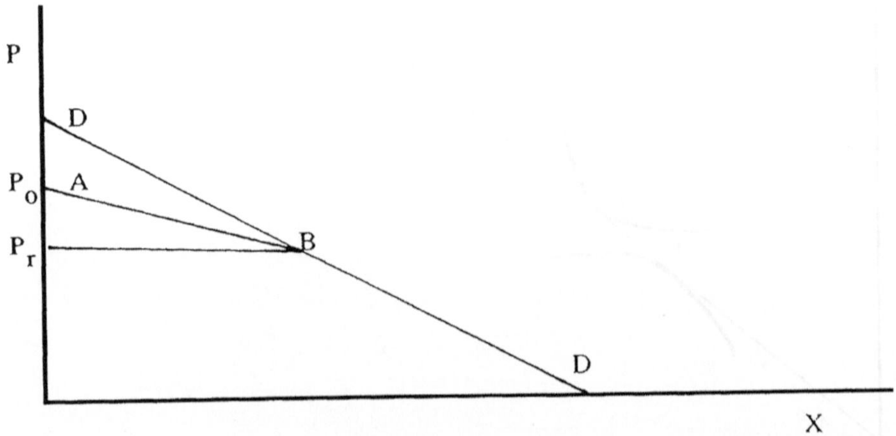

Fig. 3. X Volume of international traffic
P Price of international traffic
DD Demand curve
P_o Operator's original price
P_r Reseller's price
ABD Operator's demand curve when reseller charges P_r, reflecting customer's cost of switching to the reseller.

because of switching costs it can still retain some market share by charging a price above that charged by the reseller. The effect of this is to introduce a kink at P_r into the demand curve facing the operator. For prices below P_r the relevant demand curve for the operator is the market demand curve, as it has the whole market. For prices above P_r, the demand curve facing the operator is determined by the extent of

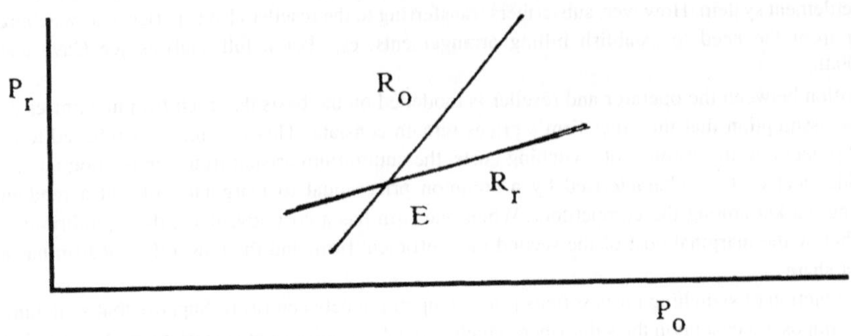

Fig. 4. P_o Operator's tariff
P_r Reseller's tariff
R_o Operator's reaction curve (to reseller's tariff)
R_r Reseller's reaction curve (to operator's tariff)
E Equilibrium

switching costs affecting customers. Thus the operator faces a kinked demand curve shown by ABD and switching costs provide the operator with some market power.

Because of switching costs the operator can in certain circumstances maintain a price above that charged by the reseller. Thus it can remain in business, and even make positive profits, even when its marginal costs exceed those of the reseller. This is one of a number of possible outcomes which depend upon the level of switching costs and the reseller's cost advantage. If the former are low and the latter great, the reseller may gain the whole market.

The market sharing outcome is represented in Fig. 4, where the curves show each firm's reaction curves, charting its response to the other's price. The equilibrium at A satisfies the condition that neither firm will wish to change its price, given the other's price decision. At A, the operator charges more than the reseller, reflecting the costs incurred by its customers if they switch to the reseller.

References

BTCE, 1993, *International Telecommunications: An Australian Perspective, Bureau of Transport and Communications Economics*, Report 82 (AGPS, Canberra).

Cave, M., Donnelly, M.P., 1994a. *The Pricing of International Telecommunications Services by Monopoly Operators*, Brunel University, Mimeo.

Cave, M., Donnelly, M.P., 1994b. *The Positive and Normative Effects of Competition in the Provision of International Telecommunications Services*, Brunel University.

Eward, R., 1985. *The Deregulation of International Telecommunications.* (RAND, Santa Monica).

Hakim, S.R., Lu, D., 1993. Monopolistic settlement agreements in international telecommunications agreements. *Information Economics and Policy*, 5:145–157.

WIK, 1994, *Network Interconnection in the Domain of ONP*. Study for DGXIII of the European Commission (WIK, Bad Honnef, Germany).

Beyond Competition: The Future of Telecommunications.
Edited by D. Lamberton.

CHAPTER 15

The regulation of unnatural oligopoly: appropriate criteria for regulators where the goals of regulation are economic progress

John Nightingale
Department of Economics, University of New England, Armidale, New South Wales, Australia

An evolutionary perspective on firm and market behaviour suggests a framework of analysis in which price-cap regulation has the natural outcome of stimulating the rate of productivity change and innovation. This was explicitly addressed at least as far back as the work of Jack Downie (*The Competitive Process*, London, 1958), a civil servant who died far too young in 1963 while head of the economics department of the OECD in Paris. But the idea can clearly be inferred in Schumpeter's *Capitalism, Socialism and Democracy*. My task is to develop the framework with a view to its application to public policy making. The present paper amounts to an argument for a different perspective on price caps, and a proposal for developing an evolutionary model of regulation for implementation in the telecommunications context.

The focus of orthodox theory on the two static efficiencies, allocative and productive, flows from the nature of that theory. Neoclassical analysis of allocative efficiency suggests that price caps may improve allocation by enforcing at least the second best of Ramsey prices. Principal-agent theory extends this analysis into the field of asymmetric information with optimising principals and agents. It therefore has a focus on the achievement of productive efficiency by manipulation of the incentives faced by agents. Both, perforce, ignore or include only by way of ad hoc and arbitrary auxiliary assumptions the changes over time of productive efficiency. Observers of technically advanced industries would agree that this advance over time has been a central part of the reason for the modern economy's efficiency differing from that of the preindustrial economy. An analysis which addresses this issue directly is therefore required.

The relevance to telecommunications of this perspective is very direct indeed. The rate of productivity change experienced in this industry and its input supply industries is such that great social gains are flowing now, and will continue to flow for some decades. A market based system in which rivalry amongst oligopolists dominates behaviour is likely, on the evidence of other industries, to best promote rapid innovation and its diffusion. During the transition to such an oligopoly the simulation

Address for correspondence: John Nightingale, Senior Lecturer, Department of Economics, University of New England, Armidale, NSW 2351, Australia.

of effective market rivalry by a price-cap system is a means of stimulating adoption of innovation and rapid productivity change.

1. The proper goal of public policy toward regulated industries

Regulated industries, like all industries, have a social purpose to their existence. It is to satisfy demand from society, expressed in the market place and the ballot box, for a stream of services. A costless stream, as the air we breathe used to be, requires no industry, no producer but the natural system in which we live. But most services are costly and require resources to produce. The fewer resources required, the better. The industrial revolution has been a constant progression of reducing the resources required to satisfy demand for services, and an opening up of possibilities to provide services previously unable to be produced. Telecommunications is perhaps the best example of an industry providing a stream of services now in extremely high demand but unsupplied at all until less than 150 years ago. The history of the industry has been one of expanding the services available and responding to demands for more and more elaborate services.

That being said, the social regulation of industry has as its goal the facilitation of the supply of its services at the least cost over time. Most industries are mainly regulated by the market, some require oversight by government, but both are subject to the rules of the game. By this I mean the structure of legal principles and conventions expressed in both legislation, and, in countries of the common law system, in judicial decisions. The present paper is not concerned with this aspect, though the argument has been applied to it in the literature on competition policy in particular. My concern is with public regulation of industries for which the market is believed to be inadequate as a social control.

1.1. Orthodox views of regulatory policy

Debate about the design of regulatory policy has been dominated by two issues. The first is how to ensure that prices set by firms with market power approximate the prices which would be set in a perfectly competitive industry. The second is how to ensure that managers in possession of privileged information may be induced to minimise social cost of production. This last can be interpreted as a problem in optimal investment in cost reducing activities. But this is as close as orthodox regulatory theory comes to dealing with economic progress.

Besanko & Sappington (1987) review this literature. Their review emphasises the complex and often contradictory nature of optimal regulation. In the presence of moral hazard and adverse selection, "policies that do a good job of dealing with one problem (e.g., prices in excess of marginal cost in the case of adverse selection) may be quite inappropriate when the other problem is present" (*ibid*, p. 39). This problem is mitigated by repeated application in which the regulator can learn (in a Bayesian manner) the firm's real costs, at least when the firm's discount rate is low enough to allow the future to be sufficiently important. The various models are characterised by special conditions defining the nature of costs, of the environment, and of the regulator's abilities to be informed. The technology, input markets and output demand are all in existence and known to the agents, the managers of the regulated firm, but curtained off from the regulator in one or another manner. Cost improving investments yield results with certainty up to a probability distribution, enabling the firm to choose such investment rationally (first best, the marginal benefits = marginal costs). The models are therefore to be understood as embodying no bounds on substantive rationality. The regulator has to engage in Bayesian learning to enjoy full knowledge, while the manager has that knowledge and may attempt to lie to the regulator. Despite these highly restrictive conditions, the inferences drawn are disappointing. Each case has its own solution, intuition is confounded by many of the inferences. The possibility of an equilibrium with second best properties is sometimes present, sometimes not. General properties are very limited, and the range of phenomena over which inference is drawn is limited to price, output, investment.

Laffont & Tirole (1993) present a comprehensive game theoretic version of principal agent analysis of regulation designed to inform regulators of appropriate mechanisms to achieve the two goals of allocative and productive (i.e., appropriate manager behaviour) efficiency. The plethora of special cases presented by Laffont & Tirole have in common the theoretical goal of an equilibrium solution in which the manager of the regulated firm finds both minimum X-inefficiency and prices at worst Ramsey, at best equal to marginal cost. Is it the case that this theory addresses the question of continuous productivity change and technological progressiveness? I would argue it does not, and certainly does not do so without imposing heavily restrictive conditions which detract from explanation and deliver prediction at cost of generality.

Of the 650 pages of their book, the concepts "technological progress" and "technology transfer" are found on some five pages. "Learning by doing" is found on five other pages. Technological progress makes its appearance in a model assuming full information. Chapter 8 deals with investment in the context of competing for a market, rather than ongoing operation of a firm. Chapters 9 & 10 on dynamics occupy just under 100 pages. These chapters deal with the problem of incentive over a number of periods, attempting to characterise schemes which may lead to optimal outcomes but without much success: their problem, of the revelation of costs by the firms being regulated, is dealt with very conditionally. The closest they come to dealing with productivity change is in the incentive effects on "inefficient types" of firm: the ratchet problem, of the disincentive that a binding constraint creates for the constrained firm to improve its performance. They suggest that the inefficient type may be stimulated to improved cost performance in the second period by a harsh constraint (*ibid*, p. 401).

It is the nature of knowledge and its absence assumed by the neoclassical and principal agent framework which so restricts these models and prevents them from dealing with discovery of previously unknown technology as an endogenous phenomenon.[1] For these theories, knowledge exists with certainty, and has existed since the beginning of the world. All that is required is the profitable use to draw that knowledge forth. The equilibrium solutions to these models imply that all the potential improvements that are worth employing, given demand and input prices, are automatically made, given the managerial objective function. The absence of real time and of the possibility of the emergence of new, i.e., unexpected, knowledge points these models away from an examination of the phenomenon of productivity change.

1.2. Some critique of welfare theory

My own critique of the rationale for price capping (Nightingale, 1990) was narrowly aimed at the information difficulties of applying such a policy in practice. My conclusion was that it was no simpler than the previously fashionable rate of return regulation, and embodied all the assumptions of that method, now conventionally regarded as impossibly information intensive and intrusive, and with various incorrect incentive effects.[2] My aim here is to broaden the critique to the goals of that policy, to show that the goal is partial and contrary to economic progress if it were to be carried out effectively.

The basis of static welfare economics has been attacked on many occasions. The two attacks to which I wish to draw attention are those of Jack Downie (1958) and Thomas Wilson (1984). Both these authors examine the underpinnings of the theory of firm and industry on which the Cambridge school is based. Downie argues that first, the gains from ensuring that price = marginal cost are likely to be very small. That argument has long since been confirmed empirically in the debates over Harberger triangles. His second argument is that firms actually differ greatly in efficiency and size, but that there is no strongly positive relation between these variables, and thus simple economies of scale cannot account for them. Were all firms to be of the best efficiency the gains to society would be far greater than those suggested by static efficiency improvement. Thirdly, he argues, the method of analysis represented by the neoclassical theory of the firm ignores those differences, and provides no means of asking why firms differ in their efficiency. Finally, he argues that the evidence for both U-shaped cost curves and downward sloping demand curves facing firms is absent, the contrary being the case. The indications for the situation examined by orthodox theory are absent: were the theory to be useful, it would be in a world comprised of noncompetitive oligopolies and monopolies. This is clearly not the world in which we live. An analysis of the behaviour of competitive industry thus lacks any plausible theory.

Wilson's critique has some similarities with that of Downie, but his paper encompasses the familiar critique of the compatibility of markets with Pareto efficiency (information problems, distributional problems, the inconsistency of market outcomes with Pareto requirements in the normal case of downward sloping supply functions, however, defined), before examining the difficulties of neoclassical price theory under oligopoly. He argues that "price competition" as expounded in the orthodox treatises is quite different from the price competition which is observed. He sees oligopolists as often engaged in fierce competition involving much more than mere price, but including price as a complementary competitive weapon. The appearance of oligopoly as not being price competitive is often merely that, an appearance created by the lens of orthodox neoclassical theory which refuses to see as competitive markets in which explicit price cutting is absent. This critique may be better accepted today than it was as recently as the early 1980s, but the means to understand the competition which is observed is still not to hand within the orthodox toolbox. That toolbox still has a presumption that oligopoly will be uncompetitive until proved otherwise, without any tools for that proof to be made.

But, as with Downie's critique, it is the inconsistency of theory with the phenomena at hand which is Wilson's greatest problem for economic theory. The most competitively structured industries are those most heavily regulated (agriculture), due not simply to political pressure, but to the real information problems of atomistic industries (Richardson, 1960). How can a producer decide on investment and production until all other, or a substantial subset of all other, producers have made their decisions? The familiar fluctuations of outputs and prices in atomistic industries can be ascribed to this problem. As Richardson pointed out (*ibid*, pp. 252–254 summarises) an equilibrium is possible only within a framework of sequences of decisions by few decision makers, such that information required to sustain those decisions exists. The same problem is now seen most clearly in decisions about locations of investments between currency areas. A volatile currency itself restrains foreign investment because it is not possible to know whether the currency will be substantially higher in the future in which the revenues, especially export revenues, will be earned and profits repatriated by investors and in which the costs of imported inputs will be denominated.

But even if a market is not price competitive, and inefficient firms can persist, it will be the efficient firms which prosper and so lead to improvement in efficiency through their greater rate of growth through investment of funds. This point is central to Downie's analysis of the competitive process (see below). Orthodox theory cannot comprehend this argument, as it cannot allow differing efficiencies and costs between competing firms. But the conditions for rapid advance in productivity may require such diversity, as Downie pointed out at length. In attempting to put this theory to practical effect, either as explanation, prediction or policy advice, applied economics has subsisted on arbitrary and ad hoc relaxations to cope with the reality of great variation in efficiencies and costs between firms competing in the same market.

A regulatory regime aimed solely at static efficiency, at ensuring that prices and costs were related as if they were in an equilibrium as near as possible to the perfect competition ideal of theoretical welfare economics, would be ignoring that aspect of efficiency which has given us the great gains in income and wealth that industrial societies have enjoyed over the past 250 years. If it is indeed the market which gave vent to the forces for progress, then a regulatory regime which mimics the market should be attempting to give those forces vent in the regulated industries. But this is what orthodox economics denies by the nature of its theory and neglects in its practice. My argument is that regulation as discussed by economists is aimed at static efficiency, allocative and productive. The policies which do aim at stimulating technological progress are those other, and usually very problematic, policies labelled "industry policy" or "technology policy".

These industry or technology policies typically lack any firm basis in theory or in social consensus. They are neither strongly expected to work, nor seen as desirable by many of their intended beneficiaries, neither firms nor their workers or their customers. Industry policies often appear as the outcome of special pleading by sectional industry groups, or by a constituency which can evince enthusiasm for only those industries which are currently "in favour" by virtue of favourable image. These policies risk failing to "pick winners", and instead support investment in quixotic ventures far removed from national interest.

A simplistic application of the "targets & instruments" theory of economic policy would regard a policy which aimed at more than a single goal as doomed to failure: to have a well defined (nonsingular, just determined) model of an economic system under manipulation by policy requires there to be just as many targets as instruments, and for the instruments to be fixed to their respective targets as allocated by the model of the economic system adhered to by the authorities in charge of policy. The theory behind this policy approach has many qualifications and questions which have never been satisfactorily answered (Webb & Wilson, 1982). The discussion and development of this theory in attempting to take account of the uncertainties and interactions between instruments that have become obvious has led toward a more pragmatic and less doctrinaire application: systems change over time, and change with, for example, the level of economic activity and the state of the external world in an open economy. Thus it can be argued that in a mainly decentralised economy, social controls such as the market in the legal and other institutional context, fulfil a number of social goals simultaneously, with considerable variation from time to time, and sector to sector, as to which of those goals is more handsomely served, in just the manner in which centralised policy instruments have been found to work (*ibid*). The crude targets & instruments model of economic policy, associated with econometric models in which the economy is described by a system of equations, can be set against the framework being used here: social control of economic activity is accomplished by many different institutions; democratic society tends to prefer decentralised controls to centralised ones; government policy is an alternative to decentralised social control. The market system is a complex of decentralised controls within the legal and conventional framework; regulatory policy is analogous to at least some of the market functions. The different normative functions of the market include both the static efficiency function and the function of best increasing the productivity of resources used in provision of services in demand. Regulatory policy then has a role in substituting for both aspects of the absent market's functions. And if, as I am arguing, the static efficiency aspect is far less significant for productivity than is the rate of change of productivity, then attention should be focussed on the latter in proportion to its importance (ironically, this comes down to the sound neoclassical principle of applying effort at the margin to various goals until the marginal results to expending effort are the same for all).

Analysis of the gains to be made by ensuring static efficiency cannot be further improved has generally revealed a disappointingly small improvement in economic well being. The review of that literature in Scherer & Ross (1990) concludes that were these inefficiencies to be removed there would be a once only gain of between 1.5 and 6% of GDP. The ORANI general equilibrium model of the Australian economy has similarly been used to demonstrate (for what CGE is worth) that the gains from successfully implementing all the microeconomic reforms currently in hand might be some 5% of Australia's GDP (Forsyth, 1992, pp. 14–16 and pp. 197–203)).[3] Filmer & Dao (1994) present an attempt at estimating the potential dynamic gains from microeconomic reform in Australia using another computable general equilibrium model. The potential gains to these reforms seen by Filmer & Dao are of the order of 12–15% of GDP by the turn of the century, or an increase of the rate of growth of GDP from 2.5 to 3.5%. These include many gains that are really due to improved X efficiency, not well measured in the CGE framework which assumes perfect competition to start with. They are puzzled by the gains from potential innovation and learning (*ibid*, p. xii). Yet many of the gains they discuss are inherently gains from improved rates of productivity change that become possible when competitive incentives and punishments are let loose. The continuing nature of these gains is ignored by the equilibrium assumption of the model. Even where that equilibrium is not expected to be attained for some to many years in the future, it is still the case that there is an asymptotic approach to it, making them expect that the gains will be greater in the earlier years and lesser in the later (*ibid*, p. 36). The expectation that equilibrium is approached limits the future rate at which GDP growth may continue to benefit from the structural changes made by the reforms. But it is the essence of a more flexible and competitive system that it responds to stimulus, carrots and sticks alike, by creative change which has unpredictable but nonnegative outcomes. This is known to people directly involved in managing and creating, but cannot be entered into the restrictive models used to show the interrelationships between sectors. It is thus entirely possible that they are underestimating the potential gains by ignoring these factors, though it must also be possible that they ignore potential costs by dismissing possible gains from government support for infant activities whose potential for creating a comparative advantage is lost by a parsimonious and very conservative assistance policy.

Is it inappropriate to suggest that regulatory policy should be aimed at the goal which promises the greatest gains to society? In the absence of a theoretical framework to suggest the implications of regulation for the rate of technological progress it certainly would be. I would hope to be able to show that there is a framework which can be applied to the question, and that that framework bears on price-cap regulation in particular.

2. The goals of policy as seen by policy makers

The conventional view of regulators' policy goals is that they are to prevent monopolists from charging prices that are too high, to prevent excessive profits from being made, to ensure they play fair in their dealings with their captive markets, and don't adopt unscrupulous tactics to rid themselves of unwanted competitors. More recently regulators have been charged with overseeing the introduction of competition into once monopoly industries. By "competition" they mean competitors for the once monopolist. This change began with telecommunications, with Judge Green's break-up of AT&T, and the UK government's privatisation of British Telecom and the gradual opening of different sectors of telecommunications to entry. And while these two actions stemmed from very different traditions, they have had similar outcomes and intended purposes.

Gravelle (1982) in his early application of principal agent theory to regulation noted a tendency on the part of some economists, including Littlechild, to reject regulation as a viable policy tool. Instead they advocated the market as regulator of industries traditionally hedged off from market forces. What Gravelle saw as a few economists evincing lack of faith in conventional regulation, we now see as the dominant viewpoint of policy makers and within the profession, though without public regulation being abandoned by any means. Policy makers' enthusiasm for market process has grown with change in the role of regulation. Instead of the permanent institutions of regulation as in the American regulatory commissions, the new regulators are overseers of competition in its early stages, and simulators of competition where real competition cannot be reasonably introduced. The by-words of the modern regulator are such phrases as "opening to entry", "breaking down the barriers to market forces" and "enforcing the pace of change". And in Australia at least "facilitating the ongoing processes of microeconomic reform" is the bureaucratic role to be most envied.

2.1. The Australian example

The Australian government's policy framework for telecommunications shows clearly the two issues central to this paper. The first is that regulation is not seen as a permanent alternative to competitive market forces, but as a means of introducing and fostering those forces. The second is that one of the goals of regulation during the transition is specifically to "encourage productivity improvements" (Beazley, 1990, p. 6).[4]

The first assertion can be supported by reference to any of a large number of statements. It has become part of the background of the policy agenda known in Australia generically as microeconomic reform. As applied to telecommunications the views of AUSTEL on its own role begin with the assumption that the path to "increased efficiency and responsiveness" (AUSTEL, 1994, p. 1) was by "increasing the exposure of what had been a monopoly industry to the forces of competition...(The industry is subject to regulation) both as ...operating in a monopoly setting and in the transition to a competitive environment" (*ibid*). Part of the regulatory role is stated by Beazley (*ibid*) to be "to prevent the dominant carrier from engaging in predatory pricing". This is restated more circumspectly[5] by AUSTEL in terms of regulation imposing "a range of safeguards to ensure that consumers and competitors are not adversely affected" by the incumbent's market power (AUSTEL, *ibid*). The then Minister's statement goes on to say that the 1990 changes to AUSTEL's remit were, *inter alia*, "a specific mandate to promote competition within the framework established by the government's decisions" (Beazley, *ibid*, p. 7). A consequence of that mandate was to make one of AUSTEL's members a part time Commissioner of the Trade Practices Commission.

Both the Minister and AUSTEL are sensitive to the problems of potential entrants evaluating their chances of success in a market dominated by a single incumbent. The question of easing the condition of entry is important in the regulatory regime. AUSTEL sees another role of the price-cap system it administers being to limit the power of the incumbent to use its relative efficiency, let alone its resources of secure noncompetitive markets, as a source of funds to meet entrants with considerable aggression. "The incumbent carrier could abuse its market power by offsetting its reductions in charges for competitive services with increases in the charges for the less competitive services. The second price cap (1992–95) has been designed to limit the extent to which the dominant carrier can abuse its market power in this way while allowing the carrier to rebalance progressively its prices toward costs and thereby reduce the historical distortions in the exiting price structure" (*ibid*, p. 5). Thus the policy of introducing competition is being pursued beyond the mere opening of the legal possibility, entrants are promised "a fair go" (a very Aussie attitude) as encouragement to promote the success of the policy.

The second element, the significance of productivity change, is supported not only by the Minister's statement, but also by the tendency within both documents being cited here to adduce both productivity change as a benefit of price-cap regulation specifically and the concept "competitive" as close to synonymous with the promotion of productivity change. For example, the Minister, having made productivity improvement the first stated aim of carrier pricing arrangements, goes on to determine that, as an implication of the aims of policy, price caps are "to ensure that customers get the benefits of additional competition", and that "the value of "X" in the overall price cap is to be increased to reflect the efficiency gains achievable through the introduction of competition" (*ibid*). AUSTEL's interpretation of this and later statements include the idea that regulation of monopoly fails as a completely satisfactory device "to either increase technical efficiency and the quality of service or charge the lowest possible price" (AUSTEL, 1994, p. 2). They go on to explicitly equate competition with a regime that is more efficacious in these regards, if lacking in others:

> The government still regards the social objectives of ensuring that telephone service remains affordable to all Australians as important, however, the imperative to reduce costs to business and customer costs as well as increase innovation in telecommunications, has led the government to rely on more competitive approaches and find alternative means of achieving the social objectives. This has involved placing certain requirements on carriers through licence conditions and in legislation (*ibid*).

AUSTEL's own view of the price-cap regime is that it is a means of enforcing a minimum rate of productivity change on Telecom: "The ceiling, or cap, is designed to ensure that consumers receive some of the benefit from the increase in productivity and efficiency in the form of lower prices. It is expected that with the use of increasingly sophisticated technology, the telecommunications industry should be able to achieve higher increases in productivity than the wider economy" (*ibid*, p. 3). "Price-cap regulation provides firms with direct incentives to improve their efficiency, since the firm can keep the efficiency gains made over and above the "X" in the form of retained profit. This is one of the key incentive properties of the price-cap approach... It has also been recognised that while the price cap may encourage improvements in technical efficiency, the constraints imposed on price changes may result in certain allocative distortions over time" (*ibid*, p. 4).

Other investigators of Australia's price-cap regime have seen the same concerns. Xavier (1994) notes that "the primary attraction of the price-cap approach was that it contained the potential to sustain incentives for economic efficiency... The scheme is expected to provide *sustained*[6] incentives for improved productivity" (*ibid*, p. 10). "Within the overall price cap, innovations in product/service pricing (consistent with dynamic efficiency) would also be permitted, indeed, encouraged" (*ibid*). "The regulated firm would also share in efficiency gains since it is allowed to retain any additional profits it generates as a result of better-than-expected performance in regard to efficiency and technological improvements" (*ibid*, p. 11). The better-than-expected refers to expectations of the policy maker.

The criteria he uses to evaluate the performance of the price-cap scheme number a dozen. Of these, six have relevance to the dynamic factors which I am arguing are of central importance to economic welfare. But perhaps of more importance to my argument is the way in which the value of the inflation discount factor is determined by the policy maker. The expected rate of total factor productivity improvement is the major factor (*ibid*, p. 12). The firm has an interest in minimising this expectation, so as to minimise the value of "X", while attempting to maximise actual performance early in the period following imposition of the cap (*ibid*, p. 13). Policy makers in fact used international benchmarking techniques to inform their guesses about how rapidly Telecom Australia could improve its performance, rather than relying on Telecom's own estimates of its potential (p. 14). The significance of productivity expectations is that a large part of productivity improvement comes routinely from incremental learning, outcomes of investment in new techniques and the enjoyment of economies associated with growth, i.e., scale economies, vintage effects as growth means investment in relatively high productivity yielding capital items, and gains from redeployment of "difficult to sack" resources (not only labour) from low productivity activities to higher productivity activities. All of these are ongoing features of industrial development rather than one off gains from rebalancing allocation of resources. In other words, they are gains from the search for continuing profits, the outcome of entrepreneurial competitive and rivalrous behaviour.

A final, and perhaps unnecessary, piece of evidence on the direction of policy goals is in a publication of the Bureau of Transport and Communications Economics.

In their (1993) they survey the literature on market dominance from the perspective of a once monopoly situation becoming competitive. They begin with the legislative basis of current policy. For purposes of my argument this is the significant aspect of their paper (*ibid*, pp. 1–9). The Telecommunications Act 1991 set out the rules under which a competitive industry would be fostered. The government's belief expressed in the Act was that vigorous competition would protect the long-term interests of Australian consumers. Indeed the term "competition" is left undefined as it is in the Trade Practices Act 1974 as amended. However, its use, the adjectives used with it, indicate that government believes competition is the key to good economic performance in all the areas of allocation of resources and economic progress. Specific mention of international competitiveness of service and facilities providers suggests that dynamic factors are considered, if only because the rhetoric of "international competitiveness" includes for the policy maker and economic journalist alike the race to match the progress made by those countries, especially East Asia, which focus "competitive threat" concerns. No further substantiation of this point will be offered in this paper.

The conclusion to be drawn from this exploration of policy attitudes is that the price-cap regulatory regime is designed to both transform the monopoly into a competitive oligopoly and speed the progress of technological change. The regime directly stimulates the latter, while for the more distant future, the presence of rivals for the now dominant incumbent is confidently expected to continue the path of technological progress as one important consequence of the process of competition. What both the government's policy goals and AUSTEL's carriage of policy lack is a sound theoretical basis within which those goals may be seen to follow from the analysis. It is that to which my suggestions are intended to direct attention.

3. Where did the price-cap formula come from?

Stephen Littlechild, currently the regulator of the privatised British electricity industry, was commissioned in 1982 to make recommendations on the method of regulation of the soon to be privatised British Telecom. His report (Littlechild, 1983) contains the first clear statement of the basic principles of price-cap regulation. Littlechild was then, and presumably remains to this day, a "radical Subjectivist", a follower of the theories of Hayek and von Mises (Littlechild, 1979). We would expect him to believe that the future is unknowable and is at least partly created by the actions of individual agents. We would also expect him to be sceptical of the efficacy of regulatory intervention, given the Austrian view of the role of the entrepreneur in achieving a general equilibrium by seeking opportunities to profit by imbalances in markets (Kirzner, 1973). In his (1979) we find him arguing, first, against orthodox welfare economics as being based on static equilibrium conceptions remote from the world of business behaviour, and second, conceding some justification for a particular form of intervention on the grounds that it improves the efficacy of the competitive process. "I quite like the Monopolies Commission because in its early days at least, people like Professor G.C. Allen were critical of welfare economics; he tended to take a view which was in sympathy with the view I have taken here that competition is a process. He was much more concerned that there should be an opportunity for other firms to compete" (*ibid*, p. 18). And indeed we find Littlechild (1983) considering the consequences of "no explicit constraints on profits", judging it superior to all other proposals except on "Protection against Monopoly", the criteria which he admitted had "been identified as the prime purpose of regulation" (Littlechild, 1983, p. 10).

The remit Littlechild had from the Department of Industry was to examine possible methods of regulating profit and pricing behaviour of the privatised British Telecom in an increasingly liberalised and competitive environment. The criteria Littlechild worked with were determined by the government. His interpretation of those criteria led him to set out five headings, of which the first was "identified as the prime purpose of regulation" (*ibid*). They were:

1. Protection against Monopoly.
2. Efficiency and Innovation.
3. Burden of Regulation.
4. Promotion of Competition.
5. Proceeds and Prospects (*ibid*, pp. 10–11).

His analysis of the two proposals brought forward by the department brought him to the conclusion that neither scheme (maximum rate of return, and an output related profits levy) could be recommended despite the inference that both did provide some protection against monopoly power, and no explicit constraint did not. "If attention is confined to ORPL vs. the MRR scheme, the choice would be between a scheme that is largely unknown and a variant of one that is known to be unsatisfactory. The absence of constraints is superior in almost all respects, and could be strongly recommended if it were not for the apparent short-term failure to limit monopoly power" (*ibid*, p. 31).

His alternative approach was to suggest a price cap. "As argued in Section 4, ultimately the only effective way of combating monopoly power is by competition" (*ibid*, p. 34). In the absence of competition... "it is possible to insert conditions in BT's licence to prevent prices of monopoly services being increased faster than the rate of inflation, or perhaps to reduce them in real terms" (*ibid*, p. 34).

His final ranking of schemes had the price-cap scheme first on Protection against Monopoly, equal first with "no explicit constraint" on Efficiency and Innovation and on Proceeds and Prospects, and second to "no explicit constraints" on Promotion of Competition (*ibid*, p. 36). The direct effect of the price-cap scheme would be to improve consumers' position, but may reduce the incentive to new entrants and perhaps expansion of services. But the effects on Innovation and Efficiency would be nonnegative, as cost reductions exceeding target are retained by the firm, so improved efficiency yields profits to shareholders and funds for investment (*ibid*). He did see a problem in that entry could be impaired, due to the effects on prices and thus on expected profitability of competitors on regulated services. However, Littlechild argued strongly for increasing competition in all areas of BT's operations, argument that has since carried the day with governments and regulators (*ibid*, pp. 38–39). Given this opening to competition, the price-cap scheme would be a reasonable compromise in reducing monopoly power while encouraging innovation and efficiency while competitive forces were fostered.

From this is can be said that Littlechild saw some of the significance of his price-cap proposal for the dynamical aspects of economic performance, but was mainly concerned to satisfy the public demand for constraints on monopoly power, without compromising other aspects of performance. He would have preferred no explicit constraints on BT's profits, or, less, on its prices. Given that constraints were required he settled on the price-cap idea because it would not compromise the other goals of policy, and may even improve pressure for efficiency until competition from other firms had strengthened sufficiently. He saw the price-cap rule as the least harmful type of regulation of market power.

4. Models of economic progress

Perhaps the best way to approach the model of the economic system which I believe to be most appropriate is to take the seminal ideas of those writers whose influence is growing amongst today's economists. The first will be familiar to most readers, that of J.A. Schumpeter. The others may be less familiar, but in another decade their ideas will be commonplace to the economically literate. I will attempt to outline the ideas and then draw together the significance they have for regulatory policy.

4.1. Schumpeter's perennial gale of creative destruction

Joseph Alois Schumpeter is now seen as a seminal thinker in the field of evolutionary economic theory. His reputation has grown since his death in 1950. The "International JA Schumpeter Society", founded in 1986, associated with the learned *Journal of Evolutionary Economics*, is devoted to the development of evolutionary economic theory.

The clearest and simplest statement of Schumpeter's ideas on the capitalist firm and its markets is to be found in Part II of his *Capitalism, Socialism and Democracy* (1943). His immediate concern in those years following the Great Depression was with the survival of capitalism. Most of his fears of stagnation and lack of technological progress have not been realised. The ideas which are now used, and doubtless abused, are stated clearly in a mere two chapters, "The Process of Creative Destruction" and "Monopolistic Practices". It is the title of the first which has captured the imagination of scholars.

Schumpeter saw that Big Business was not all bad, indeed, barely bad at all. Unlike the theories of Robinson and Chamberlin which attempted to analyse competition which did not fulfil the conditions for perfection, and were thus "imperfect" and socially undesirable in their effects, Schumpeter's theory explained why the growth of monopoly and "imperfections" went along with great increases of wealth and income for the whole of capitalist society. In Schumpeter's view it was not quantitative change but qualitative change that was of significance for capitalist progress. "Capitalism, then, is by nature a form or method of economic change and not only never is but never can be stationary... The fundamental impulse that sets and keeps the capitalist engine in motion comes from the new consumers' goods, the new methods of production or transportation, the new markets, the new forms of industrial organisation that capitalist enterprise creates" (Schumpeter, 1943, pp. 82–83). Further, he infers that a system that operates "optimally" at every point in time may pass over possibilities for transformation that could make a system that does not operate "optimally" grow wealthier by those transformations which a static system avoids. This transformation from within is his Process of Creative Destruction. This Process is that by which the firm, the entrepreneur, creates opportunities by innovation in one or more of the modes mentioned in the above quotation, carves out a profitable niche and draws demand away from the existing suppliers. In the extreme cases which have become identified as "Schumpeterian" whole industries and societies are destroyed in the creation of new ones. The transformation of the cotton industry in the century from the late 18th to the mid-19th, destroying the work and life of the hand spinners and weavers and giving birth to the factory system, is one such example. Another is the transformation of sea and land transport with the steam engine, destroying coaching, canals, sailing vessels and their economic and social concomitants.

These processes may be on the grand scale of the above examples, or on a more limited scale in which an industry is merely given an alternative and more productive method of scheduling production (e.g., Just-In-Time methods). They operate not by firms optimising known methods in existing markets, as in orthodox theory, but by their search for the new. The spur is the hope of profits beyond the normal, beyond those available to the firm which accepts the constraints of supply and demand as they appear. The competition that orthodox theory sees as binding is, to Schumpeter, a trivial constraint on a firm's performance, and a trivial indicator of economic welfare. The difference between so-called imperfect competition and perfect competition in terms of consumer welfare was nothing compared to the gains to be made by the "gale of creative destruction" (*ibid*, p. 84).

For Schumpeter's process to work required a number of conditions, at least some of which he was pessimistic about for the future. The first is the entrepreneur, a person who can see possibilities for profit and is willing to bear the uncertainty about whether the possibility will be turned into actuality. The entrepreneur is a catalyst as well as a creator, drawing with him or her the people needed to fulfil the dream. The ability to command finance is necessary along with the charisma of leadership. This function cannot, on Schumpeter's view, be carried out by a committee. The bureaucratic firm is not a vehicle for the entrepreneur. His fear for the future of capitalism was precisely that innovation would become a bureaucratic function, and thereby atrophy. One can think of a number of examples of firms which have suffered from a failure of entrepreneurship. IBM was an entrepreneurial firm in the earlier waves of computerisation, but lost that character over the following decades. It seems clear in hindsight that IBM's bureaucracy has missed out on large profits due to its entrepreneurial failures. But fortunately, and contrary to Schumpeter's fears, many other firms maintained an entrepreneurial stance. The large and unenterprising firms were surrounded by smaller enterprising ones, even in the mature capitalism of the post war period.

The second is the ability to create monopoly power. Without the prospect of monopoly profits for a significant period of time the uncertainty into which vast funds were to be emptied would make financiers unwilling to venture those funds. In order that economic progress and the broader competitive process take place, competition must not be so immediate that profits would be competed away by imitators. Patents, copyright, grants of exclusive rights, all these and other means apart from the legal system (indeed, some that are clearly illegal (*ibid*, p. 89)) are not to be seen as restrictions on competition, but as necessary conditions for the process of creative destruction that alone can bring economic progress. "The main value to a concern of a single seller position that is secured by patent or monopolistic strategy does not consist so much in the opportunity to behave temporarily according to the monopolist schema (of orthodox theory), as in the protection it affords against temporary disorganisation of the market and the space it secures for long-range planning" (*ibid*, pp. 102–103). For this reason Schumpeter was not enthusiastic about public policy to limit monopoly power. He saw such policy as coming from the orthodox theory which was against big business, neglecting the great benefits which such business had given precisely because it was able to enjoy the fruits of monopoly.

A Schumpeterian economy can be pictured as one which is constantly changing, but in which the changes are not all of the same significance. Small revolutions are continuously occurring in one or another industry. These episodes of creative destruction are of importance to each industry concerned, and make discontinuities in each industry's progress. Firms rise to prominence with innovation in production or marketing, dethroning the great performers of the previous regime. But these in turn are overthrown by another round of innovation. Current examples in the world of retailing are the rise of the discount department store at the cost of the traditional department store and the rise of the car dependent regional shopping centre at the cost of the central business district's shops. An example from the world of microelectronics is the rise of the software houses, especially Microsoft, to a position of dominance over the hardware producers who increasingly are unable to differentiate their products from those of their rivals.

But while the shopping and the home computer revolutions may be of less than revolutionary significance to society as whole, and be seen as social evolution over a number of decades, other transformations are of greater impact on society. The changes wrought can indeed be seen as disturbing the continuity of change. Examples from the past include the railway, perhaps the most disturbing of all the changes, making distance so much less important, allowing cities to grow to present scales, allowing people and goods to be transported hundreds of kilometres overnight between centres, breaking down regional markets and societies. In current times at least a couple of revolutions are under way. In telecommunications the old monopolies have been broken down not merely by governments but by technological change. The results of this burgeoning of telecommunications may not be quite as dramatic as that of the transport revolutions, but they will be considerable. The picture in broadcasting is much more dramatic, with the availability of satellites shifting control of broadcasting out of the hands of governments and giving it to the entrepreneurially creative media moguls. Rupert Murdoch does not have to bother getting licences to broadcast anywhere his satellite transponders allow, and his audience has barely to advertise their addiction as the size of dish is shrinking to dinner plate dimensions and smaller.

Schumpeter's picture of the capitalist economy is thus one which gives centre stage to the processes of economic change and to the power of the profit motive on the enterprising individual to create that change. What is the relevance of this idea to the function of the price-cap rule? The obvious relevance is that the rule acts as a surrogate for competition emerging some time after innovation has transformed the firm's profitability. The regulated entrepreneur seeks to beat the regulation by innovation. Given a specific price cap with a regimen of services covered, the firm will seek both innovation to reduce the cost of providing those services and innovation of new services which are not within the cap but also may attract demand away from the capped services.

Further relevance is in the notion that the regulated firm may be an easier target of an Schumpeterian entrant bearing services unregulated but in effective competition with the regulated firm. Impregnable monopoly, usually statutory, is no environment for the next generation of Schumpeterian innovators. The price cap is a means of keeping the regulated firm vulnerable. Indeed this is one of the complaints of Telstra in Australia and has been a complaint of the regulated Bells in the USA. These may be unintended consequences as far as the architects of regulation are concerned, but they are of much greater significance over the decades than are the levels of prices relative to marginal costs. The gains from "rebalancing" are taken but once, the gains from continual innovation are continuing. The fact that regulation of once monopolists progressing towards competition (or at least to an oligopolistic structure which is intendedly rivalrous) is being accomplished by the use of price capping amongst other tools is an indication of the role which policy makers see it having. It is intended to make entry and survival for the initial period more certain for potential and actual entrants.

However, it must be said that Schumpeter's own writings give little space to the need for continuous active competition for a market. That is an issue taken up by the next author.

4.2. Downie's competitive process

Jack Downie, a Civil Servant in the UK Treasury, was concerned at the disparities in productivity between firms, the reasons the competitive system allowed such disparities to remain, and at how the rate of productivity change could be enhanced by public policy, in particular policy against monopoly and restrictive practices. In order to analyse this set of problems he found himself having to discard the microeconomics of the firm and the market that had emerged from the Cambridge debates of the 1930s (Robinson's imperfect competition, Chamberlin's monopolistic competition, and the synthesis of Fritz Machlup). Instead he developed a completely different theory based on ideas from Schumpeter and from the growth models associated with Keynesian macroeconomics which began with the realisation that investment created productive capacity (Harrod's and Domar's models in particular). The model he developed had some similarities with the work of Josef Steindl who was in the Kaleckian tradition of post-Keynesian analysis.

In Downie's model, differences between firms contesting a market could be summarised in their relative efficiency. This was defined, broadly, as what is now called Total Factor Productivity. Efficiency was closely related to, but not the same as, profitability. For our purposes the difference can be ignored. A firm's efficiency is related to both its level of costs and its ability to sell its outputs at a margin over costs. A proportion of the profits that result from its activities are reinvested in productive capacity. He assumes this new capacity has the same efficiency as the pre-existing capacity, though his argument is strengthened if new capacity is more efficient. The propensity to reinvest depends on the rate of market growth, in turn dependent on the rate of growth of the national economy. Those firms with higher rates of profit than the average increase their market share at the expense of those with lower than average profits. Even if the most efficient firms are initially smaller than the less efficient, the latter will eventually be overtaken by the once smaller but more efficient. The average level of efficiency will increase over time as the less efficient lose market share and the more efficient gain. Market structure will also change, becoming more concentrated as the more efficient consolidate their position, and if just a very few firms are most efficient the market may evolve into the highly concentrated forms that are observed for many manufacturing industries. If economies of size are part of the cause of relative efficiency, this tendency will be exaggerated.

This process of structural change and average efficiency improvement Downie calls the Transfer Mechanism. It has been shown (Metcalfe, 1988) that this process obeys Fisher's Law. This states that the rate of change of the average of the selection variable (efficiency in this case) is proportional to the variance of the population on that variable. In other words, the greater the dispersion, the faster will productivity improve. The smaller the dispersion, the slower will productivity improve. But is this the whole story of industrial productivity change? It certainly is not. The Transfer Mechanism can take into account the learning process which will generally ensure that new investment results in production capacity that is more productive than old, but it ignores innovation. The putting into practice of inventions in all of productive processes, marketing methods, organisational methods, new markets and new products is a central element in capitalist progress. This is the capitalism emphasised by Schumpeter in his "perennial gale of creative destruction". And Downie entered a version of this into his model as the Innovation Mechanism. Anticipating Cyert & March (1963), he postulated a theory that firms who perceived themselves failing to meet targets for such performance variables as profit, market share, rate of growth, would seek a means to escape certain death. One of those means was innovation to transform the firm's efficiency. By no means all attempts at innovation would be successful, but some would be. This would change the relations on which the Transfer Mechanism acted. In more extreme cases this would overturn the existing order in the market in the way that Schumpeter envisaged. A firm's innovation need not leave the firm contesting the same markets as previously, and occasionally a firm would change its industry, by introducing a new or at least different technology.

Downie's Innovation Mechanism provides an exit from the uniformly depressing implications (monopoly and stasis) of his Transfer Mechanism, and a means of explaining the volatility of market shares in the long period of a decade or more. The market for computing services is a case in point. In the first era of this market it was served by mainframe computers running proprietary software. IBM quickly came to dominate this market, establishing a seemingly impregnable position by the early 1970s in markets of the industrialised world. A number of innovations were made by rivals in the market. The minicomputer was one, the microcomputer another. Software owned by firms not involved in hardware production was another. IBM never succeeded in dominating any of the three innovating forces. It has not been greatly successful at selling minicomputers; its microcomputer was a great success for the industry but not for IBM, because it allowed control of the software to slip out of its grasp, as well as failing to maintain its leadership in hardware platforms for that software. The microcomputer has become an easily copied technology, the producers of software earning the profits that in the previous generation of the market had gone to producers and marketers of the hardware/software package. In the process, the cost of computing services has fallen more rapidly than virtually any other business input over the same period. The mature market for mainframe computers may or may not stagnate, but that is of no relevance to the broader computing services market.

Downie's model thus encapsulates the central elements of an evolutionary model. There is a population of firms. Individual members of the population have a relatively fixed characteristic over which members of the population vary. In his case the characteristic is productive efficiency. Different values on that characteristic cause individuals to grow or shrink within the population and specific ranges of values allow new members to enter the population. Existing members may adapt by transforming to new values in a mutation process and new members may also exhibit such mutation. The demographic structure of the population on the variable at any point in time can be explained by these processes of selection and mutation. It is a phylogenetic theory, as are the Darwinian and all the neo-Darwinian theories of biological evolution: the rules governing change in the population (genetic rules) change with feedback from population changes themselves. There is no equilibrium at the "end" of the process, as there is no end to it.

Where is the relevance to a regulated monopolist in this model of a population of firms? In the absence of competitors there would be no Innovation Mechanism, no incentive to adopt or invent. Nor would there be any improvement of average or best practice productivity, as the monopolist has the entire market and no reason to innovate. This rather extreme implication of the model is borne out about as far as any generalisation in industrial economics by the evidence on rate of technical progress and concentration. Scherer & Ross (1990) review the evidence in their chapter 17 and conclude that very high concentration goes with low rates of technological change while somewhat lower concentration sees the highest rates, thereafter falling to low with very unconcentrated markets. "(Schumpeter's) less cautious followers were wrong when they implied that powerful monopolies and tightly knit cartels had... (a strong claim to dynamic efficiency)... What is needed for rapid technological progress is a subtle blend of competition and monopoly, with more emphasis in general on the former than the latter, and with the role of monopolistic elements diminishing when rich technological opportunities exist" (*ibid*, p. 660). If the regulated monopolist is to be persuaded to be progressive, it must be via a form of regulation which can be a surrogate for the competitive pressures on firms competing with rivals. How might we see regulation working in this context?

First, the bearing down of the price cap on the margin of profit is a productivity enhancement tool: the firm must respond as if it were subject to the Transfer Mechanism. Just as the firm subject to competition from more efficient firms is forced into failing ranges on its performance indicators, so the price capped monopolist will eventually be forced into failing ranges. As margins tighten under the cap, the only way to survive is to search successfully for productivity enhancement, i.e., the Innovation Mechanism.

Second, the price-cap constraint can be seen from within the multidivisional, or at least multiproduct, firm as a surrogate Transfer Mechanism. The rebalancing process, which is seen as most important by orthodox economics, can be interpreted dynamically as a force directing investment towards those activities which are yielding higher margins and profits. If the divisions invest partly as a result of their recent performance then the more profitable, and thus, usually, more productive, will expand relative to the less productive. The average productivity will therefore improve over time, and would converge on best productivity if the divisions' outputs were fungible, which they are not. In contrast to the static rebalancing notion, where the price relatives of the firm are adjusted to reflect relative marginal cost/marginal revenue relations, the dynamic interpretation sees the firm's investment, production capacity and output balance changing to reflect profit margins[7] and productive efficiency. It may be argued that this is rather semantic. But add in the Innovation Mechanism: the price-cap bearing down and enforcing productivity improving investment; and the once static rebalancing concept reappears to take advantage of the new relative productivities following the innovation. And as the firm will be innovating more or less continually, as the evidence clearly shows in the telecommunications industries, the two effects will be working continually to increase both the firm's average productivity and its leading division's productivity.

There is, however, a contraindication from the Transfer Mechanism when it is remembered that price capping is being used as a handicap on the once monopolist in the competitive struggle with unregulated entrants. The lowered profitability of the capped firm reduces its ability to invest directly from profits and indirectly by capital raising. As argued above (last paragraph on Schumpeter), entry has been fostered by regulatory policy. In fact, price caps and other handicaps imposed have had the effect of shifting market share toward more expensive providers. Industry average productivity is thereby decreased relative to what it could have been in the absence of the handicap. Under what conditions will the incentive effects of the price cap as competition surrogate outweigh the negative effects on investment of the price cap as restriction on profits? The orthodox literature attacks this as the ratchet problem of the optimal response of self interested management. The present framework sees the conflict as one between the investment routine and failure to achieve marketing targets which lead to the investment routine being transformed.

Can the belief that competition is better than monopoly, albeit regulated, be grounded in some ideas about the fostering of economic progress? Downie's Innovation Mechanism is a very simple way of arguing this point. However, it must be said that there is more to it than that. This is the issue taken up by the authors of the next set of ideas I want to examine.

4.3. Diversity as the central requirement of progress

The consequences of long established monopoly in telecommunications and the postal service for that matter were clearly seen in the last two decades. Those countries which had competing service providers tended to have a better performance outcome than the monopoly providers. The rapid transformation of industry structures over the past 15 years, with fundamental policy revision in the same direction spreading from the US and UK to the rest of the industrialised world and beyond, must be strong evidence that real gains were made, at least for governments and their powerful supporters and lobbyists. In explaining this thoroughgoing change orthodox economics is nearly silent. Slogans about competition and the side issues of static efficiency are the extent of its contribution.

To find an explanation for the vitality of markets served by once monopolist incumbents and relatively small scale entrants we must turn to the ideas of Stan Metcalfe and his colleagues whose interest is in the process of technological competition and the selection of variety in the market place. Metcalfe's work lies directly in a line with that of Downie. He makes his contribution in refining the concept of variety and adding the element of directed learning by the firm. In doing so he adds considerably to the ideas implicit in Downie's Transfer Mechanism, and sets the Innovation Mechanism in a more continuous role, integrated with the Transfer Mechanism as a routine part of market interaction.

Firms make up the population over which selection[8] takes place due to each firm's commitment to a "technological regime". This commitment comes from the knowledge base of the firm: each firm is "an island of privileged information, with a knowledge base and an organisational capacity largely determined by its past history" (Metcalfe, 1986, p289). This restricts a firm to a small set of "design configurations" which it can use as the basis for its capital asset stock, and the other resource requirements implied by that regime.[9] The differences between firms resulting from their various technological and organisational commitments results in variation in economic performance, differing profit margins in particular. Selection processes also require a relationship between profit margin and market share growth or decline. This is provided by the translation of profit into investment, via internal funds and access to external funds. If both these are broadly proportional to profit margin, the selection mechanism is in place. A third process is the ability to innovate. Metcalfe suggests that firms differ in their ability to innovate, citing the extensive literature on innovation and its diffusion (*ibid*, p. 290). These three processes result in both selection and mutation, in which firms grow or decline over time. Modelling the mutation element requires further assumptions about the differential ability to innovate, or more generally, to improve productivity. At this broadest level it is not the firm that is being selected by the competitive struggle, but the technological regime it has chosen. Each technological regime has theoretical possibilities, known with varying certainty by user firms. The firm's potential for survival and prosperity depend on the trajectory of productivity improvement that that regime allows relative to the alternatives which have been chosen by rival firms. The trajectory, given its possibilities, is largely an artefact of the internal learning process. Whether the possibilities are certain or not, the user firms' efforts at discovering improvements will gradually reveal those possibilities. The incentive to these discoveries is in the competitive pressures imposed by what is in effect the Transfer Mechanism: investment of profits creates new capacity for profitable firms, prices are squeezed by increased capacity production on the market. Capacity to innovate depends both on current output, which allows learning effects to reduce the average cost of production, and on investment in new capacity to produce and in R&D, both of which depend on profits. The resulting competitive process will usually be one in which the successful grow, and continue to be successful, so mature technologies are the province of relatively concentrated industries. However, the incentive to innovate also presses upon firms that are shrinking relatively as well as absolutely. Their attempts to innovate their way out of the vicious circle of declining profit and investment provide a counter to the increasing concentration of the maturing industry, and a stimulus to development of new technological regimes. In the even longer term, the development of the industry sees the selection of technological regimes which enjoy advantages over alternatives, and the growth of those firms which adopted those regimes rather than others.

But the process also is dependent on externalities of adoption, first adduced by Alfred Marshall in his *Principles*.[10] These external economies to the technology relate to such factors as learning by other firms' adoption and use via spillover effects, the upstream industry effects of increased demand for inputs, infrastructure provision for the industry[11] as it grows, reductions in the cost of finance as the financial services industry becomes more confident of its judgements of risk in the industry. More recently these have been rediscovered by W. Brian Arthur (1989) as technological lock-in by adoption: an early adopted technology will usually have an advantage over a later one because it will have been developed further by learning and the other economies of adoption and experience which Marshall mentions. Technological advantage then depends partly on inherent advantages of the regime, and partly on the advantages of early development of those advantages. These may indeed, as Arthur suggests, lock out alternatives which may have been more efficient had they had the chance to begin before the early starter. So while we might expect the selected technological regimes to be those with superior theoretical prospects for future productivity, the contrary case is not unexpected, and indeed may be the subject of public policy (though not one with which I am concerned in this paper).

The competitive industry then is more complex than Downie indicated. It has variety in both technologies and products. The Transfer Mechanism is not simply the growing of firms that happen to have chosen a cheap production technology relative to those which have not. It is now a race between differing combinations of product offerings produced by alternative technologies which are improving in their productivity at rates which depend on both past outputs (learning and adoption effects) and current rates of output (economies of scale). Instead of a discrete Innovation Mechanism stimulated by a need to solve an immediate problem, the innovation process is embedded in the production process which yields the various productivity enhancing effects of learning by doing and learning by using. Competition is implied by variety of both product and technology. Can the absence of competitive pressure be alleviated by a simulated competition, despite the absence of a variety of technological possibilities? The final section of the paper attempts to apply a model based on the principles outlined above, and suggests that price caps can play an important part in stimulating economic progress in the absence of genuine and full blooded market rivalry.

5. Price caps in the context of an evolving and progressing industry

Perhaps the easiest may to see how price caps effect dynamic performance in an industry becoming open to competition is to examine a recent example of a model of technically progressive industry, in effect a development of the ideas expressed by Schumpeter, Downie and Metcalfe. The model chosen for this exercise is that of Silverberg and Verspagen (1994), again, a Schumpeterian model with many explicit links to Metcalfe's concepts on technological change. As with the models outlined above it assumes considerable bounds on managerial knowledge, and a problemistic search mechanism for resolving inconsistencies imposed by external market factors and failures to meet business performance expectations. This will then be set against the analysis of Australia's price-cap experience of Patrick Xavier (1994).

The model consists of three blocks, the first describes the process of convergence on the most profitable techniques of production, the second the introduction of new techniques and new firms, the third, the evolution of new rules for generating new techniques. Evolutionary processes are seen in all blocks: the first is a pure selection process which is not unlike the heuristic used by neoclassical and Austrian writers to justify their faith in an equilibrium outcome to market processes. Writers such as J.M. Gowdy have explicitly called this "neoclassical" (Gowdy, 1992). The second block is a mutation process based on an R&D strategy which yields improvements in techniques, while the third block describes how firms learn to change their R&D strategy, mutation at another level in the system.

This first block can be described by three differential equations:

1. A linear Phillips Curve, real wage growth depends positively on employment;
2. A capital accumulation function: investment in a specific form of capital depends on current profits of that activity and on the difference between that activity's profits and the firm's average rate of profit, such that higher than average profit activities are favoured by investment of the profits from activities earning less than average profits. Capital stocks have their productivity embodied at time of investment, but labour productivity increases over time by a learning process.
3. The share of total labour employed on a particular activity depends on the share of each capital stock in the total, so the growth of any activity's share depends on its capital investment.

These three equations (Silverberg & Verspagen, 1994, pp. 211–212) are a set which, given initial conditions and parameters, describe the asymptotic approach to a steady state in which all production uses the most profitable, but given, technology. The application of this block in the present context is the process of rebalancing, where the firm or firms invest in the more profitable activities, shifting their resources in response to these profits. Firms failing to make profits give up resources in favour of those who do; those who select the more profitable activities grow more rapidly than those who do not.

The second block describes how the R&D strategy generates new techniques, and allows new firms to enter and existing ones to leave. Resources are devoted to R&D, this effort yields results stochastically. A new technique discovered begins the growth of a new kind of capital stock. Each firm may therefore be using a number of different capital stocks of varying profitability. The probability of R&D leading to an innovation in a given time period is modelled as one or another type of Poisson process, but to avoid leading to monopoly, size of firm is eliminated by making the innovation potential a relative element, dependent on the ratio of R&D to the sum of its capital stocks. There are also two kinds of spillovers, one from the general level of R&D in the system at large, the other from the product of the firm's effort and the systems, on the argument that the firm has to be capable of assimilating knowledge (Penrose, 1959 develops this idea at length). Finally, block two includes imitation by laggard firms of more advanced firms' practices, still a costly activity, but less so than discovering new techniques. Entry and exit is rather artificially dealt with by assuming that an exiting firm, one whose employment level falls below a fixed small level, is replaced by an entrant which is assigned a single technology and twice the critical small level of employment of those exiting.

This block, then, embodies a process of continual incremental improvement of best practice technology, but leaves open the question of how the R&D strategy is decided. It is in this sense without a market justification, without a competitive stick, relying on the carrot of improved profits and growth rate. In the third block of the system we find the "stick" of competitive pressure.

Two mechanisms are posited as the means whereby the R&D spending strategy is modified: the first is a draw from a normal distribution which occurs with an equal probability for each firm. This draw is for the proportion of profit to be spent on R&D. But all firms except the most profitable also have a chance to imitate the R&D spending proportion of more profitable firms. This probability varies from higher to lower as the firm's profitability increases toward that of the most profitable firm. This reflects the pressure of competition to adopt the strategy of the more profitable.

This mechanism makes the industry's structure less predictable over time, in the same way that Downie's Innovation Mechanism prevented an asymptotic approach to monopoly as the more efficient inexorably gained at the expense of the less efficient as explained by his Transfer Mechanism. Differences in profitability under the mechanism of block 1, where capital stock expansion depends on profits, lead toward dominance by the most profitable firms. Here in the third block, the potential for dominance by those most profitable firms is reduced by the efforts of the less profitable to emulate the more profitable.

The seemingly inexorable progress toward monopoly implied by Block 1 is thwarted by the shocks imposed on the system from the uncertain outcomes of R&D and the uncertain propensity to engage in R&D by different competing firms. The selection pressures of Block 1 are modified by the mutations of Blocks 2 & 3.

In the context of price capping of monopolists or dominant oligopolists this model has a number of interesting suggestions, and could clearly be implemented via a fairly simple modification of the simulation program used. For example, the firm faced not by competition from more profitable firms, but by an increasingly constraining price cap, and minor but unregulated entrants, is modelled as a set of variously profitable divisions of the regulated firm competing for funding. Each division invests in R&D according to its own strategy, and changes that strategy with fixed probability. Each division has a variable chance of imitating the strategy of a more profitable division, that chance increasing with the binding of the price-cap constraint, and consequent lowering of profitability if no action is taken. Thus the tighter the cap binds, the greater the effort to innovate out of the bind.

Thus the price-capped multidivisional firm can be characterised in terms of the three blocks of the Silverberg & Verspagen model:

Block 1: Divisions of the regulated firm, and competing unregulated (but presumably small) firms, accumulate capital and market share or market size by investing in the more profitable technologies, and running down the less profitable. The price-cap system substitutes for the market in some of these divisions, strengthening the force of the selection mechanism which results in the evolution of technologies and firms' shares of total capacity to produce.

Block 2: R&D results stochastically in the innovation of new technologies; once a technology exists it will grow according to reinvestment as in Block 1, and thus over time old technologies will be competed away. The price-cap regulation adds nothing more here than is implied from Block 1; i.e., whether the pressure of competition is supplied by normal market forces or by the artificial force of regulation, the outcome for generation of new technologies will be the same.

Block 3: R&D strategy is modified both by a general level of search, and by search imposed by the constraints on profitability imposed by the price cap and normal market forces. Outcomes of search are uncertain, and can be modelled in many different ways suitable for computer simulation. The results feed back into the prior Blocks of the model.

The model has therefore a prima facie relevance to the present problem and deserves implementation to test that relevance.

Simulations carried out by Silverberg & Verspagen demonstrate a number of features found in industrial economies. First, they show approximations to steady state growth with technical change and R&D investment; second, their simulations exhibit rapid phase transitions in the directions observed in the history of industrial development, of increasing R&D and fluctuations in industrial concentration, from the previous steady state levels through much lower levels which in turn give way to rather higher levels in the ensuing steady state; third, the following steady states are usually characterised by higher levels of R&D, technical change and investment than the previous steady states (Silverberg & Verspagen, 1994, pp. 223–224).

Their conclusion, that "an evolutionary approach appears to offer an attractive alternative explanation of how an economy can "bootstrap" itself in historical time through a succession of growth phases and market structures" is of relevance to the present concern. Their model, using a framework of bounded rationality with profit seeking in a market environment of competitive constraint, makes technical change endogenous and the central effect of the competitive constraint. This is what we currently observe in price-capped regulated telecommunications firms. The rate of change in R&D, investment and productivity improvement in firms newly exposed to competitive pressures, partly market, partly price cap, is characteristic of the Silverberg & Verspagen phase change period, or of initial conditions which include high levels of R&D.

All that remains is to show that Australia's price-cap regime has been consistent with the pictures drawn earlier in this paper. For this I will be relying on the work of Patrick Xavier, examining the practice of price capping since 1989. My summary of his conclusions merely points up the consistency of policy outcomes with my theoretical perspective.

6. The results of the price cap and conclusions

The level and structure of prices was the first target of the price-cap system, and the experience of the UK and Australia bears out this expectation (Xavier, 1994, pp. 35–41). It is important to note that while rebalancing is, in principle, a "one off" movement to a competitive equilibrium the changing demand and cost environments, exogenous and endogenous, imply a continuous need for adjustment of prices. The effects of these changes on pricing are part of the outcomes of the process of dynamic competition simulated by the price-cap regulation.

Xavier points out that excessive zeal for rebalancing under a rigorous price cap may lead to inadequate incentive to invest or extend the regulated part of the firm's activities, contrary to the aims of policy (*ibid*, pp. 43–44). Immediate gains in technical efficiency to satisfy such a cap would end in preventing the further growth of productivity. The close relation between productivity change and investment is clear from the capital intensive technology of telecommunications. Until all the software is driven on hardware that can accept any conceivable improvements, and until transmission media are an insignificant cost, the "putty-clay" model of capital is appropriate, and the firm with the more recent profile of vintages will have the better potential productivity at any point in history. Incentive to high rates of investment are the basis of good performance over time, and no matter how close the approximation to finely balanced Ramsey Price equilibrium, the technically less progressive firm will fail against the progressive; the regulated monopoly which does not have the incentive to invest will be perfectly static. He concludes that "rather than allocative and technical efficiency, the really important strategic dimensions of telecommunications performance and development will be in regard to dynamic efficiency, the quality of service and in regard to effectiveness in developing and marketing telecommunications equipment and services both domestically and internationally. The success of efforts in regard to these dimensions of telecommunications performance will depend importantly on whether adequate levels of internal, and external, funding are available to Telecom" (*ibid*, p. 54).

This paper represents only an early attempt to apply the evolutionary research programme to explain regulatory policy. I have tried to demonstrate the potential of models of competitive process and endogenous structural change in the toolbox of regulatory authorities. The tools are rapidly being developed, as my final example demonstrates, and will provide powerful and direct insights into the behaviour of regulated industries over time. While the beauty of analytical solutions may be abandoned, in their place will be general explanations of historical processes and predictions as to the possible outcomes of policy decisions. The false precision of neoclassical modelling of systems in which behavioural discretion is a dominant feature is replaced by simulation based on knowledge of behavioural generalisations and technological environments which themselves are largely endogenous. I do not expect that orthodox theory will retreat into irrelevance, but it will be supplemented by what will become more practically useful tools which satisfy the scientific requirement of economy of theoretical structure and empirical content in excess of that held by the alternatives.

Notes

1. It should be mentioned that neoclassical endogenous growth theory suffers from the same problem. Silverberg & Verspagen (1994, p. 208) point out that this theory still requires certainty up to a probability distribution, and thus avoids dealing with the radical uncertainty which characterises the environment facing technologically progressive industries.
2. Xavier (1994) comes to much the same conclusion after examination of the practice of price capping

in Australia.

3. Forsyth's figures are for the gains to be made from reallocating resources from current uses to those uses indicated by reformed pricing systems in the various institutions being suggested for reform. The figure does not include improvements in the quantities of resources used (lower unemployment of labour for example), nor improvements due to expected reductions of X efficiency. This is another area which the price-cap system is not designed to affect, though it must do so, as will be argued below.
4. The full statement of policy aims includes productivity improvement, distribution of benefits to customers and prevention of dominant carrier predatory pricing, in that order.
5. The incumbent may be expected to enjoy cost advantages in some services for some time, thus were it to undercut entrants' offers to customers it would be unlikely to be engaging in the usual definition of predation, which implies selling below cost, merely taking advantage of its cost position.
6. Emphasis in original.
7. This is, of course, just another way of looking at the marginal cost/marginal revenue relation so dear to static theory and Ramsey prices in particular.
8. A unit of selection defines the individual member of the population over which evolutionary selection operates. It has a continuity in time, is not easily or quickly adaptable to a changing environment and therefore, if a successful individual, passes on to the future its characteristics, as in inheritance and survival.
9. In the simplest terms this can be represented by the capital/labour ratio, but more realistically by the vector of inputs required by the technology. It is possible that different regimes might yield the same vector of physical inputs (given that the model did not make the fine distinctions needed to differentiate further): the knowledge used, the technology, would be different. Presumably survival of both would be unlikely given a competitive struggle as the one would produce output for less input, or produce a better quality of output for the same cost.
10. Marshall's "external economies" concept was reduced to a static concept by the Cambridge economists of the 1920s and 30s, in their need to find a reason for downward sloping long run supply curves in competitive industries.
11. By industry I mean a group of firms which are technologically capable of producing each others' outputs at least given time (Nightingale, 1978).

References

Arthur, W.B., 1989. Competing technologies, increasing returns and lock-in by historical events. *Economic Journal*, 99:116–131.

Beazley, K.C., 1990. *Ministerial Statement: Micro-Economic Reform: Progress Telecommunications*. Canberra: Parliament House.

Besanko, D., Sappington, D.E.M., 1987. *Designing regulatory policy with limited information*. Chur: Harwood Academic.

Bureau of transport and communications economics, 1993. *The Australian Telecommunications Market – When does Dominance Cease?* Canberra: AGPS.

Cyert, R.M., March, J.G., 1963. *A Behavioral Theory of the Firm*. Englewood Cliffs, NJ: Prentice-Hall.

Davey, R., 1993. *Economic Regulation in an Emerging Competitive Environment*. Paper read to the communications researchy forum. Bureau of transport & communications economics, Canberra.

Dosi, G. et al., 1988. *Technical Change and Economic Theory*. London: Pinter.

Downie, J., 1956. How should we control monopoly? II. *Economic Journal*, 66:573–577.

—— 1958. *The Competitive Process* London: Duckworth.

Eliasson, G., 1991. Modeling the experimentally organised economy. *Journal of Economic Behaviour and Organisation*, 16:153–182.

Filmer, R., Dao, D., 1994. *Economic effects of microeconomic reform*, Background EPAC Paper #38. Canberra: AGPS.

Forsyth, P., 1992. *Microeconomic Reform in Australia*. Sydney: Allen & Unwin.

Gowdy, J.M., 1992. Higher selection processes in evolutionary economic change. *Journal of Evolutionary Economics*, 2:1–16.

Gravelle, H.S.E., 1982. Incentives, efficiency and control in public firms. *Zeitshcrift für Nationalokonomie*, 2(suppl):79–104.

Hodgson, G.M., 1993. *Economics and Evolution: Bringing Life Back into Economics*. Oxford: Polity Press.

Kirzner, I., 1973. *Competition and Entrepreneurship*. Chicago: University of Chicago Press.

Laffont, J.-J., Tirole, J., 1993. *Incentives in Procurement and Regulation*. Cambridge, MA: MIT Press.

Littlechild, S., 1979. What should governments do? The Taming of Government. *IEA Readings*, 21:1–19.

—— 1983. *Regulation of British Telecommunications' Profitability*. London: Department of Industry.

Metcalfe, J.S., 1986. Technological innovation and the competitive process. In: Hall, P.H. (ed) *Technology Innovation and Economic Policy*, (pp. 287–306). Oxford: Philip Alan.

—— 1988. Evolution and economic change. In: Silberston, A. (ed) *Technology and Economic Progress*, (pp. 54–85). London: Macmillan.

Mokyr, J., 1990. Punctuated equilibrium and technological progress. *American Economic Review, Papers and Proceedings*, 80:350–354.

—— 1991. *The Lever of Riches*. Oxford: Oxford University Press.

Nelson, R.R., Winter, S., 1982. *An Evolutionary Theory of Economic Change*. Cambridge, MA: Harvard University Press.

Nightingale, J., 1978. On the definition of market and industry. *Journal of Industrial Economics*, 27:31–40.

—— 1990. Regulating telecommunications pricing: information requirements of the CPI – X Rule. *Prometheus*, 8:80–95.

Penrose, E.T., 1959. *The Theory of the Growth of the Firm*. Oxford: Basil Blackwell.

Richardson, G.B., 1960. *Information and Investment*. Oxford: Oxford University Press.

Scherer, F.M., Ross, D., 1990. *Industrial Market Structure and Economic Performance*. Boston: Houghton Mifflin.

Schumpeter, J.A., 1943. *Capitalism, Socialism and Democracy*. London: Allen & Unwin.

Silverberg, G. et al., 1988. Innovation, diversity and diffusion: A self organisation model. *Economic Journal*, 98:1032–1054.

Silverberg, G., Verspagen, B., 1994. Collective learning, innovation and growth in a boundedly rational, evolutionary world. *Journal of Evolutionary Economics*, 4:207–226.

Webb, L.R., Wilson, P.S., 1982. Microeconomic policy implementation: theory and practice, In: Webb, L.R., Allen, R.H. (eds) *Industrial Economics, Australian Studies*, (pp. 54–66). Sydney: Allen & Unwin.

Wilson, T., 1984. The microeconomic basis of microeconomic policy. In: Wilson, T. (ed) *Inflation, Unemployment and the Market*, (pp. 213–231). Oxford: Oxford University Press.

Xavier, P., 1994. Price-cap regulation for telecommunications – How has it performed in practice? Paper read to the *10th Biennial Conference of the International Telecommunications Society*, Sydney.

Part IV: Growth and development

Beyond Competition: The Future of Telecommunications.
Edited by D. Lamberton.

CHAPTER 16

The changing telecommunications environment in CEE

Nikola Holcer
The World Bank, Washington, DC, USA

Introduction

Since modern telecommunication infrastructure is considered to be vital to the emergence of market economies, most CEE countries have made the development of telecommunications a priority. During the past 4 years, as the Eastern European countries moved toward market economies, their telecommunications sectors have undergone dramatic changes. The reorganization of each country's sector, in attempts to rationalize operation, improve efficiency, stimulate competition, and attract foreign investment, touched upon all three segments of the industry: legal/regulatory environment, network and services modernization and expansion, and the manufacturing industry.

The structural changes in the legal environment include the separation of regulatory authority and operation, separation of the post from telecommunications as well as introducing some sort of competition, mainly in value-added services, mobile telecommunications, etc. New laws are being introduced, gradually liberalizing the sale of telecommunications equipment, the provision of services, and encouraging foreign investment, in order to set the stage for the eventual privatization of the state-owned operators and the state-owned manufacturers.

The transformation on the enterprise level is centered around the commercialization of the operators, by changing the public utility mentality to a more customer-oriented culture. At the same time, demand satisfaction and the modernization and digitalization of the analog network is a high priority. However, due to their limited financial resources and inexperience with modern telecommunications technology and management, the operators are ill-equipped for this demanding task. Foreign assistance is therefore needed.

The national manufacturing industries, which in the past produced and supplied about 80–90% of the required equipment and material, were not prepared for the dramatic changes following the political shifts at the end of 1980s. Deprived by COCOM of the latest digital technology, the telecom industry in the Eastern European countries was technologically outdated and not prepared for an emerging

Address for correspondence: Nikola Holcer, Telecommunications Division, Central Europe Department, The World Bank, 1818 H Street, NW, Washington, DC 20433, USA. Tel.: +1-202-473-5563. Fax: +1-202-477-0301.

competitive market.

In order to survive in the new competitive environment after the collapse of the Soviet market, the industry had to downsize, modernize, create joint ventures with the multinationals or be sold to them. At the same time a new breed of small, entrepreneurial start-ups were emerging to take advantage of the opportunities created by the liberalization, especially in some of the market niches.

The World Bank was among the first to provide assistance, both financial and technical, in the difficult process of sector transformation. With five loans to the four countries, three sector reviews and other technical assistance, a process of transformation has been initiated, which in the case of Hungary could well be terminated since a successful process of privatization was recently completed. Based on our understandings in the following paragraphs, we will review the strategies, achievements and future plans in the four analyzed countries in the important area of telecommunications manufacturing industry.

Changes in the legal and regulatory environment

Hungary

Hungary was the first of the CEE countries to begin with the difficult process of reforming the TC sector. The aim of the reform and liberalization was to increase the efficiency of the sector, eliminate direct government control of the operator, prepare for the operator's eventual privatization, and introduce limited competition in a multi-operator environment.

Prior to 1989, the regulation of the TC sector and operation of the network were both handled by Hungary's state owned PTT, Magyar Posta. In January, 1989 the Ministry of Transport, Communications and Water Management (MTCW) was formed and started to exercise its policy making and regulatory authority over Hungary's TC sector. MTCW's major functions were and are: to draft telecom regulations; to formulate and monitor compliance with the telecom policies and objectives; and to award operating concessions. The Postal and Telecommunications Main Inspectorate (PTMI) and the Frequency Management Institute (FMI) were two subordinated expert bodies which were later integrated into a single organization called the Chief Communications Inspectorate (CCI).

Hungary committed itself to the liberalization of the TC sector before any of its neighbors. However, the new telecommunications law had been under consideration for over 2 years, and went through numerous revisions and redrafts, before finally being passed on November 23, 1992. Further clarification of the law was provided by the Telecommunications Policy Statement issued by the MTCW in February, 1993. The new telecommunications law went into effect in July, 1993.

The new telecommunications law outlined the service segments which were to be liberalized.

1. HTC (Hungarian Telecommunications Company) will retain a monopoly over long

distance and international services for 8 years after privatization.

2. Concessions will be awarded for the operation of local networks, public mobile radio, nationwide public paging, national and regional distribution and broadcast of public radio and television programs.
3. Licenses are required for the operation of all other public services, including the provision of nationwide datacommunications networks.
4. Private networks limited to private property may be established without a license.

The enactment of the new law allowed the privatization of HTC and the liberalization of the service market to move forward. In the fall of 1993, MTC and AVRT (the state holding company) as the regulator and owner of HTC issued a tender for the privatization of 30% of HTC's equity. The idea was to award a national concession to the new company, which was to be owned partly by a strategic investor. In December, 1993 the award was announced and the respective documents signed. The winner was a consortium of Deutsche Telecom of Germany and Ameritech of USA. The new company MAGYARCOM was formed.

For a price of US$875 million (of which about US$400 million went to HTC while the rest was the concession fee and AVRT's share), the concession given to MAGYARCOM provided for:

1. An international and domestic long distance public telephone service, and to provide local public telephone service in 29 primary areas on an exclusive basis for 8 years, and on a nonexclusive basis for a further 17 years;
2. A possible extension of a National Concession by a further 12.5 years;
3. A minimum direct exchange line growth of 15.5% annually for the first 6 years, subject to actual demand;
4. From 1997, meeting 90% of subscriber demand within 6 months in each primary area and all demand within 12 months of receipt of applications;
5. Minimum quality of service standards;
6. Payment of an annual concession fee of 0.1% of revenue;
7. A minimum Hungarian content in products and services will be 25% from the commencement of the National Concession, rising to 50% from 1997 onward; and
8. Transaction documents to provide for investor influence over the day-to-day operation of HTC.

Simultaneously with the announcement of the privatization, the MTCW invited municipalities to submit a request for concessions to be granted in their primary areas. Twenty-five of Hungary's 54 primary regions met the MTCW's requirements for the issuance of local operating concessions. Twenty-nine primary areas were reserved for HTC. From the 25 areas put up for formal tender, 10 contracts have been awarded as of March 1, 1994 to HTC, and 15 to independent (private) companies. The process could not be delayed too much as the telecom law required that HTC operate on a concession basis after April 30, 1994.

In conclusion, it can be said that Hungary's approach to liberalization was somewhat slow but much more structured than in other CEE countries. Hungary accomplished the opening of its telecom sector with extensive and explicit regulatory activity by issuing detailed regulations and rules for building out, service obligations, licensing terms, tariffs, interconnection arrangements, etc. There are, therefore, reasons to expect progress in private sector investment and some sort of limited competition in the next few years which should result in further sector development.

However, some concern may be raised by the following:

1. The Telecommunications Policy has not yet been adopted by the Parliament;
2. The Telecommunications Reconciliation Council and the Telecommunications Engineering Qualifications Committee have not yet been formed; and
3. There is no independent regulatory authority, and the task given to the Ministry as a regulatory body by the Telecommunications Act would need experienced staff, which is at present at short supply. The shortage of human resources could hinder the fulfillment of this important regulatory function.

Poland

Through a structural reform process in 1987, a self-financing corporate entity Polska Poczta, Telegraf i Telefon (PPTT) was formed within the Ministry of Transport, Navigation and Communication (MTNC). At the same time operation was separated from the policy and regulatory functions of the Government. In 1989 the policy making and regulatory authority for post and telecommunications was transferred from the MTNC to the newly created Ministry of Post and Telecommunications (MPT).

The new Ministry drafted a new communications law which was presented to the Parliament in November, 1990. With the new Post and Telecommunications Act (PTA) passed by the Polish Parliament on November 1990 and effective as of January 15, 1991, Poland become the first country in the CEE region to liberalize the telecom services and equipment market and open its telecom sector to private participation. Under the new law:

1. TPSA retained monopoly service over the international communications market.
2. The long distance market is liberalized but foreign participation is limited to 49%.
3. Local network services are to be completely liberalized and any company may build and operate a local network. Local operators may be 100% owned by foreign companies.
4. Mobile communications, VANS and information services are all to be liberalized.

In 1991 two new regulatory bodies were established; the State Radiocommunications Agency (PAR) and the State Telecommunications Inspectorate (PIT).

On January 1, 1992 the PPTT was split into two state owned companies: Poczta Polska (Polish Post), a public utility company, and Telekomunikacja Polska s.a. (TPSA), a joint stock company owned by the State Treasury. Although Poland has moved rapidly and was ahead of its neighbors in liberalizing telecom services, the privatization of the state-owned operator TPSA has not been progressing as many foreign interested parties had expected. The actual implementation of the PTA has been slowed down by a number of unresolved regulatory issues and a lack of financing. Indeed, it was inevitable that the PTA should leave some issues unresolved since it was rushed through parliament in order to quickly stimulate network development and the demonopolization of the PPTT. A rigorous regulatory mechanism was not instituted to control the awarding of the new licenses, tariffs, and interconnection policy, and to adjudicate disputes. As a result, the performance of the new licensees has been disappointing and numerous difficulties remain inadequately addressed. In response to these concerns MPT is working on some immediate amendments of the PTA, as well as on the preparation on a new draft law, which is scheduled to be presented to the parliament by late 1995.

With regard to the long distance market, MPT has not issued any licenses for the operation of the long distance network, and probably will not until TPSA has completed the construction of a modern digital overlay long distance network, which is underway, and has established itself as a profitable and modern telephone company. This is a development expected to emerge sometime in the second half of the 1990s.

As far as the local telephone operating companies are concerned, some 40 licenses were issued (for some 100 applicants). Approximately 20 licenses were issued for paging, VSAT and VANS. The actual implementation of all private operator networks has not been very successful in the past few years. TPSA and the MPT's explanation is that the new companies are not financially strong enough to invest in local operations. The operators, however, claim that the lack of an interconnections arrangement, and especially the revenue-sharing arrangements with TPSA for the long distance services, is the main reason for hesitancy in investing in local independent operations. MPT is taking urgent measures to rectify this deficiency by devising a compromise solution acceptable to all for revenue-sharing on a temporary basis. The future will show how the further development of local operators contributes to Poland's telecommunications development.

The other liberalized telecom service sector in Poland, which does not require extensive regulations, i.e., that part including mobile communications, paging, VANS, VSAT and information services, is the most dynamic in CEE, attracting sizable foreign investment and spawning new Polish concerns.

To summarize, we can say that despite the fact that Poland initiated liberalization and changes in the legal environment more rapidly, and invited privatization earlier, results so far have been disappointing due to the less structured and regulated market, and the expected development of additional services has not taken place. It remains to be seen, however, how the looser Polish approach will perform in the near future. There are promising signs, with the MPT taking aggressive corrective measures to alleviate past obstacles to the development of local operators. Similarly, more realistic development goals have been set for TPSA's development for the rest of the century (20–25% penetration by year 2000), which should be achievable.

The Czech Republic

The Federal Ministry of Post and Telecommunications (FMPT) of the former Czechoslovak Federation was the statutory authority for ownership, regulations, and administration based on a law from 1964. In 1990 FMPT ceded a substantial amount of authority to the actual operators in the two republics, Czech and Slovak, i.e., to their SPTs in Prague and Bratislava respectively. In March, 1992 a Telecommunications Law was passed. This was an amendment to the existing 1964 Law. The new law gave the Republics the power to license and regulate local telecommunications operators but empowered the FMPT to regulate the frequency spectrum, standards, to license long distance and interrepublic operators, and to conduct international relations. The law took the essential steps of: (i) ending the mandatory monopoly of the state (and its enterprises, the SPTs) and permitting the licensing of other operators; and (ii) distinguishing regulatory authority from operations of the TC network.

After the separation of the two republics on January 1, 1993, the Government of the Czech Republic adopted a general law which maintains the previous (Federal) legal environment, including that for telecommunications, until new Republican legislation can be implemented.

The new Ministry of Economy is responsible for the policy and regulatory aspects of telecommunications. There is a consensus that the Ministry should confine itself to policy and regulatory matters. However, the MOE will need to develop more expertise in the central issues that are going to dominate its regulatory functions in the coming years, i.e., (i) regulating a dominant, increasingly independent and commercially oriented TC monopoly; and (ii) encouraging competitive forces while channeling them in directions which are constructive and valuable. The health and the growth of the sector will depend on the nature and quality of regulation as much as on the telecommunications management itself.

Further development of the sector can be expected along the following lines:

1. The MOE will prepare a Telecommunications Policy to be accepted by the Government, which will then be the basis for the preparations either for the amendments to the existing law, or for the preparation of a new law, planned for enactment sometime in 1995.
2. The MOE together with the MOF will work out a Tariff Policy, including the tariff setting and changing mechanism, which will rely heavily on SPT Telecom's cost analysis, and will promote a price-cap system.
3. In the second wave of privatization, vouchers will be issued for 26% of SPT Telecom shares, the company having restructured itself on January 1, 1994 to a one-shareholder stock company. The voucher privatization of SPT Telecom is expected to be completed in November 1994.
4. The MOE together with the National Property Fund, as the nominal owner of SPT Telecom, will invite strategic investors to purchase 27% of the equity of SPT Telecom, hoping to complete the deal in early 1995. The conditions of the deal will be defined in the Information Memorandum to be issued in the spring of 1994.

In conclusion, it is obvious that the intentions of the Czech Republic are to follow the Hungarian experience, naturally in a Czech variant and hopefully avoiding the shortcomings apparent in the previous case. The big question remains, however, whether the strategic investor will feel comfortable without a functioning Telecommunications law. Furthermore, in the Czech case the huge tax incentive is missing, which was an important consideration in Hungary. All will probably depend on the two forthcoming documents: (i) the Telecommunications Policy document, which will define the sector structure, monopoly, competition, etc.; and (ii) the Tariff Policy, which should define the future tariff policy and tariff changing mechanism, including revenue sharing regulations. However, the absence of the new law and all the relevant regulations may well turn out to have a detrimental and delaying effect on hopes for a fast privatization.

Slovakia

Up to the separation of the two republics on January 1, 1993 the regulatory situation in Slovakia was the same as in the Czech Republic under the previous Federal Government. The Slovak Government also adopted a law to enable the maintenance of the previous Federal legal environment until new Republican legislation can be implemented. The Law of 1964, with amendments from 1992, is therefore still in force. The new Ministry of Post, Telecommunications and Public Work (MPTPW) is the responsible ministry. Slovak Telecommunication (ST) is the newly formed operating entity created by the integration of the former SPT Bratislava and SR Bratislava, the Slovak Radiocommunications part of the former Federal SR. The Slovak Office of Telecommunications, under the MPTPW, is responsible for the regulatory activities, including frequency management. The preparation of the new law is contemplated but no firm timetable has been set. The privatization of the sector is not imminent as the Slovak concept is to first complete the First Telecommunications Project (financed partly by the three IFIs, i.e., WB, EIB and EBRD). The Slovak approach seems to be more pragmatic, as they try to catch up with the development they missed out on in the past.

Conclusion

With a varying degrees of urgency, all four countries have addressed legislative reform of the TC sector in their respective countries. Poland was the fastest in adopting and enacting a new law in January 1991, but due to the urgency in preparing and pushing it through the legislature in order to expediently initiate changes, some weaknesses emerged later which necessitated corrective actions. Plans are present at hand to: (i) introduce some corrections in the present act by amendments; and (ii) prepare a new law, which when enacted should correct the present shortcomings. The present target date for this is the second half of 1995.

Hungary took a lot of time, more than 2 years, to adopt and enact a new law. However, many more regulations, including a decision about the tariff-setting mechanism and revenue-sharing between operators, were prepared. These actions were strongly stimulated by the political decision to privatize a part of the HTC. The simultaneous action to invite bids for local concessions to build and operate local telephone services in the country, which were awarded in March 1994, was aimed at creating a multioperator environment with some degree of competition.

In the Czech Republic, the old law was amended while the Czechoslovak Federation still existed, and work is going on to prepare, adapt and implement a new law. A fixed timetable has not yet been announced but is expected sometime during 1995. A semiregulatory body, which is a part of the Ministry of Economy, is preparing detailed regulations. The intention of the Government to partly privatize SPT Telecom is a strong incentive to speed up the completion of the legislative agenda and the creation of a modern regulatory environment, which is a prerequisite of successful privatization.

The situation in Slovakia is somewhat similar to that of the Czech Republic with regard to the preparation of a new law to replace the old Federal Law on Telecommunication. The timetable for enactment has not yet been fixed. The regulatory authority is the newly formed Office of Slovak Telecommunications, which is under the direction of the Slovak Ministry of Post Telecommunications and Public Works. For the time being, there has been no announcement, nor does there seem to be any intention, concerning the privatization of ST.

Several interesting commonalities may be observed from the above conclusions:

1. In order to successfully sell a part of the state-owned operator, a minimum set of legal/regulatory normative acts must be in place in order to offer a sufficient comfort level to the foreign buyer-investor. These include: security of ownership, repatriation of profits in foreign currencies, tariff setting and changing mechanism, minimal legal/regulatory environment regarding operations, revenue-sharing and dispute resolution. Furthermore, serious intention on the part of a Government to privatize is a strong incentive to enact and maintain a proper legal/regulatory environment. This is clearly demonstrated in Hungary's HTC sale of shares.
2. The inadequacy and insufficiency of legal/regulatory normatives have a strong detrimental effect on sector restructuring, competition and privatization. Where the law and its associated regulations are not giving a sufficient comfort level to foreign investors, private initiative is hampered, at least in that market segment which requires a clear tariff-setting mechanism, revenue-sharing and interconnections regulations. Corrective actions will be needed to rectify these shortcomings in order to encourage private capital participation in network development and service offerings from new operators. An example of this sort is Poland.
3. In the analyzed countries, the expertise in telecommunications legal/regulatory matters is either nonexistent or rudimentary. This is understandable as no market environment in the sector has previously existed. The human resources needed for this kind of expertise are hard to come by, and it takes a long time to develop them. This was, and still is, an important cause of slow action in this area. Unfortunately, the existing expertise and experience in other developed countries, when available, is very expensive, and is not always relevant. The direct adaptation and/or introduction of foreign solutions are not possible. A lot of changes have to be made, based on an understanding and knowledge of the home country's legislature. A good example is the Czech Republic.

Telecommunications manufacturing industry

General

Historically, before 1988/89, the TCMI in CEE countries was supplying and satisfying around 80–90% of the equipment and material requirements of their respective operators. It is well known and documented that the pace of network development in the past was around 4–5% annually in DELs. Such a limited growth, despite high existing demand, was the consequence of the determination of the centrally planned economies and the role of telecommunications in the society at that time. Up to the late 1970s, the discrepancies in the technology applied in CEE countries and in the rest of Europe were not so staggering. Neither was the difference in penetration so dramatic as today. However, with the introduction of digital technology in the developed countries, and the fast development of penetration and coverage, the differences became more and more visible. After the political changes in CEE countries, the problems of demand satisfaction, and through them the technological inadequacies, surfaced forcefully as the new governments started to realize the importance of TC development for their reform programs leading towards market economy. As the functioning and further development of TCMI was an important part of the TC sector development strategy, the important question was asked: what to do with the TCMI? Overdimensioned and overstaffed for local needs, technologically not adapted to the production of modern digital equipment, and organizationally and management-wise not prepared for market economy and competition, the industry was handicapped to compete with the multinational giants. Scrapping the TCMI completely was out of question, as this had been a star performer in the past; it was the industry best equipped for eventual changes since the highly qualified workforce was already in place. Another reason for sustaining the industry was the belief that the fast development of the TC sector without a domestic TCMI is not possible since these countries do not have the foreign exchange funds required to procure modern equipment abroad necessary for a massive expansion – an expansion which is needed and expected. In the following paragraphs we will briefly describe the strategies adopted by the different countries and the results achieved so far.

Hungary

Hungary had previously exported to the CEM countries and primarily to the former Soviet Union more then 60% of its total production of telecommunications equipment. The TCMI was a star performer supported by strong R&D facilities and an excellent, trained work force. Abruptly, after 1998 all of this stopped. The question before the authorities was: what to do with this oversized, vertically integrated, technologically outdated industry? Hopes to catch up with technology and keep a market share abroad were dim. The answer was painful: the industry must restructure, downsize and implement new technology in order to satisfy primarily the domestic market, and then hope for some successes in export in some niche markets. The vitally interested parties: the industry, the operator and the Government arrived, after lengthy studies, discussions and expert assistance to a pragmatic TC industry policy with the following main elements:

1. Downsize the industry and sell the viable parts to any buyers or make joint ventures with foreign partners to create viable enterprises. Give tax holidays and other incentives.
2. Restructure technologically the industry to produce marketable goods required domestically and if possible for export. Organize the industry in a commercial, competitive, and market-oriented fashion.
3. Save the R&D facilities, valuable human resources and industrial traditions, support competitiveness and commercial orientation of the R&D activities.
4. Engage in the process HTC as the main customer of the TCMI. Reserve a market share for local production.

The result of the above policy was felt in the last few years in the following way:

1. In 1989 HTC invited an international tender for system selection of its switching equipment. Two winners were to be selected, and local production of the equipment was expected. After the second year, the payment would be in local currency only. The frame contract would be for 5 years and the market shares and prices were determined and limited. The winners were Siemens of Germany with Telefongyar and L.M.Ericsson of Sweden with Muszertehnika. The obvious looser was Northern Telecom of Canada together with AT of Austria, a partnership of Kapsh and Shrak of Austria which had already installed ADS 100 digital switching equipment in northern Hungary, of more than 100,000 lines capacity, produced under license from NT. AT had a partnership with BHG, the biggest and strongest switch manufacturer in Hungary. The interesting features of the bidding were the requirement that the frame contract was to be executed by annual competition between the two winners, that the price achieved in the frame contract cannot be increased, and that the second manufacturer, who is not the winner in the annual competition must accept the winner's lower price in order to secure a contract.
2. As regards the other manufacturers, most importantly the copper cable factory and the transmission equipment manufacturers, the first was sold to Siemens of Germany, and the latter fared with different success. Some of the industries created joint ventures, other were downsized and sold to the management and workers. Some of the industry was simply closed and liquidated.
3. Several small entrepreneurial start-ups are emerging on the market, taking advantage of: (i) the opportunities, created by the liberalization of the TCMI and (ii) especially some of the market niches such as PBX and small digital exchange production, and small series of microwave equipment production. Similarly a new breed of private installation companies are emerging to serve mainly the new private operating companies but also HTC.

Poland

Poland had in the past the biggest TC manufacturing industry in the region. A sizable portion (more than 59%) was exported to CEM countries and to the former Soviet Union. After the political changes in the country and in the region, the industry under the direction of the Ministry of Industry found itself in a difficult situation, similar to that in Hungary. There was no market, the technology was outdated, and the national operator, the Polish PTT wanted modern digital equipment immediately, of good quality and at moderate prices. This was a task impossible to fulfill. However, the new Government was determined to see to it that the industry survived. After 2 years of deliberations, the concept of an industrial policy in telecommunication was jointly devised by three Ministries (Industry, Privatization and Telecommunications), and approved by the Government. The concept's main elements are as follows:

1. The government will use the buying power of TPSA to induce multinationals to produce modern digital equipment in Poland by decreeing that the Polish TC sector may use only equipment produced in Poland with a minimum 50% local content.
2. Through international competitive bidding, the Ministry of Privatization will sell the main Polish TC manufacturers to foreigners, requesting that at least 80% of the value should be bought by the best offerers.
3. No market share will be granted to the three selected switch manufacturers: instead, annual competition will determine the market share and the two second highest price bidders will have to accept the lowest price of the winning bid. This arrangement is supposed to be in effect for 6 years, after which the market will be open for competition to everybody.

As a result of the above strategy, the following three manufacturer/suppliers were selected:

1. Alcatel SESA procured Teletra and PZT, for a price of $42.7 million invested capital and $85 million in investment commitments.
2. AT&T procured Telfa, Bydgoszcz for $33.6 million capital invested and $56 million in investment commitments.
3. Siemens procured ZWUT, Warsaw and ELWRO, Wroclaw for $38.5 million invested capital and $57 million in investment commitments.

By solving the problem of the selection of suppliers, the MPT and TPSA have so far solved only half of their problems since the sources of financing the huge development program are not yet secured. TPSA by itself is able to finance only approximately half of the cost of future development. The other half is to come from multilateral lending institutions such as the World Bank, commercial credits or supplier financing. However, the Government is not prepared to issue sovereign guaranties at the level required by the borrowing needed for TPSA to be able to execute the development program. If no new, innovative ways of financing can be found, TPSA's fast development program could be delayed, if not jeopardized.

The Czech Republic and Slovakia

The selection of the local manufacturing partners for the switching equipment in the former Czechoslovak Federal Republic followed the Hungarian model. Using international competitive selection, two partners were selected: Siemens of Germany, forming a joint venture with Tesla Karlin located in Prague, and Alcatel/Sel of Germany, forming a joint venture with Tesla Liptovsky Hradok in Slovakia. The frame contract which divided the market practically in half, is for 5 years ending in 1997. As in Hungary, the contracts are to be signed annually, and were in the beginning supported by credits from KfW, the German agency for reconstruction. After the separation of the two republics, the arrangements stayed in force and the supply and installation is financed by EIB loans to the two republics. As regards the other suppliers, some of the local manufacturing plants were privatized through the first wave of the privatization plan, and more are expected to be privatized during 1994 through the second wave of the privatization. Besides the two joint ventures for switching equipment mentioned earlier, ATT and Philips formed joint ventures to market and produce some transmission and radio equipment.

Conclusion

Although achieved by different methods, i.e., in Poland through selling the existing industry to multinationals, and in Hungary and the former Czechoslovak Federation through international bidding for a switching system selection, the manufacture of the main TC equipment has been preserved and sustained. For the respective governments, such solutions were to be preferred as they preserved the employment of at least a portion of the work force in a difficult time. For the main operator the solution was convenient as it secured the standardization of the switching equipment with domestic production for internationally competitive prices, and the payment could be affected in local currency. The overall viability of the solutions may be questioned on the basis of the possibly uneconomic size of production due to the limited market and to the many players in the field. However, the decision to invest and make joint ventures was made by the multinationals and they will bear the consequences. In any case for them, the main objective is to secure a lucrative market segment in the respective countries in the long term.

Foreign suppliers, be it by direct import or by local manufacture through joint venture, have already supplanted local manufacturers in the switching, and PBX market.

As regards other than switching TC equipment, the situation is more complex and different solutions in different countries have evolved from straight buy-outs, through joint ventures to unchanged ownership and technology. Only time will tell what will be the resulting effect of the forces of the new market environment. Two important trends are already emerging: (i) falling average prices for added DELs in the respective countries. It is too early to conclude that this is solely the effect of equipment prices since the main operators in the region are undergoing serious organizational changes, which are surely contributing to cost consciousness; and (ii) a new breed of small, entrepreneurial telecom start-ups are emerging to take advantage of opportunities created by the newly liberalized telecom market. By addressing a variety of market niches, these companies are active in the PBX market, production of smaller digital exchanges and some transmission equipment. With the emergence of local operating companies as potential buyers and suppliers to the private network market, these new companies have a chance to survive and eventually flourish. All these activities have a definitive influence on the said trends toward cost reductions of investments and diversification of the suppliers.

Service provision and network development plans

Introduction

Service provision and network development was not among the priorities of the former countries of CEE. We are not going to analyze here the reasons or the functioning of the command economies. It will suffice to state that the average annual growth of the countries in question was around 4–5% in the 1980s. The penetration rate was between six and 12 main lines per 100 population (see Table 1). After the political changes at the end of the 1980s, a hugh unsatisfied demand was present. Especially critical was the demand for telecommunication services by new emerging businesses as a prerequisite for normal operation. Furthermore, for many reasons, the existing network was not operating satisfactorily, in the international and long distance traffic congestion was enormous, many manual exchanges were in operation and considerable areas of the countries were with only rudimentary telecommunications services. After the political changes and the realization that telecommunications development is critical for economic development, new strategies for telecom development emerged. With the help of the World Bank a general framework was devised and accepted for fast development of network and services consisting of the following elements:

1. An accelerated development of network and services are needed to quickly satisfy the so far neglected demand.
2. Simultaneously modernization of network must be implemented by introducing digital technology.
3. Commercial principles and modern management must be introduced in the operations, as soon as possible.

The main elements of a broadly similar strategy for network development accepted by the four countries were as follows:

1. Rehabilitation and decongestion of the existing network with minimal immediate investments.
2. Expansion and digitalization of the international network.
3. Construction of a long distance overlay digital network (DON).

Table 1. Main lines per 100 population, 1998–1997 and 2000, actual and forecast.

Actual							Forecast				
Year/Country	1988	1989	1990	1991	1992	1993	1994	1995	1996	1997	2000
Czech Rep.	15.0	15.1	15.7	16.6	17.6	19.1	21.1	22.2	25.0	27.0	35.0
Hungary	7.7	8.6	9.3	10.7	12.5	14.3	17.0	20.0	24.0	30.0	45.0
Poland	7.5	8.2	8.5	9.0	10.0	11.2	12.3	13.6	15.0	17.0	25.0
Slovakia	12.0	12.6	13.3	14.1	15.1	16.3	18.0	19.7	21.0	23.0	27.0
CEE	10.5	11.3	11.7	12.6	13.8	15.2	17.0	18.9	21.0	23.5	30.3

The expected results of such a strategy were to:

1. Relieve congestion on the most congested portion of the old network (the long distance system) and thereby improve service.
2. Immediately improve service for business by providing them with modern digital service including international access.
3. Lay a proper foundation for a fully modern digital network in the future.
4. Increase the revenue generating capacity of the system.

The following paragraphs review the plans, and the results so far, country by country.

Hungary

Hungary was the first to start with an aggressive network and services development already in 1987. Supported by two World Bank loans it doubled its penetration rate in 6 years (see Table 1), constructed a modern digital overlay network which reaches with digital nodes and modern digital transmission equipment all its 54 primary centers. Simultaneously it expanded its international network with new fiber optic transmission media, digital international gateway exchanges and Intelsat and Eutelsat satellite communications. Beside the development of the DON and international communications, local network development in urban areas was targeted for fast development. As the politically sensitive development of the less developed areas could not be neglected, a village telephone development program was developed and implemented albeit with some delay, mostly due to the limited financial resources. HTC's historic growth over the past 30 years was some 5.6% annually. The Three Year Plan (1991–94), a US$1.4 billion program was supported by the WB's First Project (US$150 million), EIB (US$100 million), and EBRD (US$130 million).

The program concentrated on the construction of the DON which connects the 45 digital switching modes in the primary areas, extension of the international network and the expansion of the local network. The actual growth target for local lines was 14.5% per annum, higher than the 13.0% planned. While between 1980–1990 an average between 25–50,000 lines were connected, some 500,000 lines were connected between 1991 and 1993 (130,000 in 1991, 150,000 in 1992 and more than 200,000 in 1993). Although the DON was completed by some 6 months delay, it was a precondition to a functioning long distance network and a high speed interconnection network for the future local operators. A remarkable improvement in quality was achieved by increasing the long distance call completion rate from 70% in 1990 to 85% in 1993. As regards VANS – Hungary was the leader in CEE, analog cellular network was cut over in November 1990, Packet switched service (Capital 25) was available from 1989 and four VSAT providers are in operation. Videotex was available from 1989, e-mail from 1990, and paging services from 1984.

Already before the privatization HTC developed its long-term development strategy with an Accelerated Development Plan under which the target of 30% penetration for year 2000 would be achieved in 1997. This would entail a growth rate of around 22% per annum. With the resources obtained through the privatization, this target seems obtainable. Detailed plans have yet to be announced by HTC and Magyar-Com consortium, but preliminary indications for the targets for the year 1994 are already known, i.e., 250,000 new main lines.

The overall long-term strategy worked out by HTC just before privatization was: by 1997 achieve the quantitative leap by closing the gap between demand and supply; in year 2000 achieve the qualitative leap by providing services (quality wise and versatility) comparable to the European average. These goals seem more achievable with the strategic partner, who with capital, management and technology know-how, will support the otherwise successful HTC in future ambitious development plans.

As a conclusion we may say that the Hungarian telecommunications have achieved an extraordinary development result in a short time. The future looks bright, as with the help of a strong strategic partner – the Magyar-Com consortium, fast quantitative and qualitative improvements can be expected. Simultaneously, an experiment, to be watched by many began, by awarding some 14 local licenses to independent private operators. The curious observer will have to wait a few years before judgement can be made on the results of such development: will it contribute to the disintegration of a so far homogenous integrated telecom system or will it contribute to the actual and faster telecom development in the country?

Poland

Poland is the largest country in CEE and the most dynamic both economically and as an evolving telecom market. Poland's rapid liberalization of telecom services and industry has attracted foreign investment and initiated a dynamic entrepreneurial sector. Prior to 1988 telecommunications development was not a high priority of the Government Policy so between 1981–1985 only about 600,000 new main lines were connected and between 1986–1990 750,000 were planned, however, only some 670,000 new lines were connected, and this in an environment where more than two million pending applications were registered.

After political changes, as the market economy emerged, Poland suddenly awakened to the requirements of rapid expansion and modernization. With the help of the World Bank a strategy was formulated in line with the one described earlier. An emergency development plan for 1991–93, was agreed and a First Telecom Project to the tune of 578 million was devised and financed by the World Bank

(US$120 million) and EIB (US$90 million). The aim was the decongestion of the international and long distance network by building a long distance digital overlay network (DON) and expanding the international network. Simultaneously the development of services aimed at business users was targeted. At the same period some 900,000 new main lines have been planned to be added to the network, which was slightly over achieved.

As regards new plans, MPT prepared in mid-1993 a development strategy for years 1993–2000. The Plan envisages reaching in 2000 a penetration of 25% of which one-quarter is expected to come from independent local operators.

To help implement such an ambitious but feasible and certainly necessary development, Poland asked the World Bank and EIB to prepare with TPSA a Second Telecom project for the years 1995–1997 costing some US$2.1 billion (with US$800 million foreign exchange cost). The Plan will concentrate on local network expansion (as the DON will be completed in 1995) by adding some 1.5 million new main lines to the network, of which 25% should be reserved for business and high traffic subscribers, to partially secure the necessary cash flow for such a high investment. One of the main problems of past development of Polish telecommunications were financial constraints. In the pre-1988 period, telecommunications was not allowed to develop fast as the profit generated was used for other purposes (covering the cost of postal services, etc.). After 1988, the economy and the state budget was in no position to finance faster development. In the mid 1990s the GOP is limiting sovereign guarantees to sectors such as telecom considered financially able to develop on their own. Unfortunately, commercial borrowing in Poland is limited and abroad is not easy to come by. MPT and TPSA are intensively exploring alternative means of funding development, including BOT's, joint ventures, some sort of privatization, etc. The Bank was requested to help in arranging foreign and local commercial financing without sovereign guaranties. The resolution of this question will be a decisive factor in the realization of the MPT's Strategic Plan in the second half of the 1990s.

In conclusion, one may say that the dynamic development of the Polish telecom sector starting with 1990 has electrified both the private entrepreneurs and TPSA. The present complex political and economic situation influences changes in the sector which are positive, however, the results will depend on resolving the financial constraints, which could, if not alleviated, slow down the sector development.

Czech Republic

Before the separation of the Czech and Slovak Republics on January 1, 1993, the Czechoslovak Federations network and service provision was the most developed among the four analyzed countries (see Table 1). Furthermore the Czech Republic was more developed than the Slovak, already before the split. Similarly SPT Praha (the operator in the Czech Republic) was one of the more efficient operators in the region with 13.7 staff per 1000 DELs. Growth rate was, however, modest with some 5.2% annually between 1989–1992. With the help of the IFI's (WB, EIB and EBRD) a development plan was prepared and was ready to be financed. However, the political changes postponed the implementation of the plans so that the First Telecom Project for the year 1993–1995 was approved only after the separation in September of 1993. The US$891.5 million Project was financed by the WB (US$80 million), EIB (US$80 million), and EBRD (US$80 million) and with SPT Telecom's own resources (US$651.5 million). The project focuses on the construction of DON, international network and local network expansion with the emphasis of satisfying business customers. The early project implementation is satisfactory. By mid-1994 seven new digital transit exchanges will be installed and 156 local digital exchange will be added by the end of 1995. It is expected that in the 3 years some 310,000 new main lines would be connected of which 65,000 will be business customers. The quality of services will be improved and new service introduced. The sizable Institutional Development Program would assist the Czech experts in mastering the new digital technology and introduce commercial practices in operations and modern management. The intended privatization will further the development. With the help of a strategic partner and part owner, SPTs will be able to close the demand supply gap, increase penetration and coverage with simultaneous production of new services so that in early years of the next century the Czech Republic will catch up with Europe in the field of telecommunications.

Slovak Republic

The development of network and service provision, before the split of the two republics in 1993 was similar with the one of the Czech Republic (see Table 1). The growth rate of main lines in the Slovak Republic between 1989 and 1992 was 6.0% annually. The Slovak Republic after the split of the two republics completed the preparation of its First Telecom Project ahead of the Czech Republic and the WB approved it in July 1993. The First Project of the Slovak Telecommunications (ST) costing US$462 million is to be financed by the 3 IFI's (each with US$55 million) and by ST's own resources (US$296.8 million). The Project will build the DON, with a new international gateway exchange and four transit exchanges, connect some 252,000 new subscribers of which 40,000 will be business subscribers. The project comprises a strong Institutional Development Program to help the Slovak experts to master the digital technology, modern management and commercial operations. Although Slovakia started from lower penetration and do not enjoy the economic advantages of the Czech Republic, we are convinced that they will catch up fast with the rest of Europe albeit with a few years delay.

Conclusion

In two of the four countries, i.e., Hungary and Poland substantial achievements in service provision and network development are already visible, as a result of the changing telecommunications environment in the last couple of years. The process in the Czech and Slovak Republics just started and the results are expected in the coming years. In all four countries the trend is toward the closing of the gap between the demand and supply in telecom services, improvements in the quality of services and introduction of new services.

In Hungary, as a result of partial privatization further acceleration of telecom development is expected to be followed with qualitative improvement and versatility in services delivered. The aim is to reach approximately the average supply and quality level of telecom services in Europe in the early years of the next century. The technical and practical solution as well as human and material resources to achieve these goals are presently at hand. In Poland, although the development has substantially accelerated and visible improvements had been achieved, the key to final solution of the problem of demand satisfaction and material resources mobilization, it seems to us, is not yet fully at hand. Two important ingredients for successful fast development are missing so far, the financing mechanism for massive expansion and quality and service improvements and the related regulatory mechanism. Both MPT and TPSA management is intensively seeking pragmatic solutions, promised to be announced in the second half of the year 1994. However, probably due to lack of continuity in top management, modern management methods, despite remarkable human capital, are not forthcoming fast enough. The situation in the Czech Republic is good as the development is accelerating, and the base on which it is built was better than in Hungary and Poland. Similarly the economic environment is more conducive for fast telecom development. Planned next step is partial privatization, which if implemented, as planned, could positively impact further fast development and quality increase and the introduction of new services. Although the whole process started later than in Hungary, the Czech telecom sector may reach its development goals, which are similar to the Hungarians, at the same time, i.e., in the early 2000s.

Unfortunately the situation in the Slovak Republic is slightly different from the Czech in several respects although great similarities exist at the same time. The penetration base was somewhat lower, the economic situation is less favorable than in the Czech Republic and the human resource side is somewhat weaker. However, this does not mean that the development is not accelerating and that the development goals would not be achieved albeit with delays of a few years relative to the Czech situation.

General conclusions

The overall conclusion resulting from the above overview could be summarized as follows:

1. The legal/regulatory environment in the four analyzed countries underwent substantial changes, toward the European model, following the "green papers" recommendations. With more or less successes and completeness, the emerging new legal/regulatory environment is supporting the needed acceleration of the service provision and network development, commensurate to the respective

transitional market economy environment. The changes which are dependent on the political will and parliamentary procedures are more difficult and slower than anybody expected, due mainly to the inexperience and lack of human resources needed for expert design and implementation of the legal/regulatory changes.

2. The network development and service provision in the four countries, although different in speed, quantity and quality point toward a clear trend of closing the demand supply gap, improving quality of services and achieving the necessary versatility of services in a multioperator competitive environment. It can be expected that by the beginning of the next century the service provisions will not be much behind Europe's average. The digitalization and expansion of the networks are progressing satisfactorily. Major problems are in the implementation of modern management practices especially in the area of finances.
3. The telecom manufacturing industry underwent dramatic changes in the last four to 5 years. A substantial part survived, be it by sales to or joint ventures with multinational industry giants like Siemens, Alcatel, LM Ericson ATT and Philips, or by restructuring and down sizing combined with management and workers by-out and/or similar arrangements. New technology and commercial practices were introduced/adopted as a precondition to survival. Orientation toward primarily supplying the local market is the order of the day. Furthermore, a new breed of small entrepreneurial telecom start-ups emerged and took advantage of opportunities created by the newly liberalized telecom market. Most importantly, valuable human resources were preserved and industrial tradition upheld in the manufacturing industry.
4. The World Bank (and the international organizations such as EIB, EBRD, PHARE, USTDA, etc.) has contributed to the sector development by meaningful and important assistance in the region. Technical assistance and sound advise in formulating the early development strategies in the digitalization of the network, initiation of sector reforms and support for restructuring, financial support to the implementation of the agreed strategies, catalytic role in resource mobilization, and introduction of ICB with its many advantages are only some of the contributions. We may consider that Hungary already graduated from the Bank assistance in the telecom sector, by completing the two projects successfully, building the DON and a successful partial privatization of HTC. The assistance of the WB is not needed anymore, except for IFC support for private investment. We hope it will be the same with the Czech Republic in a year or two and in a few years more with Poland and Slovak Republic. The changing environment in the four countries point toward positive achievements in the sector, which is already benefiting the economies and will do it more in the near future.

Acknowledgements

This paper is based on work performed for the World Bank. Permission to publish is gratefully acknowledged. The findings, interpretations and conclusions expressed in this paper are those of the author and should not be attributed to the World Bank or to its affiliated organizations.

References

The World Bank, 1990. *Hungary Staff Appraisal Report, Second Telecommunications Project*. No. 8751–HU, September 12.

The World Bank, 1991. *Staff Appraisal Report, Poland First Telecommunications Project*. No. 9272–POL, January 31.

The World Bank, 1992. *Staff Appraisal Report, Czechoslovakia, First Telecommunications Project*. No. 10449–CS, June 1.

The World Bank, 1993. *Staff Appraisal Report, Slovak Republic, Telecommunications Project*. No. 11628–SLK, May 28.

The World Bank, 1993. *Staff Appraisal Report, Czech Republic, Telecommunications Project*. No. 11627–CZ, August 16.

Ivanek, F., Nulty, T., Holcer, N., 1991. Manufacturing telecommunications equipment in newly industrializing countries – the effect of technological progress. *World Bank Technical Paper*, No. 145, August.

Ivanek, F., Holcer, N., Nulty, T., 1992. The challenges of manufacturing telecommunications equipment in newly industrializing countries. *ITS Conference*, June 14–17, Sophia Antipolis, France.

Beyond Competition: The Future of Telecommunications.
Edited by D. Lamberton.

CHAPTER 17

Telephone penetration: industrial countries vs. LDCs

David Bowles
BellSouth Corporation, Atlanta, Georgia, USA

1. Introduction

Telephone penetration rates vary widely throughout the world. Sweden, a relatively small country in both land area and population, has nearly 70 phone lines per 100 people, the world's highest penetration rate. China, with the world's largest population and a sizeable territory, has less than two phone lines for every 100 people, one of the lowest telephones to population ratios in the world.

If the telephone was invented to provide a means of rapid communication over great distance, why do countries like Sweden, with relatively few people and little geography, have so many more telephones than China, with its much larger audience and presumably greater need? The question is not new. The simple answer is – the Swedes can afford telephone service, the Chinese cannot. The simple answer, however, is less and less a complete, and therefore satisfactory, answer. This paper shows why and makes a contribution to a more complete answer.

As a first pass at explaining a country's telephone penetration rate, it is customary to look at some measure of the country's income (Fig. 1), because income is known to be highly correlated with the demand for telephone service. In the not-too-distant past when plain old (voice) telephone service (POTS) defined the extent of the industry's offering, a customer only had to know a language and have money to pay the phone bill in order to make full use of the telephone. In that era it was nearly enough to know only income statistics to gauge telephone demand. Universal service goals notwithstanding, individuals with relatively high incomes had a telephone, many with low incomes did not.

The telephone today, however, is viewed as a necessity by many and is much more versatile than it once was. The sophistication of both the network and its users has increased dramatically in recent years. And unlike the days when it was POTS or nothing, special skills or equipment are often needed to produce and take advantage of many of the telephone's new capabilities. Computers and fax machines have greatly stimulated the use of the telephone. In businesses in industrial countries, no matter how small, more than one telephone is becoming the rule, and it is not

Address for correspondence: David Bowles, BellSouth Corporation, Room 12G 06, 1155 Peachtree Street N.E., Atlanta, GA 30367-600, USA. Tel.: +1-404-249-3351. Fax: +1-404-249-4735.

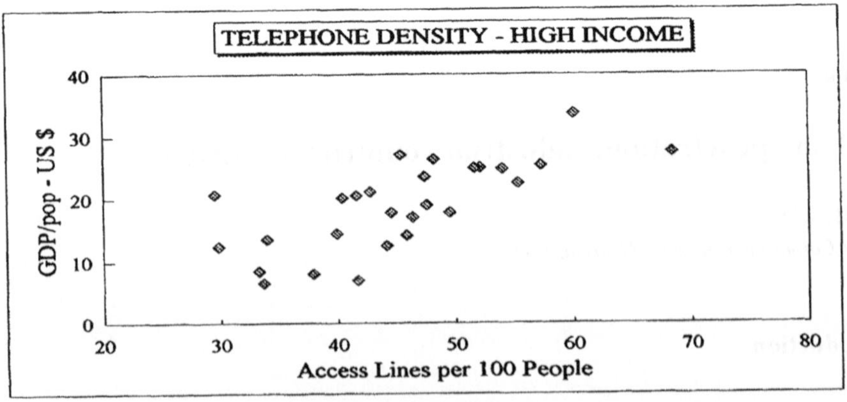

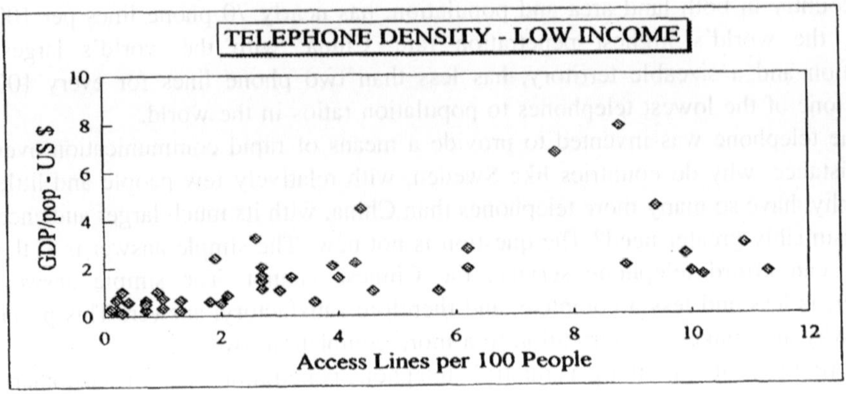

Fig. 1. High Income: Australia, Austria, Belgium, Cyprus, Denmark, Finland, France, Germany, Greece, Hong Kong, Iceland, Ireland, Italy, Japan, Korea, Luxembourg, Netherlands, New Zealand, Norway, Singapore, Spain, Sweden, Switzerland, UAE, UK, US. *Low Income:* Albania, Bangladesh, Botswana, Cameroon, Costa Rica, Cuba, Ecuador, Egypt, El Salvador, Fiji, Guatemala, Guyana, Honduras, Hungary, India, Indonesia, Iran, Jordan, Lao D.R., Lebanon, Libya, Malaysia, Mali, Mauritania, Mexico, Mongolia, Morocco, Myanmar, Nepal, Oman, Pakistan, Panama, Papua New Guinea, Paraguay, Peru, Philippines, Romania, Saudi Arabia, Sri Lanka, Suriname, Syria, Thailand, Tunisia, Zimbabwe.

uncommon to find more than one line per household. While voice messages still dominate use of the network, the data and video transmission capabilities of the system have become essential links in the operations of many businesses.

In today's more complex world, then, it seems reasonable that economic and demographic factors besides income will have significant effects on the demand for telephone service. Such relationships have, in fact, been documented. Studies by Perl (1975, 1978, 1983), Kridel (1987), Taylor and Kridel (1985), and Bodnar et al. (1988) using cross-sectional data on households in the U.S. and Canada are among many that find that socioeconomic and demographic variables affect the demand for residential telephone service. Formal work investigating the drivers of business demand for telephone service has been much more scarce.[1] Works by Antonelli (1990) and Cronin et al. (1993) on this subject find significant links between telecommunications demand and factor prices, firm size, production levels, and technology across a range of industries.

This study examines the relationship between economic and demographic variables and telephone penetration rates using cross-sectional data from 72 developing and industrial countries. It is more general than the studies cited in that it incorporates variables that might influence either residential or business telephone use, and it also considers the effect of supply constraints on telephone penetration, which can be critical in developing countries.[2]

The objectives of the study are 2-fold. The first is to learn the extent to which variables other than income have measurable effects on telephone penetration rates. The second is to determine whether these influences are consistent across economies that are at different levels of development. The answers to these questions are steps to understanding the broader issues of how infrastructure requirements differ across countries and how a country's and the world's scarce telecommunications resources should be allocated.

The remainder of the paper is divided into four parts. The next section builds a theoretical foundation of the demand for telephone service that guides the selection of the variables that are used in the empirical work. Section 3 contains a description of the variables in the demand and supply equations. Results of empirical tests of these equations are in Section 4. Conclusions are drawn in the final section.

2. The demand for and supply of telephone service

At a core level, the demand for telephone service is a manifestation of the demand for access to information.[3] The telephone itself is not an instrument of production or an object of consumption; rather, its value lies in its ability to provide information, which, in turn, increases the utility of the user. A full description of the forces shaping telephone penetration involves consideration of elements of information theory, the economics of uncertainty, and the economics of information.

Information theory makes it clear that the complexity and quality, as well as quantity, of information affect the demand for and supply of telephone service. Not all information is equally useful to all users. Error-free data transmission, for example, has no value to customers who do not have modems or do not use data. Of the information that is usable, not all is of equal quality. Some conversations, data, or documents contain mistakes or are misinformation. Usable bits of information of equal quality are not necessarily valued equally by a single user, nor are the bits of information that are of equal value to a single user valued equally by all users. A telephone network needs to be tailored in light of all these dimensions if it is to serve its purpose.

The value of the information that comes over the network depends ultimately on the extent to which it increases the utility of the user. A price quoted over the telephone could complete a transaction that might otherwise fall through. Or news of a favorable government decision might reduce the risk of an investment. In each case, information given over the telephone increases utility, but whether or not telephone service will be demanded depends on whether the utility gains outweigh the cost of service.[4]

The tradeoffs that the potential user faces can be highlighted in a slightly more formal utility maximization framework. Let U be the utility of an individual,[5] which is derived from consumption, C,

$$U = U(C) \tag{1}$$

The individual's consumption level is the result of decisions that are made with imperfect information. The telephone can raise the user's utility by providing information that either expands the individual's set of opportunities or alters the probabilities of events.

To take account of uncertainty and the benefits of the telephone, consider the consumption levels of an individual in two states of the world – with a telephone and without a telephone. Without a telephone, the relatively uninformed individual has a probability $0 < p < 1$ of attaining consumption level C_1, and, if he happens to be in the right place at the right time, a probability (1–p) of a higher consumption level, $(C_1 + C_2)$.

The individual with a telephone has superior information that allows him to make better decisions or, in effect, makes it possible for him to be in the right place at the right time. The informed individual's utility, therefore, is $(C_1 + C_2)$ with probability p, and $(C_1 + C_2)$ with probability (1–p).

The decision to purchase telephone service or not will depend on the value of the utility gain relative to the cost of information. The relevant calculation can be made by comparing the certainty equivalents of the two choices.

To do this, let t be the cost of telephone service, T denote the telephone user, and NT the individual without a telephone. The expected utilities of the user and nonuser, respectively, are:

$$EU_T = p[(C_1 + C_2) - t] + (1 - p)[(C_1 + C_2) - t] \tag{2}$$

and

$$EU_{NT} = p(C_1) + (1 - p)(C_1 + C_2) \tag{3}$$

The corresponding certainty equivalence expressions for (2) and (3) are:

$$CE_T = (C_1 + C_2) - t \tag{4}$$

and

$$CE_{NT} = (C_1 + C_2) - p(C_2) - \lambda/2(C_2)^2 p(1-p) \tag{5}$$

where λ is a measure of aversion to risk.[6]

The amount one would be willing to pay for telephone service is the amount that equates the two certainty equivalence expressions. Setting (4) equal to (5) and solving for t gives:

$$t = p(C_2) + \lambda/2(C_2)^2 p(1-p) \tag{6}$$

Equation (6) makes explicit, yet in a simple manner, the main elements of demand for access to the telephone network, namely, the potential gain associated with use of information obtained through the network (C_2), the probability of realizing the gain (p), and attitudes towards risk (λ).

The derivatives of (6) show the effects on telephone demand of changes in each of the elements.[7]

$$\frac{dt}{dp} = (C_2) + \lambda/2(C_2)^2 - \lambda(C_2)^2 p > 0 \text{ for } p < 1/2 \tag{7}$$

$$\frac{dt}{d(C_2)} = p + \lambda(C_2)p(1-p) > 0 \tag{8}$$

$$\frac{dt}{d\lambda} = \frac{(C_2)^2 p(1-p)}{2} > 0 \tag{9}$$

Equations (7), (8), and (9) show that the demand for telephone service increases as:
1. The probability of the state of the world in which gains are likely increases.
2. The size of the potential gain increases.
3. Aversion to risk increases.

In more concrete terms, the results say that users who stand to benefit the most, because of richer opportunities or higher probabilities of events that yield increased utility, and those who fear losses the most will have the greatest demand for telephone service. Equation (10) expresses this demand function in a generic form.

$$\text{tel} = t^d \text{(probalilities of expanded opportunities, benefits of expanded opportunities, attitudes towards risk, price of service)} \tag{10}$$

Specification of the supply of telephone service is more straightforward, with a few exceptions. In general, the supply of telephone service reflects the availability, prices, and productivity of capital, labor, and other inputs. Telephone supply parts company with the standard supply function to the extent that the decisions of the owners of the enterprise are not driven by profits or because suppliers have a monopoly.[8] These two conditions, plus regulation, which is usually present in some form, may distort supply/price combinations.

The features that make the telephone valuable to its users – rapid communication and the ability to provide information – may also enter the supply calculus. Some governments, for example, might view communication and the spread of information as threats to stability that should be suppressed, while others promote them in the interest of economic growth and freedom. Telephone supply, therefore, might be restricted or promoted by the government for reasons that have little to do with cost, efficiency, or resource availability.

Equation (11) is a generic form of the supply function that includes all of these elements.

$$\text{tel} = t^s \text{(price of service, capital, labor, productivity, other)} \tag{11}$$

3. Model specification

The quantitative relationships between telephone penetration and the supply and demand variables were estimated with equations (10) and (11) using data collected from World Bank, International Telecommunications Union, and OECD sources.[9]

The dependent variable in each equation is telephone penetration, which is commonly measured as the number of main access lines per 100 people.[10] In many developing countries, however, there is excess demand for telephone service, as demonstrated by the large number of unmet applications for service. Reported telephone penetration rates, therefore, are not market-clearing values.

To account for excess demand, the country's telephone waiting list was added to recorded access lines to get a truer, albeit still imperfect, measure of demand (totdem).[11] Telephone penetration in the sample ranged from a low of 0.1 lines per 100 people in Mali to a high of 68 in Sweden. Excess demand was slight, usually less than 1 percent of total expressed demand, in industrial countries, but it averaged nearly 30 percent of expressed demand in the developing countries.

The price of telephone service (price) is in both the demand and supply equations. In the empirical model, the price variable is an average of one-half of the initial connection fee plus 1 year's worth of monthly recurring charges. It is weighted to reflect both business and residential rates. In order to standardize prices to make them comparable across countries, all values are converted to U.S. dollars at market exchange rates.[12] Prices average about $180 (just over 1 percent of GDP per capita) in the industrial countries and $95 (more than 10 percent of per capita GDP) in the developing countries.

On the demand side, the price enters into the user's comparison of the expected value of the utility of information provided by the telephone and its cost. The supply price, technicalities regarding monopoly and the objective function of the provider notwithstanding, should reflect the cost of providing service. The expected signs of the coefficients of the price variables are the usual – negative in the demand equation and positive in the supply equation.

Other than the price of service there are no direct measures of the variables in (10), so, where possible, a set of instruments must be found. A usable instrument for attitude towards risk does not exist. The size of a country's stock market relative to its bond market or the popularity of gambling may yield some insight into risk preferences in a country, but there are no consistent data on these and other measures of aggregate risk across a wide range of countries. The other two generic variables in the demand equation have better candidates for instruments.

Probabilities of gains, for example, can be altered most easily by those who have the tools to make use of information. Otherwise valuable information may seem worthless to those who do not have the skill or means to interpret it. The ability to make use of the telephone to alter probabilities should be greatest in countries whose workers are well trained. Many of the uses of today's advanced telephone network require either special equipment (fax, modem, computer) or support jobs (consulting, banking, health and legal services) that are associated with higher than average levels of formal schooling. Workers' skills are measured here by an estimate of the share of the work force with secondary schooling (sch), which is a broad indicator of the sophistication of the users of telephone service.[13] The share of workers with secondary school training averages 75 percent in the industrial countries of the sample and 22 percent in the developing countries.

The value of expanded opportunities should be related to the wealth of the economy. If all participants in the market are poor and transactions are small, there is little to be gained from having information, and the incentive to acquire it will be small. Gains related to better information should also be positively correlated with the amount of economic activity in information-intensive sectors of the economy.

Considerations of wealth and the competency of workers make GDP per capita (gpd/pop) a natural demand-side variable. Income per head is a signal of the size of transactions and opportunities for gains and that the stakes are higher in an uncertain environment. These considerations suggest a strong positive relationship between telephone penetration and income per capita.

Differences in the potential gains related to the structure of the economy are evident from input-output analysis, which shows a wide range in the intensity of telephone use across industries, from light in agriculture to heavy in some types of services.[14] These differences presumably reflect the relative benefits of telephone use in each industry. The structure variables in the model are the shares of the labor force in agriculture (ag) and services (ser), the two sectors that seem to have the most distinct telephone needs.[15] Agriculture accounts for 9 percent of workers in the industrial countries in the sample and more than 40 percent in developing countries. The service sector employs an average of 52 percent of the workers in industrial countries and 28 percent in developing countries. The agriculture variable should have a negative coefficient and services a positive coefficient in the demand equation.

Foreign trade also stands to benefit substantially from the larger markets made available through telecommunication. Economies that depend heavily of foreign trade should have a higher demand for telephone service than those that have small foreign sectors. In the model, trade (trde) is defined as imports plus exports of goods and services as a share of GDP. By this definition, trade averages 100 percent of GDP in industrial countries and 80 percent in developing countries.

A final variable in the demand equation is average household size (pophh). The demand for information often occurs at the household or business unit level. The number of access lines per person, as the dependent variable is specified, would tend to fall as the average household size rises. A similar, though less transparent, relationship exists for businesses, but unfortunately the required business data do not exist for a large number of countries.

Using the instruments described above in place of the generic variables in (10), the demand equation becomes:

$$\text{totdem} = \text{d}(\underset{+}{\text{gdp/pop}}, \underset{-}{\text{price}}, \underset{+}{\text{sch}}, \underset{-}{\text{ag}}, \underset{+}{\text{ser}}, \underset{+}{\text{trde}}, \underset{+}{\text{pophh}}) \tag{12}$$

The expected signs of the coefficients are shown below each independent variable.

The translation of the variables in the generic supply equation into instruments that can be estimated is fairly straightforward. The supply equation has two variables in common with demand – labor skills, measured by schooling, and the price of service. Schooling should influence the quality of the workforce, part of which installs, maintains, and manages the telephone system. Maintenance of the system requires a reasonably high (and increasing) degree of training and technical capability. The positive relationship between potential benefits from telephone service and schooling and between efficient "production" of the network and work force quality suggests positive coefficients for the schooling variable in both the demand and supply equations.

The price of service may or may not have much to do with the quantity supplied, if the provider is not maximizing profits or competing. In most countries, for example, the telephone network is government owned, and social welfare maximization, rather than maximization of the wealth of the owners, may be the objective. Prices may include large subsidies or may be artificially high so that profits from telecommunications can subsidize other government enterprises. As a result, the output/price combinations are often different from those made by a publicly owned firm. When the state is not the owner, regulation often dictates how much service will be offered at what price. In either case, the relationship between price and quantity supplied may not be well defined, and the usual positive relationship between price and quantity may be weakly exhibited, if at all.[16]

Beyond the special price considerations, the supply of telecommunications should be much like that of any other good or service — it should be a function of the amount and quality of resources allocated to production. The labor input in the model is the share of the labor force in telecommunications (empsh). This share averages 0.6 percent in the industrial countries, and about one-half that in the developing countries.

A direct measure of the amount of capital in the telecommunications sector is impossible to obtain for a reasonable number of countries. Even if a capital stock measure could be found, aggregation problems of differences in accounting methods, measurements, vintages, and the like across countries would call its interpretation and meaningfulness into question. Because telecom production is capital intensive, however, some sort of capital variable needs to be included in the model.

One possibility is to use telecommunications investment spending. Investment data are available for a large number of countries, but are not continuous over a period long enough to construct a capital stock series. Many of the aggregation problems noted above also plague the investment data.

The alternative used here, which is admittedly a second-best solution, is access lines per employee (lnsstff), which indirectly shows the effect of the capital stock on telephone supply. An advantage of this variable is that it is consistently measured in a large number of countries. A drawback is that it reflects the influence of the quality of labor as well as the amount and quality of capital. The number of access lines per telephone company employee averages 180 in the industrial countries and only 40 in the developing countries.

Just as differences in the quality of capital and labor affect the amount of output that can be produced with a given level of inputs, network construction costs, which increase with the distance traversed, influence the amount of telephone service that can be supplied for a given level of expenditure. A larger number of customers can be connected to a network in a densely populated area than in a rural area for the same cost. This suggests that supply will be greater in countries in which the population is concentrated in a few areas or in which population density per square mile (dens) is high.

A final supply-side consideration is the possibility that governments, the usual providers and regulators of telephone services, have distorted supply by placing a high or low priority on development of the telecommunications sector.[17]

The attention given to telecommunications may be linked to the desire to promote communication and the exchange of information. Authoritarian governments may deliberately stunt telecom growth to limit communication among in-country groups or with outsiders who might stir up discontent. More democratic countries, on the other hand, may encourage telecommunications as a means of supporting freedom of speech.

To account for this possibility, an index of political freedom (free) is included as an explanatory variable in the supply function. Index values range from 1 (free) to

7 (restrictive) overall.[18] The index averages 2.0 for the industrial countries in the sample and 4.9 for the developing countries.

Substituting the variables described above for the generic elements of (11), the supply equation becomes:

$$\text{lines/100} = s(\underset{+}{\text{price}}, \underset{+}{\text{sch}}, \underset{+}{\text{empsh}}, \underset{+}{\text{lnsstff}}, \underset{+}{\text{dens}}, \underset{-}{\text{free}}) \tag{13}$$

Signs beneath each variable show expected qualitative relationships between that variable and telephone supply.

4. Testing the model

Linear versions of equations (12) and (13) were tested with 1991 data using a sample of 72 industrial and developing countries. There were several econometric hurdles to jump during the estimation. First, two-stage least squares was used to overcome the simultaneity between the price and quantity variables. Second, heteroskedasticity was also a problem. A heteroskedasticity-consistent covariance matrix was used in the calculation of standard errors and t-statistics.[19] Third, Chow tests indicated structural change in the model over the sample. Separate estimates for subsamples of 42 (countries with per capita incomes less than $4,500) and 30 (per capita incomes above $6,500) countries are given where tests with dummy variables indicated statistically significant differences in the coefficients.[20] Fourth, multicollinearity was pervasive. Its suspected effects are considered in the discussions of the individual variables. The results of the regressions are shown in Table 1. All coefficients except the constant term have been converted to elasticities.

The overall results of the regressions are encouraging. For each equation as a whole, the adjusted R^2 is very good for a cross-sectional sample, indicating that the equation explains a high percentage of the variation in telephone demand and supply. The coefficients of individual variables have plausible magnitudes and, for the most part, the expected signs.

Several general conclusions stand out. First, the results of the demand equation suggest that, in addition to price and income, the average household size and the quality of the workforce are important determinants of the demand for telephone service. On the supply side, the level and skills of workers are very important. In both equations, there are significant differences in the explanatory power of specific variables across industrial and developing countries.

The first column of Table 1 shows coefficients from the demand equation in which penetration is defined as access lines plus applications for access lines per 100 people (totdem). It shows that the relationship between telephone penetration and income remains strong. There is no statistically discernible difference in this income elasticity across high and low income countries.

The 0.4 estimate of income elasticity is in the middle of the range of income

Table 1.

Eq no.	12			13		
Dep Var	Totdem			Lns100		
Sample	High income	Total	Low income	High income	Total	Low income
price	–0.14		–0.82[a]	0.47[b]		–0.61
	(1.0)		(1.9)	(4.9)		(1.1)
GDP/pop		0.37[b]				
		(4.2)				
ag emp		–0.06				
		(0.7)				
ser emp		0.10				
		(1.3)				
schooling	0.45[a]		0.22	0.43[b]		0.23[b]
	(2.0)		(0.9)	(2.8)		(2.1)
trade	0.01		0.27[a]			
	(0.3)		(1.8)			
pop/hh	–0.24[b]		–2.6[b]			
	(2.7)		(3.0)			
tel emp/lab fce				0.93[b]		0.27[a]
				(3.3)		(1.7)
lines/employee				0.74[b]		0.49[b]
				(4.3)		(4.0)
liberty index					0.15[a]	
					(1.9)	
pop density					0.02[b]	
					(4.6)	
C	13.7		25.9	–3.9		–32.1
	(2.4)		(1.4)	(1.5)		(2.9)
Mean dep var	42.2	20.8	5.5	41.7	19.9	4.4
std error	6.7	4.8	3.6	5.9	4.4	2.4
adj R2	.76	.94	.64	.81	.95	.77
sample size	30	72	42	30	72	42

Absolute values of t-statistics are in parenthesis; [a]significant at 90% level; [b]significant at 95% level.

elasticity estimates in other studies of telephone penetration. Early CCITT correlation studies reviewed in Saunders et al (1983) and Taylor and Kridel (1990) find income elasticities at or above unity. Taylor's (1985) survey puts the income elasticity at 0.5, and Bodnar et al (1988), using aftertax household income, finds an income elasticity less than 0.1. All of these studies focus on residential demand rather than the broader concept used here. Income is correlated with both structure variables, schooling, and household size, but the statistical significance of the coefficients of these variables suggests that the collinearity is harmful only with respect to the structure variables.

As expected, the price of service has a negative effect on penetration, and, unlike income, the effect is not stable across the sample. The –0.14 elasticity for the industrial countries is small but not significant, which is consistent with estimates of the price elasticity of access in US and Canadian studies. Bodnar, for example, finds a price elasticity less than –0.1, and Taylor's survey has the price elasticity of access between –0.03 (connection) and –0.1 (monthly service). Kridel and Taylor report a small but nonzero price elasticity. None of these studies is strictly comparable to the present one. The price elasticity in developing countries is significant and much higher, but not implausibly so.

Of the two structural variables – the shares of workers in agriculture and services – both have the expected signs but high standard errors. This finding lends only weak support to the hypothesis that utility gains from telephone service are greater in economies with large information intensive sectors. Regression experiments show that the result reflects multicollinearity. It may also be due to the broad definition of services.

The trade variable, another instrument for the size of potential gains from better information that does not appear to be afflicted with collinearity problems, fares somewhat better. The sign matches the expectation in both subsamples and is significant at the 90 percent level in the developing countries.

The hypothesis that better trained workers have a greater use and therefore higher demand for telephone service is supported by the positive coefficient of the schooling variable. The larger coefficient in the industrial countries, moreover, is evidence that there are potentially larger gains from telephone service in upper income economies.

The number of people per household has a significant effect on demand in both the high and low-income countries. As with prices, the difference between the industrial country coefficient and the developing country coefficient is very large. The result is strong evidence of information demand at the household rather than individual level.

In the supply equation, the resource variables – employment share, lines per employee, and schooling – all have significant effects on telephone service. The employment elasticity shows the direct effect of the number of workers in the industry on telephone supply. The schooling coefficient is a fairly pure skill-level effect. The elasticity of lines per employee reflects a mix of the effects of capital per worker and the skill level of workers in the industry. In all three cases, the coefficient of the variable is much larger in industrial countries, which suggests that an intangible such as management may be a very important to telephone supply.

The effect of price on the supply of telephone service is fairly large in high income countries but is insignificant and has an unexpected sign in low income countries. The latter outcome is consistent with the theoretical conclusion that monopoly, government ownership, and regulation, one or more of which affect most telephone suppliers, distort the normal output/price relationship.

Political freedom has a small positive effect on telephone supply, which may indicate, contrary to expectations, that authoritarian governments may be using their influence to funnel resources toward the telecom sector rather than away from it. Regression tests, however, suggest that collinearity among the index and the employment and schooling variables may be obscuring the true relationship.

The population density has the expected effect on telephone supply but a very small coefficient. A one percent increase in population density raises telephone penetration by only 0.02 percent.

5. Conclusions

It is clear from the results of the study that telephone penetration does not have a simple explanation. A large part of the variance in telephone demand across countries can still be attributed to differences in incomes, so the short explanation of why the rich have phones and the poor do not is still correct. The estimated elasticity of demand with respect to income, however, suggests that raising the average incomes of the developing countries to equal those of the industrial countries would leave the poor countries with telephone demand less than half that in the rich countries.

The results also support the hypothesis that differences in the benefits of the telephone, which, in turn, are tied to the structures of economies and abilities of users, explain telephone penetration. The varying influence of some nonincome variables – schooling, household size – moreover, says that the level of development matters and suggests that the increasing sophistication of the telephone network and its uses affects demand.

The supply equation tells a similar story. The effects of the basic variables on the supply of telephone service – the price of service and input quantities – also differ across countries. And a seemingly secondary variable, the quality of the labor force in this case, is another important part of the explanation of supply across countries at dissimilar stages of development. Where telephone service is in short supply, resources need to be directed to improving labor quality as well as the amount of labor and capital to ensure efficient production. Throwing more workers at inadequate supply may solve only a small part of the problem if those workers are poorly educated or managed.

In general, the results say that there should be no expectation that two countries whose incomes and levels of development are nearly identical should have similar telephone penetrations. Shooting for such an outcome, moreover, could result in an under or overbuilt network. Suppliers should take into account the wealth, abilities, and economic structure and diversity of a country to ensure that the network does not offer too little or too much.

One final observation: more and better telecommunications and economic data are sorely needed.

Notes

Helpful comments were received from Emmanuel Gyawu, Rod Maddock, Bill Marxsen, Tom Schaap, and Jim Young. The paper does not necessarily represent the views of BellSouth Corporation.

1. In his review of *Telecommunications Demand Modeling*, Wolak (1993) notes the lack of work on

business demand for telecommunications. Although Wolak reviews a monograph containing papers written in the mid-1980s, little research on this topic seems to have been done since that time.

2. Although there are data separating residential and business access lines for many countries, actual business and residential use is not clear cut. Many business calls, for example, have a social reason behind them, and many residential calls have a business component. In some countries, moreover, higher tariffs on business lines give subscribers the incentive to get business phones classified as residential. Grouping the two types of line together and estimating an overall demand for service, therefore, seems reasonable.
3. The demand for information has been widely discussed in the literature. See Lamberton (1971) (introduction and papers by Boulding and Marschak) and Arrow (1984) for general views of the drivers of the demand for information.
4. Although the two examples given here apply to business situations, it is easy to think of similar examples that apply to individuals. Social use of the telephone is an exchange of information that raises the utility of individual users. Many calls from residences are for business reasons – placing orders, news gathering, emergencies – that also raise utility.
5. The problem can easily be written from a business perspective, in which case the object would be profit rather than consumption.
6. In this case, λ is simply assumed to increase as risk aversion rises. For a more comprehensive treatment of risk aversion, see Pratt (1964).

   Equations (4) and (5) are specific versions of the general certainty equivalence expression $CE = E(U_{T,NT}) - \lambda/2\sigma^2(U_{T,NT})$. $\sigma^2(T) = 0$. From the formula for the variance of a sum of two random variables, $\sigma^2(NT) = p(C_1-C_1-C_2+pC_2)^2 + (1-p)(C_1+C_2-C_1-C_2+pC_2)^2 = C_2^2p(1-p)$.
7. Technically, this is the amount one would be willing to pay rather than the quantity demanded.
8. Of the 72 countries in the sample, the main telephone provider is wholly government-owned in 58 and only 8 have competition.
9. By design, the model uses only publicly available data that have some history as well as a high probability of being published in the future.
10. In industrial countries and a growing number of developing countries wireline access underestimates telephone access because of wireless (largely cellular) access. Neglect of this fact, however, should have a minimal affect on the results, because in industrial countries, cellular phones are largely a complement to the wireline system and thus do not represent incremental access. In developing countries, cellular access is often a substitute for landline access, but cellular penetration as a share of the total was still very small or nonexistent in the countries in our sample in 1991. Illegal connections to the network are probably a bigger distortion.
11. Anecdotal evidence suggests that waiting lists probably underrepresent the true level of excess demand because of those who, knowing that the wait is extremely long and that a deposit may be required, fail to register.
12. Ideally, the average price of a call should also be included as part of the price, but it is not possible to isolate a reasonably defined call price for more than a handful of countries with any degree of confidence.
13. The literacy rate is an alternative measure, but secondary schooling seems more appropriate because use and production of advanced telephone services requires more than basic reading and writing skills.
14. See Saunders et al. (1983), pp. 87–100, for an early survey of studies of telephone intensity.
15. A narrower definition of services than the one used would be preferable, but the number of countries for which such data are available is very small.
16. The question of whether price or quantity should be treated as endogenous is a tough one in regulated industries. In practice, there is a large element of predetermination in both, while in theory, they are determined simultaneously, so both are endogenous. In the supply and demand equations estimated here, prices are treated as exogenous, but the values have been created from the first stage of two-stage least squares.
17. Stone (1993) presents evidence regarding the wide variation in the size of the telecommunications

component of government budgets across countries.

18. See Scully and Slottje (1991) for a full description of the index and methods used in its calculation. The model uses L6 from Table A.1 in that study. The index might also be an instrument of trade restrictiveness.
19. See White (1980) for a description of the test and calculations.
20. Empirical studies that use aggregate macroeconomic and demographic data inevitably suffer from aggregation problems. In this case, a line is not a line is not a line, and workers, among other inputs, are by no means homogeneous across countries. The standard criticisms, therefore, ring true. But, in addition to addressing structural change, the split sample should dampen the sting of aggregation problems by grouping like countries, whose inputs and output should have similar characteristics, together.

References

Antonelli, C., 1989–1990. Information technology and the derived demand for telecommunications services in the manufacturing industry. *Information Economics and Policy*, 4:45–55.

Arrow, K., 1984. *The Economics of Information*, Cambridge, Harvard University Press.

Bodnar, J., Dilworth, P., Iacono, S., 1988. Cross-sectional analysis of residential telephone subscription in Canada. *Information Economics and Policy*, 3:359–378.

Cronin, F., Gold, M., Hebert, P., Lewitzky, S., 1993. Factor prices, factor substitution, and the relative demand for telecommunications across US industries. *Information Economics and Policy*, 5:73–86.

De Fontenay, A., Shugard, M., Sibley, D. (eds), 1990. *Telecommunications Demand Modeling*, Amsterdam, North Holland.

Hirshleifer, J., Riley, J., 1992. *The Analytics of Uncertainty and Information*, Cambridge, Cambridge University Press.

Kridel, D., 1987. *An Analysis of the Residential Demand for Access to the Telephone Network*. PhD Thesis, University of Arizona, Tucson.

Kridel, D., Lehman, D., Weisman, D., 1993. Option value, telecommunications demand, and policy. *Information Economics and Policy*, 5:125–144.

Lamberton, D. (ed), 1971. *Economics of Information and Knowledge*, Middlesex, England, Penguin Books.

Orr, D., 1981. *Applied Econometrics*, unpublished manuscript, University of Toronto.

Perl, L. J., 1975. Economic and demographic determinants of telephone availability. In: *AT&T filing before the Federal Communications Commission*, Docket No. 20003, Exhibit 21.

Perl, L. J., 1978. *Economic and Demographic Determinants of Residential Demand for Basic Telephone Service*. National Economic Research Associates, New York.

Perl, L. J., 1983. *Residential Demand for Telephone Service 1983*. National Economic Research Associates, New York.

Pratt, J., 1964. Risk aversion in the small and in the large. *Econometrica*, 32:122–136.

Saunders, R., Warford, J., Wellinius, B., 1983. *Telecommunications and Economic Development*. Baltimore, The Johns Hopkins University Press.

Scully, G., Slottje, P., 1991. Ranking economic liberty across countries, *Public Choice*, 69:121–152.

Stone, P., 1993. Public-private alliances for telecommunications development, intracorporate Baby Bells in the developing countries. *Telecommunications Policy*, 17:459–469.

Taylor, L., 1983. Problems and issues in modeling telecommunications demand. In: Courville, L., De Fontenay, A., Dobell, R. (eds) *Economic Analysis of Telecommunications*. Amsterdam, North Holland.

Taylor, L. D., Kridel, D., 1985. Residential demand for access to the telephone network using Census Tract data. Paper presented at: *BELLCORE Telecommunications Demand Modelling Conference*, New Orleans.

White, H., 1980. A heteroskedasticity-consistent covariance estimator and a direct test for heteroskedasticity, *Econometrica*, 48:817–838.
Wolak, F., 1993. Telecommunications demand modeling. *Information Economics and Policy*, 5:179–196.

Beyond Competition: The Future of Telecommunications.
Edited by D. Lamberton.

CHAPTER 18

Growth dynamics under shifting telecommunications and trade regimes in Australia

Neil Dias Karunaratne
Department of Economics, University of Queensland, St. Lucia, Q 4072, Australia

1. Introduction

The macroeconomic dynamics of an emergent information economy can be captured through key macrovariables such as growth, trade, world economic performance, international competitiveness and telecommunications. This study aims to empirically analyse the growth dynamics underpinning the Australian economy during the period 1971Q1–1992Q4. During this study period there were revolutionary changes in information technology and a liberalisation of the world trading environment. In Australia there was a major reversal of the long standing protectionist trade policies. Furthermore, after much foot dragging the regulatory framework governing Australia's telecommunications monopoly was duopolised. The key role played by telecommunications in maintaining Australia's competitive edge in the global information economy is now widely recognised. Australia's growth and competitiveness in trade depends crucially on the cost effective supply of efficient telecommunication infrastructure and services to business and industry.

The dramatic changes in telecommunications technology (microelectronics, fibre optics, satellites, computers, microwave radio, cellular mobile telephony, digital transmission of voice, text, image and multimedia) have reduced costs and increased transmission capacities and versatility of services and products manifold. The new telecommunications technologies can bypass established networks and offer scale and scope economies making state imposed monopolistic barriers contestable (Baumol, 1982). The universal provision of POTS (plain old telephone services) by cross-subsidisation of thin routes from thick route revenues to achieve equity through government monopolies has lost its economic rationale. The rapid deregulation and establishment of competitive market structures in the world's leading information economies (USA, UK and Japan) have led to dramatic price reductions and improved customer services. In the USA the adoption of pro-competitive policies has boosted total factor productivity and reduced labour cost per unit of telecommunications by 63% within a decade (Cronin et al., 1991). Australian labour productivity in

Address for correspondence: Prof Neil Dias Karunaratne, Department of Economics, University of Queensland, St. Lucia, Q 4072, Australia. Tel.: +61–7–3365–6570. Fax: +61–7–3365–7299.

telecommunications has been laggard and there is scope for improving it by 40% to reach the best-practice input-output utilisation bench mark (BIE, 1992).

Despite international demonstration effects of the benefits of deregulation, competition and privatisation in the provision of basic and nonbasic telecommunication services there has been a reticence in Australia to keep up with the rest of the world. Public interest theories justify the monopolistic regulation of lumpy infrastructure investments with lengthy gestation periods (Stigler, 1971; Posner, 1971). It is argued, that bereft of such monopolistic market power, the spillovers or externalities create problems of inappropriability and could lead to suboptimal investment from a social standpoint. But special interest group theories would critique such arguments as a ploy to justify monopoly rents by exploiting the free-riding psychology of the unorganised consumers (Olson, 1965).

In Australia the separation of telecommunications from postal services in 1975 led to a brief spurt in innovative activity. The Davidson Report (1982), containing recommendations for pro-competitive policies in telecommunications, was put on the back burner. The Evans Report (1988) recommendations, paid lip service to the benefits of competitive market structures and deregulated value added services and customer equipment sales, but continued to reserve for Telecom the monopoly of the basic network and associated services. The Beazley Report (1990) on microeconomic reforms paved the way for a major regime change in Australian telecommunications policy. A carrier duopoly with Telstra and Optus with Austel as the regulator was to operate until full competition in mid-1997. The Telecommunications Act 1991 provided safeguards for new entrants with regard to filing of tariffs, equal access to the PSTN (public switched telecommunication network). A brief chronology of the major changes in Australia's telecommunications policy regime is presented in Table 1.

During the study period, in the arena of trade policy a major reversal of the long standing legacy of protection was initiated. There was a significant liberalisation of protectionist policies on trade and the deregulation of the financial system culminating in the floating of the exchange in 1983Q4. The regime shift in trade policy has led to a reconfiguration in the Australia's manufacturing sector. Many firms in the TCF (textile footwear and clothing) sector and the PMV (passenger motor

Table 1. Major changes in Australian telecommunications.

Date	Regime change
1975	Separation of Telecom and Australia Post
1976	Start of International Direct Dialling
1982	Davidson Report (Recommends competition)
1988	Evans Report (Network monopoly retained)
1989	Telecommunications Act
1989	AUSTEL (Independent regulator)
1990	Beazley Report (Microeconomic reform)
1991	Telecommunications Act
1992Q1	AOTC due to merger of Telecom & OTC
1992Q4	Telstra & Optus duopoly until mid-1997

vehicle) sector have been drastically revamped to withstand the fresh winds of competition after massive reductions of tariff and nontariff barriers (Karunaratne, 1992). The liberalisation of Australia's trading environment has integrated Australia's information economy firmly with the global information economy. To maintain Australia's international competitive edge in this context requires the adoption of best practice technology and competitive pricing policies. The existence of bidirectional causality between telecommunications and economic growth has been confirmed by validating Granger-Sims causality tests for the USA. It has been shown that telecommunications investment causes an increase in total factor productivity (Cronin et al., 1993). On a broader front several studies underline that investment in public sector infrastructure such as telecommunications improve the performance of the private sector in a spectacular fashion (Auschauer, 1989; Cullison, 1993). There is widespread recognition that the competitiveness of Australian enterprises in this information age will depend on the efficient supply of state of the art telecommunications services and inputs (BIE, 1992). The convergence of telecommunications and computers and the delivery of multiproducts through ISDN (integrated services digitial networks) has blurred the distinction between goods and services. The burgeoning trade in information services and goods has created new pressures for the establishment of a GATT like framework to regulate the international trade in information services (Karunaratne, 1993). The globalisation or the growing international integration of the information economies of the world has created a new organisational and techno-economic paradigm (Dosi et al., 1991; Lipsey, 1992). In this new techno-economic paradigm the sharpening of Australia's competitive edge by adopting a pro-competitive regulatory stance in telecommunications appears to be imperative. Inappropriate policy stances in trade and telecommunications would cause irreparable damage to Australia's long-term growth dynamics.

The aim of this study is to review the dynamic interactions of growth, trade and telecommunications in the emergent Australian information economy during the period 1971Q1–1992Q4. The study is organised as follows: Section 2 analyses the stationarity and cointegration or long-run equilibrium properties of the time-series capturing dynamic interactions between growth, trade and telecommunications in the context of international competitiveness. Section 3 uses vector autoregressive (VAR) model innovation accounting to measure the strength of the dynamic interactions between the prime macro variables in the Australian economic system. Section 4 derives a conditional growth model or an autoregressive distributed lag (ADL) model to parsimoniously estimate the meta parameters underpinning Australia's growth dynamics using the general to simple methodology. Section 5 reports on the superexogeneity test results and reviews the implications of regime shifts in trade and telecommunication regulatory policies on Australia's growth dynamics. Section 5 presents the concluding observations of the study.

The data base of the study uses seasonally adjusted quarterly time-series for the period 1971Q1–1992Q4. The growth rate has been proxied by the change in GDP per capita (y_t), trade has been proxied by changes in the value of exports per capita (x_t), world production has been proxied by the changes in the OECD production index

(w_t), trade-weighted exchange rate (e_t), terms of trade (v_t) and change in telephone penetration or main lines per capita proxied expansion of network capacity and information technology generally (z_t). These macrovariables have been used in the study to capture the stochastic processes that underpin the interactions between growth, trade and telecommunications in Australia. The definitions and sources of the data used in the study are given in Appendix 1. In empirical modelling the lower case letter notates the natural logarithm of the variables that have been used in the study. The econometric analyses of the study are based on the use of several computer software packages: PC GIVE 7.00 (Doornik & Hendry, 1992); RATS 4.0 (Doan, 1992); MICROFIT 3.0 (Pesaran & Pesaran, 1991); SHAZAM 7.0 (White, 1993).

2. Unit root and multicointegration tests of growth-telecommunications dynamics

Insights on long-term properties of macroeconomic time-series require the use of stationary data. Both parametric Augmented Dickey Fuller (ADF) (Dickey & Fuller, 1981; Fuller, 1976) and nonparametric Phillips-Perron (PP) (Perron, 1988; Phillips, 1987; Phillips & Perron, 1988) tests have been used to test for stationarity or unit roots. The test results reported in Table 2 indicate that most of the time-series have unit roots and would have to be differenced to induce stationarity.

Table 2. Unit root test results.

y_t (time-series) and

$$\Delta y_t = \beta_0 + \beta y_{t-1} + \sum_{i=1}^{k} \Delta y_{t-i} + \epsilon_t$$

k = lag length required to achieve white noise residuals

Time series	ADF stat.	min AIc (k=lags)	PP stat.	Truncation lag
Growth pc: y_t	−2.47	−8.12 (k=2)	−4.08	2
Trade pc: x_t	−2.64	−6.16 (k=2)	−3.04	2
World: w_t	−3.13	−1.05 (k=1)	−3.14	1
RER: r_t	−2.39	−5.68 (k=7)	−2.45	7
Tele: z_t	−2.22	−2.22 (k=1)	−2.23	1
95% CV	−3.13			

The macrovariables proxying the growth, trade, world production, competitiveness and telecommunications dynamics can be defined by the vector $X_t = (y_t, x_t, w_t, r_t, z_t)$ and referred to as the growth dynamics vector for brevity. The unit root test results indicate that the all the time-series are nonstationary or I(1) processes. Therefore, there could be several bivariate cointegrations linking the macrovariables (Engle & Granger, 1987). However, such bivariate cointegrations are vitiated by small sample bias (Banerjee et al., 1986) and fail to provide insights on the multivariate links in the data. The Johansen procedure (Johansen, 1988; Johansen & Juselius, 1990) facilitates the testing for the number of cointegrating vectors spanning the multivariate space.

The Johansen procedure starts with a p variable VAR of order k VAR(k) which can be defined as:

$$X_t = \mu + \sum_{i=1}^{k} \pi_t X_{t-i} + \epsilon_t$$

where π_{pxp}, X_{px1} are I(1) processes. The dimensions of the matrices are shown in subscripts, and μ and ε_t are conformable vectors of constants and Gaussian errors. The ECM (error correction mechanism) representation (based on the Engle-Granger Representation Theorem) is given below:

$$\Delta X_t = \mu + \sum_{i=1}^{k-1} \Gamma_i \Delta X_{t-i} + \pi X_{k-1} + \epsilon_t$$

where,

$$\Gamma_i = -I + \sum_{i}^{k-1} \pi_i \text{ and } \pi = -I - \sum_{i}^{k} \pi_1 .$$

The ECM representation is composed of a stationary or a first difference component and a nonstationary component πX_{t-1}. If there is cointegration, the matrix π can be factorised so that: $\pi_{pxp} = \alpha_{pxr} \beta'_{rxp}$.

Under certain conditions the reduced rank r < p of the linear combinations $\beta' X$, where X denotes stationary processes, gives the number of cointegrating vectors. The maximum likelihood tests statistics: λ_{max} and Trace statistics can be computed using Microfit 3.0 software to test for the number of cointegrating vectors in a given VAR. The multicointegration test results for the five variable VAR for Australian growth dynamics is reported in Table 3. The test results suggest that there are at most two cointegrating vectors according to the λ-max-statistic and one according to Trace statistic. Therefore we conclude with 95% confidence that there is at most one long-run equilibrium or cointegrating relationship linking the variables describing Australia's growth dynamics.

Table 3. Johansen multicointegration tests.

Null (r)	λ_{max}-Stat	95% CV	Trace-stat	95% CV
(0)	71.11*	33.46	125.40*	68.52
(1)	28.79*	27.07	54.29	47.21
(2)	15.58	20.96	25.50	29.68
(3)	6.51	14.07	9.91	15.41
(4)	3.41	3.76	3.41	3.75

Eigen values in descending order: .5711, .2902, .1693, .0745, .0397.

3. VAR innovation accounting of growth-telecommunication dynamics

The VAR approach can be used to account for the effects of an innovation (shock) effect of one variable on other variables in the system. The Simultaneous Equation Modelling (SEM) methodology has been used to explore the symbiotics of trade and telecommunications in an international context pooling time-series and cross-section data (Karunaratne, 1988). However, the SEM approach has been criticised as "incredible" because it requires the imposition of zero restrictions, dichotomisation of variables into endogenous and exogenous based on a priori theory, and cumbersome estimation to overcome simultaneity bias problems (Sims, 1980). The VAR methodology overcomes some of the limitations of the SEM methodology and it is based on the philosophy that everything depends on everything else.

The VAR analysis assumes that the growth dynamic vector $X_t = (y_t, x_t, w_t, r_t, z_t)$ can be represented in an autoregressive (AR) form as: $\phi(L)X_t = \varepsilon_t$ where ε_t is a vector of Gaussian errors process and $\phi(L)$ is a matrix polynomial in degree p with L as the lag operator. The moving average (MA) process for innovation accounting is obtained as: $X_t = H(L)\varepsilon_t$, when invertibility $H(L) = \phi(L)^{-1}$ prevails (Judge et al., 1992). Innovation accounting requires orthogonalisation to rectify the contemporaneous correlation errors in the covariance matrix. This has been accomplished by using the Choleski decomposition technique available in the RATS software. The independence of the resulting error variances facilitates the forecast error variances (FEV) and impulse response function analyses.

The FEV (forecast error variance) decompositions show how forecast error variance of a specified variable can be apportioned to the different variables comprising the VAR system. The magnitude of the forecast error variance indicates the causal strength of an innovation. The large FEV magnitudes indicate a high degree of endogeneity and strong causal effects and small magnitudes indicate weak causal and a high degree of exogeneity of the variable in the innovation process. The FEV decomposition for 12 and 24 quarters ahead forecasts are reported in Table 4 for the Australian VAR model.

The first column of the above table gives the standard error of the forecast. Reading horizontally the other variables indicate the magnitude of the innovation that is explained by the variable. For the purpose of this study FEV less than 5% are regarded as weak, 5–15% moderate and over 15% strong. The 4-period ahead forecast of FEV is regarded as the short run and the 24 period ahead forecast of FEV is defined as long run.

In the case of growth (y_t) in the short run nearly 70% of the innovation is accounted for by growth itself and trade (x_t) accounts only for a moderate 6.5% and telecommunications (z_t) account for a meagre 2.3% of FEV. However, in the long run the innovation accounted for by growth itself decreases to 44% of the total FEV approximately. In the meantime the FEV accounted for by trade nearly trebles to 18.7% and that of telecommunications becomes a moderate 10% of the total FEV. Thus the contribution of trade and telecommunications to the growth innovation process increases over time starting from small beginnings. In the short run the fact

Table 4. Forecast error variance (FEV) decomposition.

Step	se	y_t	x_t	w_t	r_t	z_t
y_t4	91.8	69.7	0.5	8.8	10.9	10.1
12	127.9	50.0	8.9	6.7	10.6	23.7
24	157.2	43.6	11.7	7.6	7.2	29.9
x_t4	431.9	6.5	79.6	7.2	3.6	4.1
12	560.8	13.5	61.4	7.8	6.4	10.9
24	685.9	18.7	45.9	7.3	4.7	23.3
w_t4	3409.6	5.3	3.3	82.8	5.9	2.6
12	5202.8	3.0	13.4	68.7	9.5	5.2
24	5395.6	3.2	14.8	65.1	9.9	7.0
r_t4	600.9	15.0	1.8	7.9	74.5	0.8
12	794.6	10.0	17.4	13.6	54.0	4.3
24	862.8	10.4	16.0	17.9	51.3	4.4
z_t4	9.3	2.3	1.2	7.7	10.0	78.8
12	15.3	2.4	3.5	28.8	23.9	41.4
24	17.4	9.6	4.0	28.2	21.9	36.2

growth is propelled by itself is consistent with the new or endogenous growth theories that foreshadow that human capital and skill levels are main propellers of growth dynamics (Lucas, 1988; Roemer, 1990). However, in the long run other variables such as telecommunications technology as postulated by the neoclassical Solow growth theory seem to play an important role in Australian growth dynamics (Solow, 1956) according to the VAR results in Table 4.

In the case of trade (x_t) innovations or shocks are in the short run (80%) mostly explained by trade itself while growth and telecommunications account only for 6.5% and 4.1% respectively. However, in the long run the innovation accounted for by trade itself declines to 46% of the total FEV while the share of growth and telecommunications rises to moderate levels of 19 and 23% respectively. Furthermore, it is noteworthy that in the innovations associated with the world economy and international competitiveness trade accounts for nearly 15 and 16% of the total FEV respectively in the long run. The opening-up or liberalisation of an economy could have strengthened the causal effects on growth through access to the stock of world technology and knowledge (Grossman & Helpman, 1990). The VAR analysis indicates that the process of trade liberalisation in Australia had positive effects in relation to telecommunications and growth not only in the long run but also in the short run when effects were negligible (see Table 4).

Nearly 79% of the innovation in telecommunications (z_t) in the short term could be explained by itself while growth and trade accounted for 10 and 4.1% of the total FEV, playing only a moderate role. However, over the long term the total innovation accounted for by telecommunications itself declined by more than half accounting for only 36.2% of the total FEV, while the share of innovation accounted for by trade and growth in telecommunications innovation process rose to strong levels as 30% and over 23% of the total FEV respectively. Telecommunications arguably is a good proxy for high-tech in an information economy. Strategic trade theories contend that high-tech sectors such as telecommunications could be promoted by protecting the domestic firm against the foreign rival. However, whether such strategic postures will increase productivity and efficiency in the long run will depend on the absence of retaliation by foreign rivals. Also on whether the domestic and foreign duopolists will engage in Cournot (quantity fixing) passing on the benefits to consumers rather than Bertrand (price fixing) policies resulting in enhanced monopoly profits (Brander & Spencer, 1983; Krugman, 1986).

The VAR analysis indicates that although in the short term the innovations process is mostly accounted for by the variable in question, over the long run the other variables in the system emerge as strong explanators in the innovation process. The dynamic interactions of growth trade and telecommunications as captured by the FEV clearly indicate that the contribution of telecommunications to the growth and trade performance increases significantly over the long period. The impulse-response functions (not reported) further corroborate the increasing importance of the other variables in a specified innovation process in the long run. Furthermore, although in the short run much of the innovation appears to be accounted for by a variable's own FEV, in the long run this endogeneity seems to erode. In the long run the FEV of other variables increase dramatically indicating that innovations are caused by the increasing interplay of other exogeneous variables in the system.

The VAR analysis indicates that any innovation of the variables in the system in the long-run results in a virtuous circle wherein the contributions of other variables increase. In particular, innovation or technological change in telecommunications has the potential to make sizeable contributions to growth and trade in the long term rather than in the short term. Infrastructure investment in telecommunications has a long gestation period and therefore their dynamic multipliers grow over time as shown in the VAR analysis above (Table 4).

4. Dynamic analysis of the conditional growth process

The general model or the ADL (autoregressive distributed lag model) explaining growth in terms of: trade, world conditions, competitiveness and technology parameters, using quarterly time-series, can be formulated as:

$$y_t = \beta_0 + \sum_{i=1t}^{4} \beta_{1t} x_{t-i} + \sum_{i=1}^{4} \beta_{2t} w_{t-i} + \sum_{i=1}^{4} \beta_{3t} r_{t-i} + \sum_{i=1}^{4} \beta_{4t} z_{t-i} + \in_t$$

The long-run static solution of the above ADL model yielded:

$$y_t = 6.4560 + 0.0026x_t + 0.2947w_t - 0.0174r_t + 0.1995z_t$$

Wald $\chi^2(4) = 336.28$.

Over the long haul it appears that world economic performance, telecommunications penetration and trade, in that order of importance, exerted positive effects on growth. The appreciation of the real exchange rate or decrease in competitiveness seemed to have exerted an adverse effect on growth. The behaviour pattern of the real exchange rate and Australia's international competitiveness discerned in this study broadly accords with the findings in other studies (Blundell-Wignall et al., 1993).

Conditional growth model for Australia

The general to simple modelling strategy (Hendry & Richard, 1983) reduced the ADL model using LM (Lagrange Multiplier) restrictions to a conditional growth model to capture the growth dynamics of the DGP (data generation process) in a parsimonious manner. The unit root null on the residuals of the OLS regression of growth (y_t) on competitiveness (r_t) was rejected. This signalled the existence of a cointegrating or long-run equilibrium relationship and an ECM proxied by the lagged residuals of the OLS regression of y_t on x_t according to the Engle-Granger two-step procedure (Engle & Granger, 1987). In the first step, the cointegration regression linking the two I(1) variables were estimated as:

$$y_t = 0.9677 - 0.2848r_t$$
$$\text{se} \quad (.0752) \quad (.1727)$$

Diagnostics

$R^2 = 0.76$ RSS = 0.23 DW = 0.25 Reset F(1,85) = 0.29

In the second step the unit root null on the residuals of the OLS regression was rejected as the statistic ADF(4) = –5.28 was lower than 95% critical value of –3.41. These econometric tests are based on Microfit 3.0 software. The lagged residuals of the cointegrating relationship was the ECM that was used to model the short-run dynamics of the conditional growth model for Australia. The estimated parsimonious conditional growth model using the general to simple methodology (Hendry & Richard, 1983) is reported below.

$$\Delta y_t = -0.0109 + 0.4899\Delta y_{t-1} - 0.4608\Delta y_{t-4} - 0.2329\Delta y_{t-2} + 0.0672\Delta r_{t-2} + 0.4485\Delta z_{t-1}$$
$$(t) \quad (4.07) \quad (3.21) \quad (2.59) \quad (1.94) \quad (3.01) \quad (3.42)$$
$$-0.1689ECM_{t-1}$$
$$(2.17)$$

Diagnostics

T = 1971Q1–1992Q4	$R^2 = 0.64$ $\sigma = 0.1717$	DW = 1.59
AR(5,65) = 1.24	ARCH(4,62) = 2.55	Norm(2) = 1.55
RESET(1,69) = 0.42	Forecast $\chi^2(20)/20 = 0.8165$	
Chow(20,50) = 0.6125		

The estimation results indicate that growth dynamics depend crucially on its own past history, competitiveness, and the last period's telecommunications infrastructure. The above conditional growth model appears to pass a battery of diagnostic tests and shows no signs of misspecification. The Godfrey (1978) fifth order serial correlation denoted AR is nonsignificant. The Engle (1982) fifth order ARCH is not significant at 10% level indicating the absence of evidence against homosecedasticity. The Jacques-Bera (1980) normality test and the Ramsey (1969) test for functional mis-specification are both nonsignificant. The parameter constancy of the model has been confirmed by the fact that forecast stability and the Chow F-tests are both less than the critical value 2.

5. Methodology of superexogeneity testing and results for australian regime shifts

In this section we examine whether during the study period the changes in the trade and telecommunications policy regimes caused a paradigmatic shift undermining the meta parameters capturing the growth dynamics as estimated in the conditional growth model for Australia. The Lucas critique (1976) emphasised that shifts in policy regimes caused changes in expectational mechanisms and this changed the structure of the model. The composite superexogeneity testing which comprises testing for weak exogeneity and structural invariance (Engle & Hendry, 1993) enables us to determine whether shifts in the policy regime changes the structure of the model specified for empirical analysis. The Engle-Hendry superexogeneity testing methodology is outlined in the sequel using it to test the effects on the growth dynamics of a possible telecommunications regime shifts in Australia.

The growth process y_t and a policy process x_t such as trade could be represented as a joint normal density for an information set Θ of interest as follows:

$$D(y_t,x_t;\Theta) = N[\mu_x, \Sigma],$$

where μ_x is a vector of the conditional means and Σ is the variance-covariance matrix.

It is assumed that the joint density model can be factorised into a conditional growth model defined by parameters Θ_1 and a marginal policy model defined by parameters Θ_2 as follows:

$$D(y_t,x_t;\Theta) = D_1[y_t I x_t;\Theta_1]\ D_2[x_t;\Theta_2] \tag{1}$$

The conditional growth model in (1) can be defined as a univariate normal distribution, thus:

$$D_1[y_t I x_t;\Theta_1)] = N[\mu_y + \delta(x_t-\mu_x); \Omega] \tag{2}$$

In (2) δ is the regression coefficient of y_t on x_t and Ω is the covariance matrix, conditional on the information set Θ_1.

The behavioural regression of expected conditional growth on expected trade with z representing a vector of current and past macrovariables as follows:

$$\mu_y = \beta\mu_x+z'\tau \tag{3}$$

Substituting for μ_y from (3) in model (2) yields:

$$N[\beta\mu_x + z'\tau + (\beta-\delta)(x_t-\mu_x);\Omega] \tag{4}$$

Superexogeneity testing can be performed by substituting in (3) a linear function in moments for β and a linear function for error covariances and generating the following testable regression:

$$\mu_y = \beta_0\mu_x + z'\tau + (\delta-\beta_0)(x_t-\mu_x) + \delta_t(x_t-\mu_x) + \beta_1 E(x) + \beta_2 V(x) + \beta_3[V(x)/E(x)] \tag{5}$$

The above expression clarifies that superexogeneity requires the satisfaction of first the weak exogeneity property by $\delta = \beta_0$ and moments being nonsignificant in the regression (5). Second, the satisfaction of structural invariance requires parameter constancy of the conditional model when parameters of the marginal model change. These properties are clarified further in the sequel when empirical results of superexogeneity tests are presented.

Superexogeneity is a composite test requiring the passing of structural invariance and weak exogeneity properties. The structural invariance property is met if the dummy variable proxying the regime shift is significant in the marginal model or the IV regression representing the policy behaviour and nonsignificant in the conditional model. The weak exogeneity property is met if the residuals or the fitted values from the policy IV regression are nonsignificant when run in the conditional model as postulated in Hausman-Wu-Engle test (Wu, 1978; Hausman, 1978; Engle, 1982). Furthermore, these tests have power if the residuals and fitted values are significant when the policy variables of interest is excluded from the conditional model. This study examined specifically whether changes in the trading policy regime and the telecommunications regime during the study period caused a paradigmatic change in the growth process in Australia during the study period. The IV (instrumental variable) regressions for trade and telecommunications policy are reported below.

Regression for policy regimes

Trade

$$\Delta x_2 = \underset{(0.19)}{-0.0037} - \underset{(5.36)}{0.4415\Delta x_{t-1}} - \underset{(5.21)}{0.3956\Delta x_{t-4}} + \underset{(2.14)}{0.4482\Delta y_t} - \underset{(1.90)}{0.2298\Delta w_{t-4}}$$
$$\underset{(2.14)}{-0.0897\Delta r_t} + \underset{(1.90)}{0.5469\Delta z_{t-4}} + \underset{(4.01)}{0.0530DM83}$$

Diagnostics

$R^2 = 0.62$ $\sigma = 0.0465$ DW = 1.89   RSS = 0.1556

AR(5,67) = 3.11   ARCH(6,64) = 1.43   NORM(2) = 1.61

RESET(1,71) = 0.39

Telecommunications

$$\Delta z_t = 0.0078 - 0.675\Delta z_{t-1} + 0.1031\Delta y_{t-1} - 0.0085DUM75$$
$$\quad (2.51) \quad (8.18) \quad (1.96) \quad (1.67)$$

Diagnostics

$R^2 = 0.59$ $\sigma = 0.1235$ DW = 2.13 RSS = 0.0116
AR(5,71) = 2.17 NORM(2) = 54.02 HET(6,69) = 1.56
RESET(1,71) = 0.03

The IV regression for trade policy was based on changes in past trade, the contemporary growth conditions in the domestic economy, competitiveness, and changes in world growth, and technology lagged four periods. The dummy variable DUM83 captured the shift in the trading policy regime. For the purposes of this study the structural break is at 1983Q4 when the trade liberalisation and financial deregulation climaxed with the floating of the Australian dollar. The process of trade liberalisation was marked by significant reduction of tariff and nontariff barriers and marked a reversal of a long legacy of Australian protectionism. The intervention dummy DUM84 proxies the shift in the trading regime. It is highly significant both in the IV regression on trading policy and nonsignificant in the conditional growth model indicating that the trade regime shift did not undermine the growth dynamics by causing structural variance and giving credibility to the Lucas critique. However, the trade policy changes did cause a change in growth model parameters as the residuals and fitted values of the IV regression of trade are both significant in the conditional growth. This violates the weak exogeneity property. It should be noted that weak exogeneity test on the residuals of the IV regression lacks power (see Table 5).

The IV regression proxying shifts on telecommunications policy was based on past telecommunications penetration, lagged growth and on an intervention dummy DUM75 capturing regime shifts in the telecommunications sector. The separation of Post from Telecommunications in 1975 was followed by a brief spurt in innovation which has led to the introduction of IDD (international direct dialling). The next major regime shift occurred in 1991 when the state owned carrier monopoly was partially deregulated by the establishment of a duopoly. The complete deregulation of the telecommunications sector and full competition is envisaged in 1997. The regime shift dummy DUM75–91 is significant in the telecommunications IV regression and simultaneously nonsignificant in the conditional growth model establishing the prevalence of structural invariance. The residuals and fitted values

Table 5. Weak exogeneity and structural invariance t-tests.

Instrumental	res	fit	res'	fit'
Δx_t	-2.26^{**}	-2.26^{**}	0.90	3.04^{**}
Δz_t	0.12	−0.35	−0.29	3.37
Dummy	DUM83	DUM75		
t-value	0.12	0.35		

of the telecommunications regression are both nonsignificant in the growth model confirming the prevalence of weak exogeneity. The weak exogeneity test based on residuals (res') seem to lack power while that based on fitted values (fit') has power (See Table 5). Overall we can conclude that the telecommunications policy (z_t) is superexogenous with regard to growth and the Lucas critique is nullified. Both IV regressions for trade and telecommunications appear to be quite robust with regard to specification as they pass a battery of diagnostic tests.

6. Concluding observations

The operation of a robust telecommunications network is central to the growth performance and international competitiveness of a modern information economy. The VAR analytics of Australia's information economy clearly demonstrate that innovations in growth, trade and telecommunications exhibit endogenous behaviour in the short run, but in the long run the role of the dynamic interactions explode. In the long run the mutually symbiotic behaviour of trade and telecommunications could generate a mutually reinforcing virtuous cycle. The operation of a regulatory environment in telecommunications and trade that is conducive to the adoption of best practice technology that will enable international competitiveness of private enterprise will determine the pace of growth of the Australian economy in a fast expanding global information order. The need for far-sighted telecommunications and information policies have been identified as vital for promoting trade and economic integration of Europe (Commission of European Communities, 1993). Australia could play a leading role in establishing an Asia-Pacific free trade community by spearheading the development of a regional telecommunications network. Of course, this requires as a precondition the pursuit of more open and competitive policies domestically.

There have been revolutionary changes in information technology and drastic deregulation and privatisation of overseas telecommunication monopolies resulting in a cornucopia of service benefits for customers and an explosion of growth to the telecommunications and dependent information industries. However, the super-exogeneity tests for the change in the Australian telecommunications regime indicated structural invariance or the lack of a paradigmatic shift in growth dynamics due to reform of the regulatory framework governing telecommunications. Perhaps the short history of deregulation resulting from a long period of indecision and prevarication could explain the lack-lustre results from the regime change in Australian telecommunications. A massive injection A$3.3 billion is under way by Telstra to modernise the telecommunications infrastructure by investing in smart and intelligent technology such as SDH (synchronous digital) transmission and ADSL (asymmetric digital subscriber lines). This will provide a plethora of new information services (telecommuting, telemedicine, teleconferencing, distance education and video-phoning). The technological efficiency and range of services provided by the telecommunications sector will determine not only growth and consumer welfare but also Australia's competitive edge in the fiercely competitive international information economy. There is concern that restricting competition to a duopoly and erecting barriers to carrier entry and service provision has undermined international competitiveness and the long-term growth prospects of the nascent Australian information economy. There appears to be a strong case to accelerate the pace of deregulation and competition in both international and domestic telecommunications to overcome constraints to the development of a vibrant information economy in Australia.

Appendix 1: Definitions and data

Definitions of variables and data sources

All variables refer to seasonally adjusted time-series. The lower case letters refer to the natural logarithm of the original variable indicated in upper case form.

Y_t Per capita GDP in thousands of $A calculated using GDP expenditure at 1989/90 prices divided by Australian population (persons). [Source: Econodata & ABS Cat. 5206.0].

X_t Exports of goods and services in $A millions at 1989/90 prices $A millions seasonally adjusted. [Source: Econodata & ABS Cat. 5206.0]

W_t World production proxied by OECD index of industrial production in seasonally adjusted quarterly terms. Base 1985=100.

[Source: Econdata & OECD Quarterly National Accounts. Various issues].

R_t $= E_t x V_t$ a measure of competitiveness or the real exchange rate.

V_t Terms of trade index or the ratio of export price deflator to import price deflator. Seasonally adjusted. Base: 1989/90=100.
[Source: Econdata & ABS Cat. 5206.0]

E_t Trade-weighted index (TWI). Base. May—1970=100.
[Source Econodata & Reserve Bank of Australia Bulletin, various issues].

Z_t Telecommunications or lines per capita.
[Source: Telstra Corporation Ltd. and ABS]

DUM83 Intervention Dummy for trade regime shift, taking values 1 for 1983Q4—1992Q4 and zero otherwise.

DUM75-91 Intervention Dummy for telecommunications regime shifts, taking values 1 for 1975Q1—1975Q4 and 1991Q1—1992Q4 and zero otherwise.

Data sources

- Econodata Pty Ltd. (1994) *Time Series Data Express, 1985—1992.*
- Australian Bureau of Statistics (1988, 1991), *Australian National Accounts. Historical Estimates,* Catalogue No. 5206.0, Canberra.

- OECD (1988, 1993) *Quarterly National Accounts*, OECD Department of Economics and Statistics, Paris.
- Reserve Bank of Australia, *Bulletin* (various issues).

References

Aschauer, D.A., 1989. Is public expenditure productive? *Journal of Monetary Economics*, 23:177–200.

Banerjee, A. et al., 1986. Exploring equilibrium relationships in econometrics through static models: some Monte Carlo evidence. *Oxford Bulletin of Economics and Statistics*, 48:253–270.

Baumol, W.J., 1982. Contestable markets: an uprising in the theory of industry structure. *American Economic Review* 72(1):1–15.

Beazley, K.C. 1990. *Microeconomic Reform: Progress Telecommunications*. AGPS, Canberra.

Blundell-Wignall, A., Fahrer, J., Heath, A., 1993. Major influences on the Australian dollar exchange rate. *Conference on the Exchange Rate and the Balance of Payments*. Reserve Bank of Australia, Sydney.

Brander, J.A. and Spencer, B.J., 1983. International R&D rivalry and industrial strategy. *Review of Economic Studies*, 50:707–722.

Bureau of Transport and Communications Economics, 1993. International telecommunications an Australian perspective. *Report 82*. AGPS, Canberra.

Bureau of Industry Economics, 1992. International performance indicators: Telecommunications. *Research Report 48*. AGPS.

Commission of the European Communities, 1993. Growth, competitiveness, employment. The challenges and way forward into the 21st Century (White Paper). *The European Communities Supplement* (Bulletin) 6/93.

Cronin, F.J., Colleran, E.K., Herbert, P.L., Levitzky, S., 1993. Telecommunications and growth: the contribution of telecommunications infrastructure investment to aggregate and sectoral productivity. *Telecommunications Policy*, 17(9):677–690.

Cronin, F.J., Parker, E.B., Colleran, E.K., Gold, M.A., 1991. Telecommunications infrastructure and economic development. *Telecommunications Policy*, 17(6):415–430.

Cullison, W.E., 1993. Public investment and economic growth. *Federal Reserve Bank of Richmond Economic Quarterly*, 79(4):19–33.

Davidson, J.A., 1982. Committee of Inquiry into Telecommunications Services in Australia. Report. AGPS, Canberra.

Davidson, J.E.H., Hendry, D.F., Srba, F., Yeo, S., 1978. Econometric modelling of aggregate time-series relationship between consumers' expenditure and income in the United Kingdom. *Economic Journal*, 88:661–692.

Dickey, D.A., Fuller, W.A., 1981. Likelihood ratio statistics for autoregressive time series with a unit root. *Econometrica*, 49:1057–1072.

Doan, T.A., 1992. *RATS User's Manual*. (Version 4). Estima, Evanston, Illinois, USA.

Doornik, J.A., Hendry, D.F., 1992. An Interactive Econometric Modelling System, Institute of Economics and Statistics. *PC GIVE* (Version 7). University of Oxford, Oxford, UK.

Dosi, G.C., Freeman, R., Nelson, G., Silverberg L. Soete, (eds), 1988. *Technical Change and Economic Theory*. Pinter, New York.

Engle, R.F., Hendry, D.F. 1993. Testing superexogeneity and invariance in regression models. *Journal of Econometrics*, 56:119–139.

Engle, R.F., Hendry, D.F., Richard, J.F., 1983. Exogeneity. *Econometrica*, 20:83–104.

Engle, R.F., Granger, C.W.J., 1987. Co-integration and error correction, representation, estimation, and testing. *Econometrica*, 55:251–276.

Engle, R.F., 1982. A General Approach to Lagrange Model Diagnostics. *Journal of Econometrics*, 20:83–104.

Evans, G., 1988. Australian telecommunications services: a new framework. *Statement by the Minister of Transport and Communications*. AGPS, Canberra.

Fuller, W.A., 1976. *Introduction to statistical time series*. Wiley, New York.

Godfrey, L.G., 1978. Testing against general autoregressive and moving average error models when regressors include lagged dependent variables. *Econometrica*, 46:1293–1301.

Granger, C.W.J., 1969. Investigating causal relations by econometric Models and cross-spectral methods. *Econometrica*, 37:24–36.

Grossman, G.M., Helpman, E., 1990. Trade innovation and growth. *American Economic Review*, 80:85–91.

Hausman, J., 1978, Specification tests in econometrics, Econometrica 46, 1251–1271.

Hendry, D.F., Richard, J.F., 1983. The Econometric Analysis of Economic Time Series. *International Statistical Review*, 51:111–163.

Hendry, D.F., 1985, Monetary economics: Myth and reality, *Oxford Review of Economics and Policy*, 1:72–84.

Jacques, C.M., Bera, A.K., 1980. Efficient tests for normality, homoscedasticity and serial independence of regression residuals. *Economic Letters*, 6:255:59.

Johansen, S., Juselius, K., 1990. Maximum likelihood estimation and inference on cointegration – with applications to the demand for money. *Oxford Bulletin of Economics and Statistics*, 52:169–210.

Johansen, S., 1988. Statistical analysis of cointegrating vectors. *Journal of Economic Dynamics and Control*, 12:231–254.

Judge, G.G. et al., 1992. *The Theory and Practice of Econometrics*. Wiley, New York.

Karunaratne, N.D., 1993. Reality and the rhetoric of free trade in services – The case of telecommunications. *Intereconomics*, 28(2):95–102.

Karunaratne, N.D., 1992. Growth and trade liberalisation debate in Australia – A VAR analysis. *Rivista Internazionale di Scienze Economiche e Commerciali*, XLI(8):625–643.

Karunaratne, N.D., 1988. Symbiotics of telecommunications, trade and development. *Economia Internazionale*, XLI(1/2):3–19.

Krugman, P. (ed), 1986. *Strategic Trade Policy and the New Industrial Economics*. MIT Press, Cambridge.

Lipsey, R., 1992. Global change and economic policy. In: Stehr, N. and Ericson, R. (eds), *The Culture and Power of Knowledge*. New York, Walter de Gruyter.

Lucas, R., 1988. On the mechanics of economic development. *Journal of Monetary Economics*, 2:3–142.

Lucas, R.E., 1976. Econometric policy evaluation: A Critique. In: Brunner, K. and Meltzer, A.H. (eds). The Phillips Curve and Labour Markets. *Carniegie-Rochester Conference Series on Public Policy*, Vol. 1, (pp. 19–46), Amsterdam, North Holland.

MacKinnon, J.G., White, H., 1985. Some heteroscedastic consistent covariance matrix estimators with improve finite sample properties. *Journal of Econometrics*, 29:305–325.

Olson, M., 1965. *The Logic of Collective Action*. Harvard University Press, Cambridge, Massachusetts.

Perron, P., 1988. Trends and random walks in macroeconomic time series: Further evidence from a new approach. *Journal of Economic Dynamics and Control*, 12:297–332.

Pesaran, H.M., Pesaran, B., 1991. An interactive econometric software package, *MICROFIT 3.0*. Oxford, Oxford University Press.

Phillips, P.C.B., 1987. Time-series regression with a unit root. *Econometrica*, 55:277–301.

Phillips, P.C.B., Perron, P., 1988. Testing for a unit root in time series regression. *Biometrika*, 75:335–346.

Posner, M.V., 1961. International trade and technical change. *Oxford Economic Papers*. New Series, 13: 323–341.

Ramsey, J.B., 1969. Tests for specification errors in classical linear least squares regression analysis. *Journal of Royal Statistical Society*, (Series B) 31:350–371.

Romer, P.M., 1990. Endogenous technical change. *Journal of Political Economy*, 98:S71–S102.

Sims, C.A., 1980. Macroeconomics and reality. *Econometrica*, 48:1–48.

Solow, R., 1956. A contribution to the theory of economic growth. *Quarterly Journal of Economics*.

Stigler, G.J., 1974. Freetiders and collective action. *Bell Journal of Economics*, 5:359–365.

White, K.J., 1993. *SHAZAM* The Econometrics Computer Program Version 7.0 User's Reference Manual. McGraw-Hill Book Company, New York.

Wu, D.M., 1973. Alternative tests of independence between stochastic regressors and disturbances. *Econometrica*, 41:733–750.

Part V: Some macroeconomic aspects

Beyond Competition: The Future of Telecommunications.
Edited by D. Lamberton.

CHAPTER 19

The macroeconomic effects of new information technology, with special emphasis on telecommunications*

Gunnar Eliasson
IUI and The Royal Technical Institute (KTH), Stockholm, Sweden

1. Introduction

Economics has always had a problem with information (Stigler, 1961, Lamberton, 1971). It is difficult to measure. Much of its use occurs within the heads of individuals. Because of lacking *receiver competence* important aspects of knowledge in society cannot even be communicated. It is tacit and cannot be translated in a communicable form as information (Eliasson, 1990a). This is particularly important for economic analysis when the importance for growth of new technology and business judgement is to be taken into account. What is the state of information in markets for investment?

The number of investment opportunities residing at any time in an advanced economy is immense. The theory of economic growth has to explain how these opportunities are discovered, explored and exploited. The sheer size and complexity of the space of business opportunities and the free access of innovative agents to enter and compete with incumbents mean that anything can occur. What does this everyday situation mean for the definition, the use and the communication of information?

2. The scope of opportunities

Attempting to understand the space around us is not only a practical problem for the firm, or a scientific problem for the information specialist. Neither is it unique to economics. The notion of open vs. closed systems has long been the concern of philosophers, and it has always been considered intellectually uncomfortable to think in terms of open, nontransparent state spaces. Hume and Locke discussed the world in terms of *memory*, *logic* and *imagination*. Logic and memory were acceptable concerns for the human mind and from the two, a large number of combinations of new facts could be derived. Imagination was the rest. Leibnitz accepted no more

Address for correspondence: Gunnar Eliasson, IUI and The Royal Technical Institute (KTH), S–100 44 Stockholm, Sweden. Tel: +46–8–790–6803. Fax: +46–8–790–6741.
*This paper has been written as part of the IUI telecommunications study for Swedish Telia.

imagination than all possible combinations of logic and memory (the reducibility assumption, Gustavsson, 1991). This was also sufficient to introduce *complexity* and a limit to understanding.

Kant, however, was willing to go beyond the computable world of Leibnitz and accept the presence of imagination as a separate dimension of human awareness, i.e., the representation of something that is not derivable from existing facts, or the past. Imagination was created out of nothing, from the unknown depths of the human mind. Schumpeter (1911) used a similar notion to introduce *innovation,* as a "deus ex machina", previously incomprehensible.

That a stable structure, a language, a theory is needed to organize facts and thoughts such that information can be communicated (Eliasson, 1990a, b) took a long time to understand. Not until the 1930s, and Wittgenstein, did philosophers and linguists realize that language is a limiting factor of knowledge. Pelikan (1969) used the same idea to study the impossibilities of being fully informed, and in control of a large organization from a central point. The language limits the kind and volume of knowledge that can be communicated on code as information.

It is very obvious that going beyond an economy that limits itself to a codable representation of facts – the computable world – into *the experimentally organized* economy that allows for nonpredictable imagination or innovation, represents an entirely new experience to the analyst. We will find, in this paper, that it changes the nature of the concept of information, and that it all has to do with the size of state space and the difference in opinions, Kant, on the one hand, Leibnitz on the other.

3. The experimental nature of economic activity

Adam Smith (1776) laid down the principal design of a decentralized market economy in which division of labor made economies of scale "on a small scale" possible and the realization of large macro productivity effects feasible. He saw no limits to the possibilities of productivity advance or the potential wealth of nations. His world was essentially an open system (Loasby, 1991). The benefit of productivity advance and economic wealth, however, came at a significant cost, a fact that "modern" mathematical representations of the "invisible hand" have had difficulties incorporating. The organization of the division of labor is an instance of innovative behavior. It evolved gradually in the market. Once realized, economic activity had to be *coordinated* physically (transports) and through communication of information.

Once an innovative design, whether technical, organizational or commercial has been accomplished, competitors will be "on your doorstep" to learn (imitate). If your organization is large enough you will want to diffuse the new knowledge throughout your organization. You may also want to sell your knowledge for a profit (consulting). *Learning,* hence, becomes a general and very resource-using economic information activity.

Even very simple tasks (you soon learn) can normally be solved in a large number of ways. The higher-up, the more complex the decision problem and the larger the number of possible solutions. Some of these solutions are better than others. The problem, however, is that *you will never know until you have tried them. This is the essence of the experimentally organized economy.* The number of possible solutions defines the size of the *business opportunity set* that each agent probes into through experimental search, a search being directed by a limited vision (theory) of all possibilities (*bounded rationality,* Simon, 1955). With his or her particular vision as guidance, agents are limited in communicating with each other because of limited and differently composed *receiver competence* (Eliasson, 1990a, p. 17; 1990b). The result will be a heterogeneous structure of competence, defined by the organization of people in the economy.

Much of the knowledge put to use in a firm, especially high level competence, is difficult or impossible to communicate in coded form as information. It is *tacit.* Tacit knowledge is acquired through on-the-job learning and filters through the economy (selection) through the acquisition of whole, or parts of firms in the mergers and acquisitions market, or through the mobility of people or teams of people with competence in the labor market (Eliasson, 1991c).

I have now introduced four general, knowledge-based economic information activities; *coordination, innovation, learning,* and *selection* (see Table 1). How much of all resources used in an economy that falls within these categories depends on the economic measurement system chosen. Measurement taxonomies that make them cover all economic activity can be defined, making information processing an intellectual superstructure (the memory), that controls all other activities. If there were no limits to our possibilities of measurement, all economic activity would consist of using and communicating information. For instance, automating production in a work shop can be described as a change of the information system that coordinates the underlying physical production. In this particular case it is also very obvious that this

Table 1. The four basic economic activities in the knowledge-based information economy.

1. Coordination (organizational structure)	The invisible and visible hands at work • competition (in markets, Smith, 1776) • management (of hierarchies, Chandler, 1977)
2. Innovation (exploring state space)	Creation and exploitation of new business opportunities (Schumpeter, 1911) • innovation • entrepreneurship • technical development
3. Selection (organizational change)	Incentives for change • entry • exit • mobility
4. Learning	Knowledge transfer (Mill, 1848) • education • imitation • diffusion

Source: The Knowledge Based Information Economy, IUI, Stockholm, 1990, p. 73.

change in the information system is a change in work organization. Available statistical measurement technology, however, does not allow the economist to work with the high resolution necessary to automate factory production. The standard national accounts statistical system is, nevertheless, sufficient to demonstrate the dominance in total resource use of knowledge-based information activities. In the economic model which we will use *coordination, innovation, selection* and *learning,* will also dominate economic resource use, making the *use* and *the communication of information* the dominant production activity. We can also establish that most of what we call technical change, as observed through macro production function analysis, is really composed of changes in the technology of economic information processing, which in turn originate in *innovative, organizational change,* including innovative change in the organization of learning to accumulate new competence (Eliasson, 1992). This observation, however, demands a very broadly based definition of innovation, that is not restricted to hardware equipment in factories. Innovation occurs throughout business organizations and is particularly important at the level of the top competent team in the form of *innovative organizational change,* exercising strong leverage effects on the entire firm (Eliasson, 1995a). Such organizational change therefore becomes part of the technology of the firm. Innovative organizational change is also associated with the organization of the entire economic system and its capacity to efficiently coordinate economic activities. Particularly important for macroeconomic performance is the balancing of coordination through free markets, on the one hand, and through nonmarket, central regulation by the government on the other. Another equally important problem is the trade-off between the short and the long run.

4. Information processing the dominant production activity – some basic facts

Even though clearly observable in reality, and recognized by economists, the attempts to mathematize economic theory, beginning in the late 1900th century, had problems with knowledge and information from the start.

The academic solution to this problem was typical. Knowledge and information was simply ignored. General equilibrium-based models of Walrasian origin dominated pure economic analysis well into the 1970s. With such models information gathering and communication required no resources. The auctioneer or God made sure the economy was safely in equilibrium. Industrial organization theory has increasingly recognized information use in the form of asymmetric information and incomplete markets, and the possibility of limited understanding (bounded rationality). Observing that value added was rarely exhausted by observed factor payments, McKenzie (1959) suggested that residual profits should be seen as compensation for unmeasured knowledge inputs. Analysts carefully avoided selection mechanisms (like entry and exit) and information activities drawing large resources and/or being subjected to innovative technical change.

For economists attempting to understand the use and communication of knowledge in economic growth such simplifications are not acceptable. Significant selection mechanisms and/or large information costs that are (unavoidably) subjected to unpredictable technical change dramatically change the properties of the economic model. Above all, it will have to be recognized that *technical change in information processing in an economy occurs through organizational change*, that significantly affects the reliability of information in markets and the capacity of markets to coordinate the economy. Let me state one empirical fact to make that point, and then go on.

The fact: information costs dominate

Figures 1A and 1B present Swedish manufacturing and the entire Swedish economy on a break down in physical, manual production and various forms of service and information activities like R&D, product development, global marketing and internal education. Even with the rough categorization used in the national accounts, it is possible to show that the bulk of labor costs in Swedish manufacturing goods production consists of service production, and that most of this service production consists of information activities of some kind. The dominance of information activities would become even more pronounced if the presentation were in terms of value added, since profits are a form of compensation for financial services rendered, and/or the compensation for ownership services of various kinds, including risk taking

A

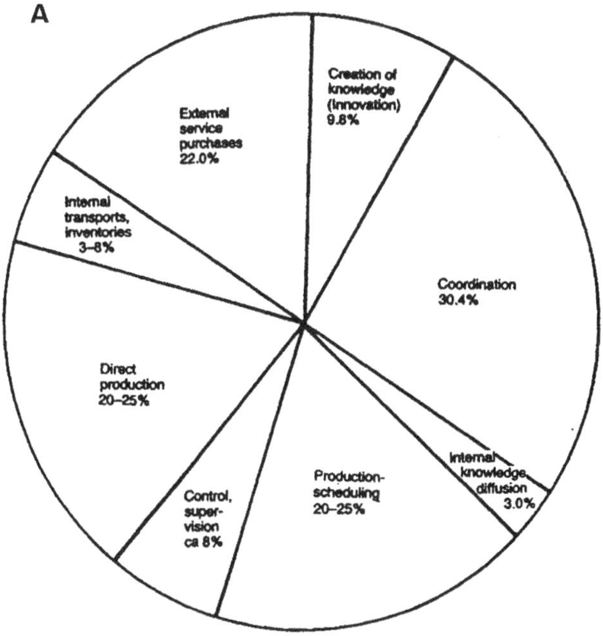

Total: 122 percent
***Source:* Eliasson (1990a, p. 68)**

Fig. 1A. Distribution of labor costs over production activities.
– large Swedish manufacturing firms
– global operations

B

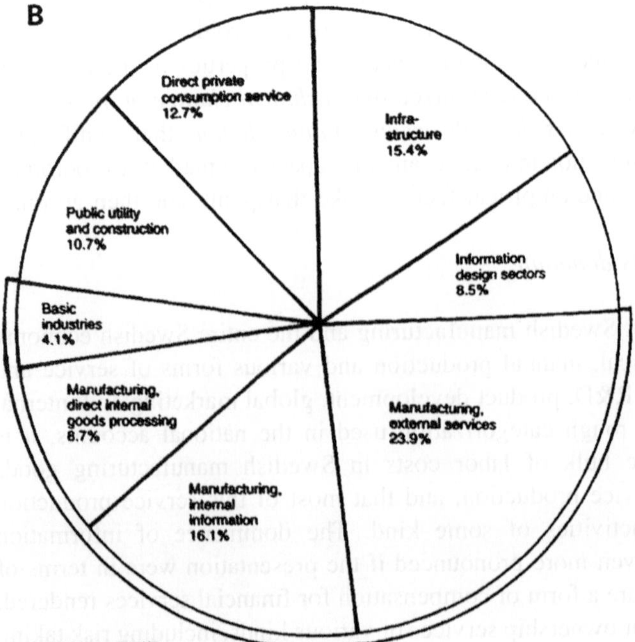

Total: is GNP.
Source: Eliasson (1990a, p. 70)

Fig. 1B. The information economy.

and competence contributions. Figure 1B gives the same picture (in terms of GNP value added contributions) for the entire economy. Again service production dominates, and among services various forms of information processing and communication activities dominate.

Against this statistical background the use and communication of knowledge in various forms becomes the dominant resource use in production, unavoidably being subjected to innovation and technical change.

Consequence one: innovative organizational change reduces information in markets and hierarchies

Information processing occurs in markets and within firms (hierarchies), the goal of agents is to be able to identify better business solutions and to support economic decisions that are superior to other decisions. With information costs zero or small and predictable and state space small and transparent, the exact nature of existing options can be identified analytically. In the type of models commonly used by economists it is a matter of routine analysis to find the optimal solution that is related to external factors that remain unaffected by the allocation decisions in the economy.

With information, the dominant resource use, costs are affected by technical change in information and communication use and not known in advance. Then *the optimal positions of the economy as well as of firms will depend critically on the actual technology of using information.* In a market, prices signal information on quantities. In static equilibrium this relation is one-to-one. Prices reflect quantities accurately, and vice versa. Out of equilibrium, as in most nonlinear economic environments signalling mechanisms are much more complex and unreliable (Eliasson, 1985, Chapter VII). If the production organization that emits price signals is subjected to innovative change the information content of prices also changes, and is reduced until the actors have learned to interpret them. With (fact one) the bulk of resource use in an economy engaged in information processing in various forms, quantities become extremely difficult to identify through interpreting market price signals. The problem is identical within hierarchies. The efficiency of internal communication top-down or bottom-up depends critically on the stability of the information content of internal statistical categories. Reorganization of firms changes the information content of such internal statistical measures and, hence, reduces the state of information in the firm (Eliasson, 1976, 1984a, 1995a). Hence, organizational change is always resisted by the middle-management hierarchy responsible for communicating information and exercising control within the firm (see below). This is also a very practical concern for designers of internal information systems in firms.

Consequence two: economic growth is organizational change

Organizational change, we have just shown, reduces the coordination capacity of markets through price signals and the coordination capacity of hierarchies through administrative (planning) techniques. Three important conclusions emerge from research carried out by Bo Carlsson. *First,* in an early IUI study, he (Carlsson et al., 1979, p. 34; Carlsson, 1980) demonstrated that, when stripped down to the level of a division, or an establishment, more than 50 percent of total factor productivity change at the manufacturing industry level originated in structural adjustment between existing establishments, most of which was due to the exit of low performing units and the transfer of resources to high-performing units. *Second,* Carlsson (1989) reports that technological change in manufacturing is generally making smaller scale production more economically viable than earlier, which is reflected in a general reduction in the average size of both plants and firms among the industrialized countries. This observation is strengthened by the relatively faster advance of private service production. Smaller scale, service-oriented, and competence-intensive production will increase the importance for macro performance of structural adjustment at the plant and establishment levels.

Third, Carlsson (1991) observes that in a 20-year perspective, total factor productivity growth is almost all a matter of reallocation of resources within and between existing plants. This tallies nicely with my own results on entry and exit (Eliasson 1991a, 1993c). Beyond the 20-year horizon, the introduction of new technology through entry and through new investment begins to exhibit sizable macroeconomic effects. Summing up, this means that (1) organizational change determines long-term productivity change and macroeconomic growth, but (2) the faster organizational change, the lower the state of information in markets and the less efficient the capacity of markets to coordinate the economy in the short term. *Organizational changes* at all levels are instances of innovation. But contrary to the exchange of tacit knowledge, the efficient transmission of (coded) information requires *organizational stability,* or a stable grammar, something that linguists and business information systems designers have long understood (Eliasson 1976, 1995a). Hence, in an economy where information processing is the dominant production activity, and where information processing is subjected to technical change, *the experimentally organized economy,* the nature of information becomes fundamentally different from that in the classic, static model of a transparent world. The question asked is how individuals and firms find, interpret and use information under such undefined circumstances. We will find that intuitive business decisions that may turn out right, and mistaken decisions that cause business failure constitute the critical, and positive dynamics of a growing economy.

Learning theory

Modern learning theory illustrates these practical problems clearly. Learning theorists like Blume, Bray and others (see Lindh, 1993) have been confronted with a kind of impossibility problem, or the impossibility (already in the classical model) of learning to be fully informed, through standard statistical methods. If you have no prior knowledge of the important structural characteristics (the specification) of your model you will be stuck with limited and biased insight. Thus the notion of "bounded rationality" or limited insight (Simon, 1955, Eliasson, 1990b) is already forced upon the analyst in the classical model. Financial economics has – on similar grounds – come up with the idea of rational "noise traders" that realize that other agents may not be well informed and/or do not "behave rationally". Therefore they cannot behave as if all are informed, or as if all agents "behave rationally", but have to make guesses about the state of information in the market and about behavior. This situation is not tractable by standard analysis. Antonov and Trofimov (1993) find that in the experimentally organized economic environment of the Swedish micro-to-macro model (MOSES) agents (firms), equipped with sophisticated, econometric learning instruments (rational statistical learning), still do not improve performance, if they do not know the structure of the economy beforehand. The structure of the economy depends, however, as we have just shown on the state of information in markets and on the nature of behavior of agents. Hence, the perfect state of full information does not exist as a feasible operating domain of the experimentally organized economy. In an experimentally organized economy, "noise-trading-like behavior" – Antonov and Trofimov (1993) show – improves long-term macroeconomic performance compared to rational statistical learning or, above all, centrally imposed "information" based on structurally specified, but misspecified econometric models. Noise-traders by chance or (to others) by incomprehensible intuition stumble on opportunities in the state space of the model that are excluded in the centrally restricted, "optimal" decision environment.

This paradoxical result can also be expressed in terms of *prohibitive complexity.* If you know the structure of the economy – expressed as the prior structural specification of an econometric model, the prior requirement of learning theory – then you can proceed to try to estimate the model. If you do not know the structure of the model you have to make prior assumptions, and run the risk of *misunderstanding through misspecification.*

If you know the structure of the economy and (hence) of your model, it may still not be estimable by standard econometric techniques. This is almost always the case with models like MOSES (see below), featuring selection mechanisms like entry and exit, and other forms of nonlinearities. Even if you would know the code you would be unable to estimate its coefficients.

The next step would be to work with simplified representations of the underlying model structure. This is what Antonov and Trofimov (1993) did. Even though they knew the code of the underlying standard model (MOSES), it did not help, since the structure of the underlying linearizable and estimable model was not stable over the entire estimation period, due to agent behavior and multimarket interactions. This is sufficient to render the standard equilibrium model dangerous for any analysis that has to do with resource allocation. And note carefully that we are still conducting our argument as if in the highly unrealistic world of *costless information processing.* The experimentally organized economy changes all that.

5. The experimentally organized economy (EOE)

The model of the experimentally organized economy (Eliasson, 1987, 1988, 1991b) confronts the analyst with the impossibility of identifying and obtaining all information needed for a fully informed economic decision and of communicating the complex of information acquired. This problem is similar to that of distinguishing (Knight 1921) between *uncertainty* and *computable risks,* a distinction that has no meaning in the classical economic model which uses risk and uncertainty as synonyms, but that carries utmost significance in the experimental model.

The main conclusion of this section is that when fully informed decisions are impossible, and when decisions are normally significantly biased, business mistakes become a normal cost for economic development. The rational expectations or efficient market model treats such mistakes as random noise. In the experimentally organized model business mistakes cannot be treated as random, insurable noise. They furthermore become the most important learning cost for economic development (Eliasson, 1991b). This proposition is best illustrated with the help of the Swedish micro-to-macro model.

A generalized Salter curve analysis of innovative learning and competition

A market, or the entire economy, can at each point in time be represented by a

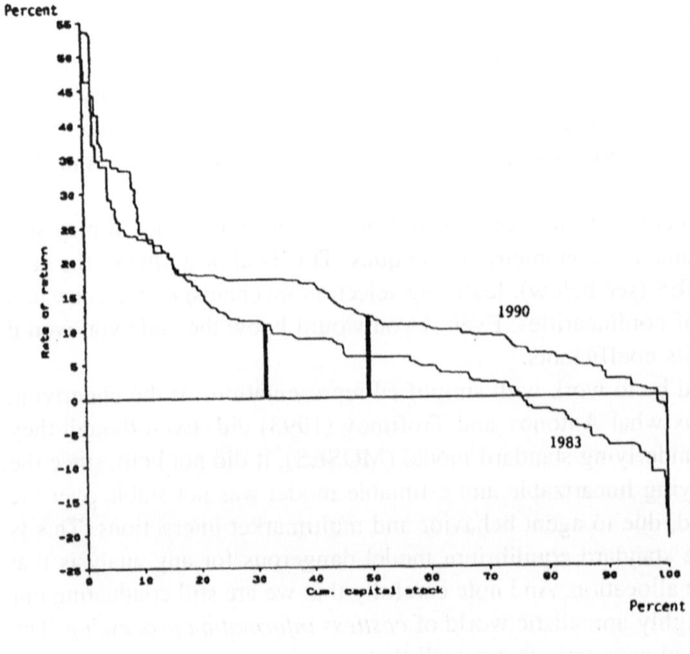

Fig. 2A. Distributions of rates of return over interest rate (=ε̂), Swedish manufacturing, 1983 and 1990.

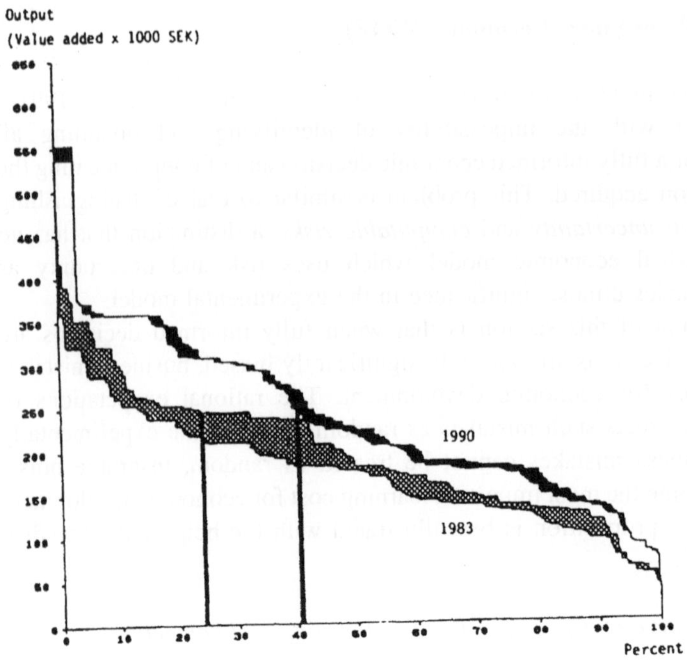

Fig. 2B. Actual and potential labor productivity distributions, Swedish manufacturing, 1983 and 1990. Note: Shaded areas denote unused labor capacity. Source: Eliasson, 1991b, p. 164.

distribution of potential performance characteristics, like the rates of return over the interest rate ($\bar{\varepsilon}$) in Fig. 2A. These types of distributions – especially if presented as productivity rankings of establishments (Fig. 2B) – are often referred to as Salter (1960) curves. Each firm is represented on this curve by a ranking on the vertical axis (the two columns), the width of the column measuring the size of the firm in percent of all other firms. Figure 2A shows that even though the firm in the model has increased its rate of return between 1982 and 1991, it has lost ranking. Figure 2B shows the same firm's labor productivity and wage cost positions. Finally, each firm is operating under its productivity frontier to position itself on the productivity and rate of return rankings (see Eliasson, 1991b, Fig. 1). This is still actual *ex post* performance 1982 and (simulated) 1991. The dynamics of markets, on the other hand, are controlled by the potential *ex ante* distributions that capture the planned action of all firms. The Salter curve can be upgraded by new entrants.

Existing firms can improve their position through *reorganization* or *rationalization.* If it does not succeed it will eventually be acquired by some other firm or exit. This leaves us with four fundamental growth investment mechanisms:
a) entry
b) reorganization
c) rationalization
d) exit
that together make reorganization and selection the fundamental mechanisms behind economic growth (Eliasson, 1993a, b).

Competition

There is a third set of Salter curves that tell how *each firm sees itself positioned relative to other firms.* The real world of the experimentally organized economy, as well as its model approximation, the Swedish micro-to-macro model, shows large divergences between actual and perceived positions.

The *ex ante* distributions show the potential for the firm to outbid all other firms in wages, or in paying a higher interest rate.

Learning about one's competitive situation – in reality or in theory – occurs in different dimensions. Prices offered in the market tell something about how other firms – notably the best firms – view their competitive situation. Competition, production, hiring, etc. can also be directly observed. The firm, finally, learns directly itself, when it enters the market. (The critical learning experience occurs when firms observe that competitors can do better. Firm management then knows that this *can be done* and that it had better improve in order not to be pushed down, right along the Salter distribution, and, perhaps, out). Similarly, when the firm finds itself at the top, or close to the top, it knows that a whole lot of "closely inferior" firms feel threatened and are taking action to better their positions through innovation or imitative learning.

The conclusion is that if potential Salter distributions are sufficiently steep and if all firms know it, firms – and especially the top left-hand group – will feel sufficiently threatened to actively aim at improving their positions on the Salter curve through innovation. If such innovative activity, notably through innovative entry in markets, is free, necessary conditions for maintaining sufficiently steep Salter distributions to move the entire economy through a self-perpetuated competitive process have been established (Eliasson, 1985, 1991a,b). These conditions become both necessary and sufficient if the opportunity set (Eliasson, 1987, 1991b) is sufficiently large. This also establishes the link between dynamic competition through the Schumpeterian (1911) entrepreneur and innovative entry, argued by Smith (1776) to be the critical mover of macroeconomic growth that perpetuates a disequilibrium economic process type Wicksell (1898). A sufficiently large and heterogeneous state space, boundedly rational behavior on the part of agents, and sufficiently free innovative entry are the small modifications of the classical model needed to release the dynamics of the experimentally organized economy.

The Swedish micro-to-macro (M-M) model exhibits these features. Dynamic competition determines entry and exit, and hence the selection process that creates a path-dependent evolution, and the nonstationary behavior that prevents classical learning.[1]

Simulating approximate static equilibrium through faster market arbitrage ($\bar{\varepsilon}$ competition)

The competitive process described above culminates in stock market competition for funds. The more intense competition the flatter $\hat{\varepsilon}$ distributions (see Fig. 2A) will be through forced exit, if new entry does not add even better performers from the left. New entry, besides keeping competition going, also preserves diversity of the performance distributions. Large positive $\hat{\varepsilon}$ in a market signal profit opportunities, and entrants enter on the basis of their expected $\hat{\varepsilon}$. Most new entrants, however, fail (Eliasson, 1991a). The classical notion of a perfect capital market equilibrium is that all agents' rents ($\hat{\varepsilon}$) have been forced into line and made equal to zero (no profits over costs). Efficient financial markets perform that function in the classical general equilibrium model. This process can be simulated in the MOSES model. We then find, however, that the more efficient financial market arbitrage (through forcing agents' $\hat{\varepsilon}$ closer to a zero profit situation) the more volatile $\hat{\varepsilon}$ development and the more unstable the real economy (Eliasson, 1983, 1984b, 1991b). The closer to "optimal" positions agents have been guided by market signals and forced by intense competition, the more unreliable price signals become, the more frequent major business mistakes and the more disorderly macroeconomic behavior; a true Heisenbergian situation. This will be an important observation for the final discussion of new instruments in financial markets.

Three different forms of learning

Three different forms of information processing or learning can take place in a dynamic or experimentally organized economy.

Agents:

1. use *analytic* or classical (costless) learning methods, like statistical learning
2. *invest* in methods or equipment to assist in the gathering of knowledge or information
3. learn through *experiments and mistakes.*

(1) is the preferred method in economic theory, but it has also been practiced widely and wildly by firms (forecasting) and by policy makers and macroeconometric model builders. The 1960s was the analytical forecasting and planning decade (Eliasson, 1976).

The *investment* method is more useful. Its typical application is in systems control like automated production etc. Such applications are surrounded by very costly investments in measurement equipment, process monitoring devices and control instrumentation. A necessary requirement for such applications is a stable underlying structure, meaning that signals emitted by the system (the measurement system) always contain exactly the same information, barring a small random error (tolerance).

Economic learning in more broad applications takes you into the third category. Business information systems are good examples. This time you lack full information, both of the interior of your system, the firm and its environment (Eliasson, 1976). The relevant decisions are *choices.* The very interesting first fact is that the typical business information system has long been, and still to a large extent is, designed on the format of category (2) or factory automation. The interesting second fact is that it should be organized (Eliasson, 1995a) as learning through experiments and mistakes, reflecting the fact that the local economic environment of a firm can now and then be completely disorderly and unpredictable.

6. The nature of the firm in the experimentally organized economy

The way you look at the economy affects the way you look at, and theorize about the firm (Eliasson, 1994). If you look at the economy in terms of the transparent static general equilibrium model you tend to see the firm[2] as a forecasting-planning machine that never makes mistakes, except random forecasting errors. Firms can plan (in principle) to achieve a state of full information, and this theoretical possibility of the classical economic model exerted significant influence on the business administration literature of the 1960s and early 1970s, which abounded with treatises on "business planning" (see Loasby, 1967, Jantsch, 1968 and Eliasson, 1976), until reality struck back in the form of the series of macroeconomic crises of the 1970s. This literature, and its promotion of formal, long-range business planning, is now gone. The problem, pointed out already by Knight (1921), is that the assumption of computable errors makes business mistakes insurable and, hence, leaves no room for business competence, which is utterly at variance with facts (Eliasson, 1990b). The EOE and reality, however, require a very different animal, with different nonanalytic competence specifications. Each firm has to reckon with the presence of many competitors aiming for its market niche through technological product competition (Eliasson, 1987). The set of business opportunities is huge and mostly nontransparent to the individual firm. A firm that wants to survive cannot wait to compute its fully informed plan of what to do. It has to weigh the costs of committing an error (type *one* mistake) against the costs of being beaten by a more daring competitor (type *two* mistake). If it does not act prematurely on a very incomplete information base, it can be sure that many of its many competitors will do and that one might score a success. Thus, top firm management had better be equipped with a good *sense of direction,* the first, dominant competence requisite for success (item 1 in Table 2). If it does not, it will fail anyway, and a business firm that wants to continue to exist has to have a business idea which its management and its owners believe in. Hence, a firm will have to demonstrate itself to outsiders as being a gambler, taking on seemingly large risks. With a good sense of direction, however, the true risk exposure to the insider management is much smaller. Thus the competence of the firm can be defined in terms of its ability to transform uncertainty (Knight, 1921, Eliasson, 1990b) into (for its own management) computable, subjective risks.

There are also learnable, administrative techniques to minimize the costs of mistakes. Techniques can be developed that make it possible to take on (reduce the risk of taking on) large risks, that is, a technique to manage in situations when the first competitive requisite (intuition) has failed. This management technique consists of two elements: *identifying* mistakes early and *correcting* mistakes immediately (Eliasson, 1990b, 1995a). Once these tests (elements 3 and 4) have been passed, another and quite different element of competence has to be clicked on: the ability to *operate* the firm *efficiently* on a day-to-day basis and to feed experience back to the top (*learning*). Here the next problem shows up, namely the limited internal transparency of a large firm. The large firm (at least) appears (Eliasson, 1976) as an organization with a very limited top down insight into its internal life and only a partial understanding of its external market environment. After the disorderly 1970s administrative techniques have currently moved away from trying to understand ("forecast") the external market environment towards improved internal monitoring and control (Eliasson, 1984a, 1995a). This organizational technique dominates when firms have found themselves in the same market for a long time and in high volume production. (The large Swedish multinationals were very successful in doing this in the 1980s). New information and communication techniques support the firm in exercising these competences, and a business firm is far from a controllable economic

Table 2. Competence specification of the experimentally organized firm

Orientation	1. Sense of direction (intuition)
	2. Willingness to take risks
Control	3. Efficient identification of mistakes
Operations	4. Efficient coordination
	6. Efficient learning feedback to (1)

Source: Eliasson (1990a).

system. Thus, intuition and arbitrary choice rule at the top. New communications technology has found successful applications at that level in making access to people and data faster and more easy. Data communication has found more applications in manufacturing process control, in global inventory systems, in production design and CAD applications etc. (see further Eliasson, 1995b). At the same time new electronics technology has contributed to improved product quality ("robots", etc). Data communication intensity increases with firm and division size. This can be observed from data collected in the so-called planning surveys carried out annually by the IUI and the Swedish Federation of Industries. Nevertheless, Taymaz (1990) found a negative correlation between both productivity and profitability and telecommunications intensity in production. Since most of telecommunications costs are for telephone communication this may not be altogether wrong. Telephone conversations do not necessarily contribute to, or be associated with new technology. Another possible interpretation of the negative correlation might be that intensive use of data communications services is mostly in the hardware-oriented, large engineering firms that service global markets for not so sophisticated products. Here efficient routinized production and inventory control matter. But the same orientation might also have prevented the firms from moving into new, high value added products and markets. Success in routine volume management may negatively affect the more important "innovative" tasks, critical for long-term survival, and even very large firms are at peril in the EOE.

People in charge of the first orientation tasks (intuition) are always organizationally separated from the operations people. Earlier information systems developers attempted to make analytical methods (forecasting) the main management tools at this top orientation level. These approaches, however, failed miserably and have now been pushed down to the operations level (Eliasson, 1995a).

7. Policy making in the experimentally organized economy

A business firm is not a well controlled economic system. Complexity and high resolution measurement make even simple production processes difficult to automate. The external environment is largely unpredictable. Much of the interior of a large business is unknown terrain for its top decision makers.

One would expect the control situation to be even more difficult for the central planner or policy maker of an entire economy. It of course is, and the policy maker has an extra problem to cope with. He normally acts against the market. His autonomy to do that has diminished in pace with the globalization of markets increasingly making each national economy an integrated part of the international economic system. New communications technology (see below) has played a decisive role in this development towards diminished policy autonomy of nations. The policy makers have used three countermeasures to remain in control. They:

1. act on a simplified notion (model) of the economy and disregard possible mistakes, that occur with a long delay,

2. lower ambitions, and/or announce less detailed policy targets,
3. restrict innovative activities that cause structural change, which reduces the state of information in the economy, but increases central control.

The *third* method has been practiced by the centrally planned economies within the former Soviet empire. By regulating unpredictable innovative activities "to a halt", economic growth was more or less eliminated. The *first* method has been increasingly typical of western industrialized countries since the Keynesian revolution. Simulation experiments on the Swedish micro-macro model (Eliasson 1993b, Eliasson and Taymaz 1992) strongly suggest that an advanced industrial economy is very robust and takes very long mistreatment before it responds with deteriorating growth performance.

The *second* method is the one chosen by business firms, after the bad experience with the forecasting planning mode (1) in the 1970s. They have increasingly relinguished central control of operational details and focused on the central monitoring of profit targets (Eliasson 1984a, 1995a).

In the experimentally organized economy firms and government alike are unable – as illustrated by Antonov and Trofimov (1993) – to reach a state of full information and systems control. The same is true of stochastic control in the efficient market model. While the little firm that mismanages soon fails in the efficient market, large firms with large financial capacity can live longer. Government has the capacity to sustain mismanagement for generations, eventually moving the economy beyond restoration. I will conclude on this note with the amusing experience we have had over the many years of modeling work. If you sit down at the computer and attempt to correct unexpected, disruptive, and "socially undesirable macro behavior" in the Swedish micro-to-macro model by using its almost full assortment of traditional policy parameters you tend to create more and stronger disruptive macro behavior of the same kind at some later period (Eliasson, 1985, pp. 78 ff; Eliasson and Taymaz, 1992).

8. The macroeconomic effects of improvements in communication technology – a research project design

The use and communication of information is the dominant resource use in a decentralized market economy. Information processes are governed by a dominant competence capital. The economic value of that competence capital depends on the ability of agents of the economy, including government to achieve certain targets. The value functions, however, will have to include not only target achievement measures but also the costs of failed decisions. The latter has normally been disregarded in analysis.

Information processing, broadly defined, includes critical tacit knowledge transfers within teams and firms. The efficient communication of tacit knowledge is a matter of organization, notably the organization of people. I have already gone through this and concluded that information in the experimentally organized economy, even though coded and technically communicable, will be based on subjective appreciations of the underlying fundamentals. Furthermore, only a fraction of the total stock of knowledge can technically be coded and communicated through artificial means. Even though the most important use and communication of knowledge occur through the interaction of people in teams, modern communication technology has enhanced the efficiency of that knowledge exchange process through improved *access* to people with knowledge and information (Eliasson, 1995a). Improved education and presentational techniques (theory) have also made previously tacit knowledge codable and communicable. However, the stock of knowledge (economic and technical) has also increased rapidly. The information processing capacity of the advanced decentralized and internationally integrated industrial economies has increased rapidly, but the share of total knowledge that can be communicated may have decreased.

The final section of this paper will present a method to evaluate the macro systems effects of this change in information technology, explaining how it is being used in an ongoing project at the IUI. Within the ongoing project telecommunications or rather information processing and communication in general is being studied empirically under three headings:

Demand

1. – from households
2. – from firms

Productivity change

3. – in firms
   – in the information and telecommunications sectors

Investment allocation and growth

4. How telecommunications-based product technology in financial markets affects the allocation of investment resources in, and the growth of the economy (dynamic systems effects).

Hence, we study three principally different macroeconomic effects of information and telecommunications use;

- *demand,* and the importance of *economies of scale,* notably in the production of telecommunications services
- the *supply* (productivity) effects of new technology at the micro level
- improved investment *allocation* techniques and economic systems stability.

The macroeconomic effects of each category can be studied separately in the Swedish micro-to-macro model. *The strongest leverage* on the economy is expected to occur through the dynamic investment allocation mechanisms in financial markets, and multimarket interaction between firms. In this last and most difficult part of the study we expect to find the most interesting cases of learning through mistaken business decisions.

The particular empirical applications, partly carried out already, but mostly to be done involve analyzing the explicit economic systems effects of the use of coded information and technical change in the production and use of particular and well defined information services.

Basic to this analysis is the production of information services under §4 above. We currently have the two information and communication sectors integrated in the MOSES model. We can make macro assumptions of technical change in these sectors, and, hence, the long-run cost determined price of communication services. In the future both macro analyses and micro firm-based analyses will be carried out. These, constantly less costly services, enter firms through §2 above and affect firm production through §3. They will all be empirically investigated. The simulation analysis on the micro-to-macro model will be based on a combination of empirical results from that study, notably about the effects of new data communications services on the relative development of capital and labor saving technical change in firms (see Eliasson, 1987). This is one sequence of well defined systems effects of the new communication technology.

The third type of effects, however, may be potentially more important. They occur through the capital market and improved allocation mechanisms (under §4). Such dynamic systems effects have been studied on the model for a long time. By improving what we call Schumpeterian innovative competition in the capital market (so-called $\mathcal{E}$ competition; see above and Eliasson, 1983, 1984b, 1988, 1991b), resources are more efficiently allocated over firms. Up to a limit faster arbitrage (through competition and more informed agents) will speed up long-term economic growth. The closer to a static equilibrium situation, however, the more disorderly market signalling, reducing the productivity effects through mistaken business decisions. The second type of allocation effects occur through the introduction of improved financial instruments, making it possible to distribute risks more efficiently. More exactly, new electronics-based financial instruments make it possible to establish separable markets for risks of the Markowitz (1952), Modigliani-Miller (1958)–Sharpe (1964) type. Even though only parts of the modelling work have been done, this is so important that we devote a separate section to this topic.

9. Improved allocation of resources through new telecommunications-based products in financial markets

Property rights are the fundamental institution in a market economy (North and Thomas, 1973). Tradability in property rights requires well defined ownership institutions. The most difficult ownership to establish is the financial contract that entitles you to future profits from intangible wealth, like the stock certificate (Eliasson, 1993a). New information and communication technology has played a remarkably efficient role in making the design of, and trading in new financial instruments feasible.

The emergence of the market for financial derivates

The services traded in financial markets are explicit (and sometimes implicit) contracts or entitlements to future profits from investments today. Such profits are generated from a pyramid of financial contracts in the form of interest and dividend payments and capital gains. At the bottom of this pyramid there should be contracts defining entitlements to the future profits from tangible and physically defined investments, like a factory or a firm (the stock certificate). But investment companies can own shares in firms that own shares in other investment companies that at the bottom hold shares in a physical operation, government bonds or other forms of financial contracts. The market for financial derivates is a pyramid of such contractual entitlements to future profits (Rybczynski, 1993), and modern information and communication technolgy has played the critical role in making the market for financial derivates feasible. The emergence of these markets has been made technically possible by new, electronically based information and communication technologies, but has also been stimulated by other incentives to capture rents in incomplete markets, *through:*

- avoiding regulation and taxes
- increasing risk sharing opportunities
- reducing transactions costs
- etc.

Globalization through the growing Euromarket occurred as a way to avoid government regulation during the 1960s. The options and financial futures markets developed in the 1970s as a means to improve the sharing of risks, making financial markets more or less complete (Allen and Gale, 1994). The markets for securitized loans and swaps, has grown enormously during the 1980s.

Has the decrease in total risks increased uncertainty in the financial system?

The more efficient distribution of risks, Allen and Gale (1994) argue has significantly increased the value of US stocks, as much as two-thirds between 1973 and 1986. The advantage for the individual of this market, as it has developed is that it allows you to take on the risks you want, and to hedge the risks you do not want to carry. This is indeed a very useful service. There are, however, three things to consider in this context. *Firstly* you have the computable risks that are being traded in the market. *Secondly,* in addition to the computable risks there may be true uncertainty associated with the underlying fundamentals. If this is the case, *thirdly,* how does the pyramid of risk trading in the market for derivates affect the total degree of *uncertainty* in the system; is it reduced or increased, or is the market destabilized?

The possibilities of taking on (buying) the risks you think you understand and passing other risks on to others for a fee, mean that you can give your portfolio of investments any risk profile you desire, making it possible for risk-averse individuals to enter typically risky markets, where competitive investors offer you risk coverage at a reasonable price. This essentially amounts to making financial markets more informed, lowering the total risk level by allocating total risks over the market, such that those who are competent to evaluate risks earn money by taking them on. Allen and Gale (1994) estimate that the increased efficiency of financial markets achieved through these improvements in markets for risks has reduced aggregate risks and thus increased the value of stock on the US stock market by as much as two thirds between 1973 and 1986.

Uncertainty, on the other hand, produces events that cannot be calculated from past observations. More particularly, if profit is your concern, and the value of your entitlements to future profits depends on the expected value of future $\hat{\varepsilon}$ (in the previous section), then if the $\hat{\varepsilon}$ could be forecasted by a stochastic equation with known parameters, no uncertainty would prevail. This is, however, not the case in the experimentally organized micro-macro model, as we have shown, and it should not be the case under the assumptions of Knight (1921). Were it the case, there would also be a case for insurers to enter the market for business risks.

The new situation that has emerged with the market for derivates is (1) that the arbitrage process has been speeded up, and (2) that a new uncertainty may have been layered on top of the fundamental uncertainty associated with business investment decisions. How do these two effects balance over time?

Separate markets for risks

Financial market arbitrage in itself (without a market for derivates) integrates (through multimarket interaction in price and quantity setting) all the markets affecting business decisions with the market for savings. For a separable market for risks (the market for derivates) to be established intermediaries trading in those risks have to be introduced. To do that risks have to be well defined, *for them,* to be tradable. As I have demonstrated above they would not be if the traders in risks were able to assess the entire business market situation. Hence, these traders have to set up their particular perceptions (models, theories) of their market niche, in the particular risks they are trading in.

This necessarily means a partial model, excluding consideration of a number of relevant market interactions. Such partial markets for risks are often based on observed stabilities in the past, like the small corner of the financial market for bonds backed by home mortgages where risks were assumed to be "hedgeable" by the empirically established "theory of market neutral mortgage investing". This market collapsed in the first week of April 1994, when movements in the thick hierarchy of derivates suddenly broke the empirical relationships upon which some heavily leveraged funds were based. Their equity value suddenly disappeared.

For a complex market economy the empirical fact is that for long periods there will be thousands of seemingly stable partial relationships that will be observed and used to trade on. Each trader will think of his observed relationship as a business secret, and invite customers to share his expected rents for a fee.

In the real world such empirical relationships build on physical and technical facts combined with less stable demand and supply conditions in markets.

The thicker financial market structures may, however, be internally quite unstable, resting as they do on a large number of observed relations between financial variables that in turn are based on the underlying value creating structures of the production system.

Stabilizing or shifty expectations?

To the extent that the market for derivates parcels out risks on institutions capable of evaluating them more efficiently than before, the market for derivates will reduce the total risk level as well as financial transaction costs in the economy. The market will also exercise a stabilizing influence.

To the extent that uncertainty prevails on top of computable risks, more traders in risks presumed to be computable might, on the one hand, reduce the total level of exposure by creating a more diverse set of assessments. This, however, is only true as long as such trading in seemingly computable risks does not increase the overall uncertainty of the economic system, subjecting it to sudden and unpredictable, expectational shifts in one direction. Such destabilizing shifts of expectations can only be avoided if individual assessments are very diverse and independent and not suddenly and dramatically moved now and then by some common underlying factor.

Empirically we know very little about that, but we can conjecture that if the information base available for more efficient allocation of risks through new financial instruments and more efficient markets, is only a part of the total information base that moves the entire economy, then the level of uncertainty in the economy at large will increase. It then also becomes very important to understand more about the potential exposure of the economic system to the uncertainty created by such shifts. We also know that they can only be studied through simulation of an economy-wide model.[3]

The long and the short term

The paradoxical result thus emerges that a growing pyramid structure of derivates effectuates the first task of diversifying risks, placing them in competent hands, thus lowering the total risk level of the economy, driving up stock prices and decreasing capital costs. There should be two types of positive effects on the macro economy:

1. decreased capital costs increase investment
2. more efficient arbitrage in financial markets allocates funds better on projects and risks.

The same effects will to some extent occur temporarily even in the case where these assessments are based on observed partial relationships in financial markets, rather than venture capital type evaluations of investment projects, through increasing diversity in the economy.

This takes us back to the earlier stability analysis. On the informational assumptions of the static or the efficient market model, risk reduction in more efficient separate markets for risk lowers capital costs and increases investment. But this time both the underlying fundamentals of the real economy and the complex pyramid of financial derivates may respond – if markets come too close to clearing – with destabilizing feedback through multimarket interactions or one-sided expectational shifts, causing significant real systems instability and even further reductions in the information content of market prices.

This analysis leaves us with an impressive structure of economic activities; a relatively thin real production system and an enormous superstructure of financial contracts, based indirectly on the real economy that speeds up allocation mechanisms in the real world through making the sum of all expectations of future profits immediately manifest in cashflows to and from real investors. The robustness of such a financially leveraged economy can only be studied through sophisticated simulation techniques based on the known and assumed behavior of agents in experimentally organized markets.

Notes

1. This is the case even though the M-M model, for all practical purposes, is deterministic. If you have the code of the M-M model, you can of course predict through a deterministic simulation. The question was, however, whether you would be able to learn the structure of the model (to perform that prediction), without access to the code, from observing the output from a large number of simulations, and with such precision that it would predict over a chosen future period, barring a predetermined and limited stochastic error. This question reduces to the problems of (1) finding an acceptable, estimable approximation of the M-M model, and (2) obtaining unbiased estimates of the parameters of that approximate model. If (3) the error terms between the M-M simulation ("reality") and the corresponding computed model values do not pass a test for randomness over any chosen simulation period, classical learning is not feasible and the particular behavioral characteristics of the firm of the EOE should exhibit themselves. The seemingly erratic behavior exhibited by the model economy, like major macro collapses that occur out of the blue (Eliasson, 1983, 1984b, 1991c) all originate in the endogenous changes of the Salter distributions, structural characteristics that are impossible to reproduce in a predictable way by known estimable modeling techniques. This is sufficient to rule out classical learning in the experimental setting of the M-M model.
2. If you can see it at all. It is atomistic.
3. See Eliasson and Taymaz (1992) where such simulation techniques have been used to determine how far policy makers can go against the market without destabilizing the macro economy.

References

Allen, F., Gale, P., 1994. *Financial Innovation and Risk Sharing*. Cambridge, MA: MIT Press.

Antonov, M., Trofimov, G., 1993. Learning through short-run macroeconomic forecasts in a micro-to-macro model. *Journal of Economic Behavior and Organization*, 21(2), June.

Carlsson, B., 1980. The content of productivity growth in Swedish manufacturing. *The Firms in the Market Economy, IUI 40 years 1939–1979*. Stockholm, Sweden: IUI.

Carlsson, B., 1989. The evolution of manufacturing technology and its impact on industrial structure: An International Study. *Small Business Economy*, 1(1):21–37.

Carlsson, B., 1991. Productivity analysis: A micro-to-macro perspective. In: E. Deiaco, E. Hörnell, G. Vickery (eds), *Technology and Investment – Crucial Issues for the 1990s*. London: Pinter.

Carlsson, B., Dahmén, E., Grufman, A., Josefsson, M., Örtengren, J. *Teknik och industristruktur – 70 – talets ekonomiska kris i historisk belysning*. (Swedish technology and industrial structure – the crisis of the seventies in historical perspective). Stockholm: IUI and IVA.

Day, R.H., Eliasson G. (eds), 1986. *The dynamics of market economies*. Amsterdam: North-Holland and Stockholm:IUI.

Day, R.H., Eliasson G., Wihlborg C. (eds), 1993. *The markets for innovation, ownership and control*. Amsterdam: North Holland and Stockholm: IUI.

Eliasson, G. 1976. *Business economic planning – theory, practice and comparison*. London: John Wiley & Sons.

Eliasson, G. 1983. On the optimal rate of structural adjustment. In: G. Eliasson, M. Sharefkin, B.-C. Ysander (eds), *Policy Making in a Disorderly World Economy*. IUI Conference Reports 1983:1. Stockholm: IUI.

Eliasson, G., 1984a. Informations- och styrsystem i stora företag (information and control systems in large firms). In: G. Eliasson, H. Fries, L. Jagrén, L. Oxelheim (eds), *Hur styrs storföretag? (How are large firms run?)*. Stockholm:IUI.

Eliasson, G., 1984b. Micro Heterogeneity of Firms and the Stability of Industrial Growth, *Journal of Economic Behavior and Organization* 5(3–4).

Eliasson, G., 1985. *The Firm and Financial Markets in the Swedish Micro-to-Macro Model – Theory, Model and Verification*. Stockholm: IUI.

Eliasson, G., 1987. *Technological Competition and Trade in the Experimentally Organized Economy*, Research Report No. 32. Stockholm: IUI.

Eliasson, G. 1988. Schumpeterian Innovation, Market Structure and the Stability of Industrial Development. In: H. Hanush (ed). *Evolutionary Economics, Applications of Schumpeter's Ideas*. Cambridge, UK: Cambridge University Press.

Eliasson, G. 1990a. The Knowledge Based Information Economy. In: G. Eliasson et al. (eds), *The Knowledge Based Information Economy*. Stockholm: IUI.

Eliasson, G. 1990b. The Firm as a Competent Team, *Journal of Economic Behavior and Organization* 13(3):275–298.

Eliasson, G. 1991a. Deregulation, innovative entry and structural diversity as a source of stable and rapid economic growth. *Journal of Evolutionary Economics* (1):49–63.

Eliasson, G. 1991b. Modeling the experimentally organized economy; complex dynamics in an empirical micro-macro model of endogenous economic growth. *Journal of Economic Behavior and Organization*, 16(1–2):153–182.

Eliasson, G. 1991c. The international firm: A vehicle for overcoming barriers to trade and a global intelligence organization diffusing the notion of a nation. In: L. Mattsson, B. Stymne (eds), *Corporate and Industry Strategies for Europe*. Amsterdam: Elsevier Science Publishers B.V.

Eliasson, G. 1992. Business competence, organizational learning and economic growth – establishing the Smith-Schumpeter-Wicksell (SSW) connection. In: F.M. Scherer, M. Perlman (eds), *Entrepreneurship, Technological Innovation, and Economic Growth. Studies in the Schumpeterian Tradition*. Ann Arbor, MI.: University of Michigan Press.

Eliasson, G., 1993a. Företagens, institutionernas och marknadernas roll i Sveriges ekonomiska kris, Nya villkor för ekonomi och politik. Ekonomikommissionens förslag. Stockholm: *SOU 1993:16*, bilaga 6 (pp. 195–233).

Eliasson, G., 1993b. *A note on privatization, contract technology and economic growth*. In: R.H. Day, G. Eliasson, C. Wihlborg (eds), 1993.

Eliasson, G., 1993c. *Endogenous Economic Growth through Selection*. Working Paper No. 397. Stockholm: IUI.

Eliasson, G., 1994. The Theory of the Firm and the Theory of Economic Growth. In: L. Magnusson (ed), *Evolutionary and Neo-Schumpeterian Approaches to Economics*. Boston/Dordrecht/London: Kluwer Academic Publishers.

Eliasson, G., 1995a. *The firm, its objectives, its controls and its organization*. Boston, Dordrecht, London: Kluwer Academic Publishers.

Eliasson, G., 1995b. Communication, information technology and the firm. Paper presented to the conference *Telecommunications, Infrastructure and the Information Economy*. Ann Arbor, Michigan, (March 10–11) 1995.

Eliasson, G., Taymaz, E., 1992. *The limits of policy making: An analysis of the consequences of boundedly rational government using the Swedish micro-to-macro model (MOSES)*. Working Paper No. 333. Stockholm: IUI.

Gustavsson, L., 1991. Om kreativitetens gränser (On the limits of creativity). In: *Den svenska modellen.* Stockholm: SAF.

Jantsch, E., 1968. Technological forecasting in corporate planning, *Long Range Planning Journal.* 1(1), Sept.

Knight, F., 1921. *Risk, uncertainty and profit.* Boston: Houghton-Mifflin.

Lamberton, D.M. (ed), 1971. *Economics of information and knowledge. selected readings.* Harmondsworth: Penguin Books.

Lindh, T. 1993. *Lessons from learning about rational expectations.* In: R.H. Day, G. Eliasson, C. Wihlborg (eds) 1993.

Loasby, B.S., 1967. Long-range formal planning in perspective. *Journal of Management Studies* 4:300–308.

Loasby, B.J. 1991. *Alfred Marshall's connecting principles of firms and markets.* Paper presented to the RES Annual Conference. Warwick: University of Warwick (April 8), 1991.

Markowitz, H.M., 1952. Portfolio selection. *Journal of Finance* 7:77–91.

Marschak, J., 1968. The economics of inquiring, communicating, deciding. *American Economic Review.* 58(2):1–18.

McKenzie, L.W., 1959. On the existence of general equilibrium for a competitive market. *Econometrica* 27(1):30–53.

Mill, J.S., 1848. *Principles of Political Economy with Some of Their Applications to Social Philosophy.* London.

Modigliani, F., Miller, M.H., 1958. The Cost of Capital, Corporation Finance and the Theory of Investment, *American Economic Review* (48), June:261–297.

North, D.C., Thomas, R.P., 1973. *The Rise of the Western World, A New Economic History.* Cambridge: Cambridge University Press.

Pelikan, P., 1969. Language as a limiting factor for centralization. *American Economic Review.* 59:625–631.

Rybczynski, T., 1993. Innovative activity and venture financing: *Access to markets and opportunities in Japan, the U.S. and Europe.* In: R.H. Day, G. Eliasson, C. Wihlborg (eds), 1993.

Salter, W.E.G., 1960. *Productivity, Technical Change.* Cambridge, UK: Cambridge University Press.

Schumpeter, J.A., 1911 (English edition 1934). The theory of economic development. *Harvard Economic Studies* XLVI. Cambridge: Harvard University Press.

Sharpe, W.F., 1964. Capital asset prices: A theory of market equilibrium under conditions of risk. *Journal of Finance,* XIX(3):425–442.

Simon, H.A. 1955. A Behavioral model of rational choice. *Quarterly Journal of Economics* 69:99–118.

Smith, A., 1776. *An inquiry into the nature and causes of the wealth of nations.* New York: Modern Library, 1937.

Stigler, G.J., 1961. The economics of information. *Journal of Political Economy,* 69:213–225.

Taymaz, E., 1990. A microsimulation analysis of manufacturing firms' demand for telecommunications services (Chapter V). In: G. Eliasson, S. Fölster, T. Lindberg, T. Pousette, E. Taymaz (eds), *The Knowledge Based Information Economy.* Stockholm: IUI.

Wicksell, K., 1898, *Geldzins und Güterpreise* (interest and prices), 1965. New York: AMK Bookseller.

Beyond Competition: The Future of Telecommunications.
Edited by D. Lamberton.

CHAPTER 20

Managing international research and development activities: the role of communications and information technologies

Peter Hagström
Graduate School of Business Administration, Harvard University, Boston, Massachusetts, USA

What is the possible contribution of international computerized information systems when it comes to coordinating geographically dispersed research and development (R&D) operations in multinational corporations (MNCs)? The relevance of the question is predicated on the observed geographical dispersal of R&D in MNCs and on the robust finding that effective R&D hinges on direct personal contact. However, the issue is largely absent in the literature on the management of international R&D.

A series of exploratory interviews in six selected MNCs and questionnaire data from 105 R&D units in Swedish MNCs reported here indicate an important and growing role for international information systems in MNCs. They have had an impact on old problems such as alleviating the "tyranny of distance" and improving integration. New problems and shortcomings have also surfaced, particularly as regards interdependent systems development where a project draws on inputs from several, specialized locations. On the other hand, many of the benefits from using international information systems evolve often in unanticipated ways. For instance, where cooperation was sought, constructive competition could result when different units fought to keep within deadlines and specifications in order to "beat" their counterparts in a joint project.

As the prevalence of the computer-mediated R&D activities tends to increase in industrial R&D, much of the R&D activities are automatically "prepackaged" for dissemination on the corporate communications network. In turn, that tends to improve knowledge-sharing in general, the working of joint projects, and local acceptance of new ideas in the MNC. In a related vein, the roles of R&D establishments emerge as both being varied and changing over time. Contrary to the common view that such roles should be assigned to different R&D units by corporate management, far-flung R&D sites were found to be much more active. Not content with implementation only, they also assumed different roles in the corporate R&D network.

This chapter first explores the challenges of managing international R&D activities in the light of existing research and frames the question of R&D coordination with

---

Address for correspondence: Peter Hagström, Harvard Business School, Soldiers Field, Baker West 188, Boston, MA 02163, USA. Tel. +1-617-495-6922. Fax. +1-617-496-2910.

regard to usage of corporate communications networks. The second section reports on the empirical findings. Lastly, some implications for the management of international R&D are discussed. Particular attention is given to the close kinship of computer-mediated R&D activities with the communications network, and to the varied and changing roles of R&D establishments.

What is the "problem"?

The *basic issue* is the tension between a) that geographically dispersed R&D *de facto* is very prevalent in today's MNCs and that this dispersal of R&D resources gives rise to nontrivial problems of coordination, and b) that the communication needed to support coordination of R&D activities continues to be almost exclusively framed as a question of direct personal contacts. Following on the seminal work of Allen (1970 and 1977), personal contacts have consistently been demonstrated to be the best form of communication for R&D operations. That insight clearly poses a problem for firms already having their R&D activities spread around the world.

Whereas patterns of communication have been extensively studied in the area of R&D management in general, that is not the case for international R&D within firms. The first, largely explorative, major academic research was carried out by Ronstadt (1977), later followed by Behrman and Fischer (1980) and Hewitt (1980). The importance of geographically dispersed R&D having been established, subsequent research has shown that there is a pervasive trend towards increased such R&D activity in MNCs (e.g., by Ronstadt, 1984; and Pearce, 1989).[1] In fact, it transpires that the 1980s saw a considerable surge in the process of internationalization of MNC R&D (cf. Granstrand et al., 1992; and Dunning, 1994).

International exploitation of knowledge – in the wide sense of the word – is in essence seen as the very *raison-d'être* of MNCs in the dominant economic theory of the MNC.[2] In the face of imperfect markets for knowledge, the firm is a superior vehicle for transfer of knowledge simply by internalizing the transactions pertaining to knowledge. Clearly, "knowledge" is a much broader concept than R&D or even technological know-how, but the international exploitation of the knowledge generated by R&D is a fundamental reason for why some firms operate across borders.

Exploitation does, however, not necessitate that R&D is performed abroad. For the firm that is basically a technology transfer problem which can be handled in a variety of ways. Maintaining international R&D operations becomes very costly by bringing difficulties of coordination, by being likely to slow down development and to lower productivity, and by possibly spreading the firm R&D resources too thin (De Meyer, 1992). The incentives for the traditional strategy of "developing at home and selling internationally" (cf. Terpstra, 1977) are still strong. Perceived economies of scale in carrying out R&D (Caves, 1982) reinforce this tendency.[3]

Explanations for geographically dispersed R&D such as access to more abundant and lower cost inputs (notably engineers), and the need for local, market-driven adaptation of products are weak at best. In particular, they seem ill-suited to account for the more recent surge in international R&D activities of MNCs. Newer approaches instead focus on the internationalization process itself. Since capabilities, opportunities and particular locational advantages are unevenly spread around the world, the ability to exploit such differences and learning from them are a powerful explanation for the continued relative success of MNCs *vis-à-vis* national firms, offsetting the relatively much higher coordination costs of MNCs.[4] Consequently, developing the required organizational capabilities for management and deployment of technological resources have commonly come to be seen as much more important than only technological prowess *per se* if the benefits of international operations are to be utilized effectively.[5] The process of internationalization has undermined any particular local or home-country monopoly on technological know-how, and the expansion of individual firms' operations abroad (notably of manufacturing and by acquisitions) has tended to bring R&D activities with it.[6] In short, conducting international R&D has both become something of a competitive requirement for, and a natural evolution of, the modern MNC.

Unable to escape the need for coordination of geographically dispersed R&D activities, MNCs have been found to have several, particularly important goals for such coordination:[7]

1. To avoid duplication of effort.
2. To reduce product variation on different geographical markets.
3. To facilitate technology transfer.
4. To ensure systems are compatible (in fact, any individualized approach is argued to be suboptimal when the technologies impact on several business areas).
5. To reap the potential benefits from combination of different logics and data residing in different parts of the MNC.
6. To capitalize on the underlying learning that can be seen to take place in international firms.

Now, dispersed R&D cannot be allowed to be disorganized R&D. Given the ends, what about the means of coordination? Whereas formal means are readily identifiable,[8] informal means are fuzzier in that they involve more of social interaction[9] and, as such, are more in line with the particular requirements of R&D management. In addition to the need for direct personal contact, R&D activities constitute a very complex set of activities not readily standardized;[10] on both counts offering particular challenges and costs. Any contribution by communications and information technology to mitigate these adverse circumstances would clearly be welcome.

In a study[11] of R&D units in 14 MNCs basically all of them were found to be using corporate international computerized information systems, but that their main contributions were as a temporary replacement for face-to-face communication and as lowering the communications threshold. A handshake was an important precondition for effectively using electronic communications.[12] Confidence is argued to have a "half-life" due to its erosion in the absence of renewed direct personal contact, but the decay of confidence could be significantly retarded by employing electronic communication. With truly international development projects, replacement of and support for personal contact must, however, be instituted (De Meyer, 1991). A second important issue retarding usage was concerns about security (De Meyer, 1992). Still, electronic communication was better at prolonging the "half-life" of confidence than either the telephone or written media (De Meyer, 1991).

Thus, usage of international computer-based information systems in MNC R&D amounts to more than applying a common functional tool. Such usage is also an integral part of the wider issue concerning coordination and, in our view, control of geographically dispersed activities. By their "infrastructural nature", international information systems support new ways of communicating, also across functions, as well as complement other means of communication.[13] In short, there are both direct and indirect, both formalized and informal ways of using international information systems in MNCs; something that can wield considerable influence on the way international R&D is carried out and managed.

Whatever the potential, there is a notable dearth of research on what usage of electronic information systems imply for international R&D in MNCs.[14] In turn, that leaves us little to compare with when turning to the empirical results at hand.

How do MNCs cope?

The purpose of the empirical investigation is largely an explorative one, given the limited prior research available. The inquiry has followed two tacks in parallel:

– First, a *questionnaire* was distributed to foreign R&D units of Swedish MNCs *in order to* establish "the lay of the land" in terms of computer-based information systems usage.[15]
– Second, a series of semistructured *interviews* was carried out in firms with advanced information systems capabilities *in order to* gain a richer understanding of the possible role of information systems in the R&D process.[16]

It is clear, both from the questionnaire and from the interviews, that using international information systems is perceived as a "must" for competitive reasons, but also that firms are still struggling "to make it happen". Evidence of such lingering problems was found even in the interviewed firms. Systems for reporting and financial management, electronic mail, logistics, product databases, and for computer-aided design and manufacturing (CAD/CAM) are well established in the firms singled out for closer scrutiny. Their collective experience tallies well with earlier findings[17] of firms starting with implementing control-related information systems and messaging systems before braving the task of introducing systems for direct (e.g., logistics) and indirect (e.g., common databases) coordination. Internationally accessed systems for development or project support purposes are still felt to be a complicated affair, in particular concerning integration across several such systems. In short, the R&D function is found at the tail-end when it comes to MNC international, coordinated computerization.

However, an authentic vision from one of the interviewed firms is that the *"critical mass[18] of an R&D department should be one person"*. Far from being outlandish, the vision was seen in the firm in question to be a realistic goal and achievable at least in theory given the on-going refinement of the corporate information systems infrastructure. The point is not one of necessarily wanting to create "one man R&D departments" but rather that it is perceived as being a quite feasible proposition in a manufacturing MNC today. A result from the survey data, on the other hand, inspires caution in that only four percent of the R&D units could be classified as avid users of video conferences[19] (see Fig. 1). The vast majority of units confess to never using video conferencing or only using it on what is likely to be a trial basis.

Nevertheless, the somewhat unusual vision above does have a bearing on how R&D may be managed once distributed. All firms in our sample of case studies – without exception – had experienced, and saw a continuing trend towards, an increased non-home-country share of corporate R&D. At the same time we found a pronounced preoccupation with the high cost of conducting R&D in many countries. That concern appeared stronger and was different than the one found in the literature of R&D in MNCs, which tends to be treated more as one of conducting R&D in different locations.[20] Cost and "critical mass" on the one hand, and *de facto* dispersed R&D on the other, then present something of an empirical enigma.

When raising this question in the interviewed firms, an important reason for dispersed R&D could be identified as "defensive" exploitation of dispersed resources/skills. These resources stem from earlier departments for local market adaptation, more recent acquisitions, and, to a lesser degree, host-country induced local R&D. Local learning effects, especially coming as a function of having a leading local customer, was also mentioned in a couple of the firms as a reason for having "a larger than normal" R&D unit in a particular country. Insights gained from that relationship can then be exploited on other markets.

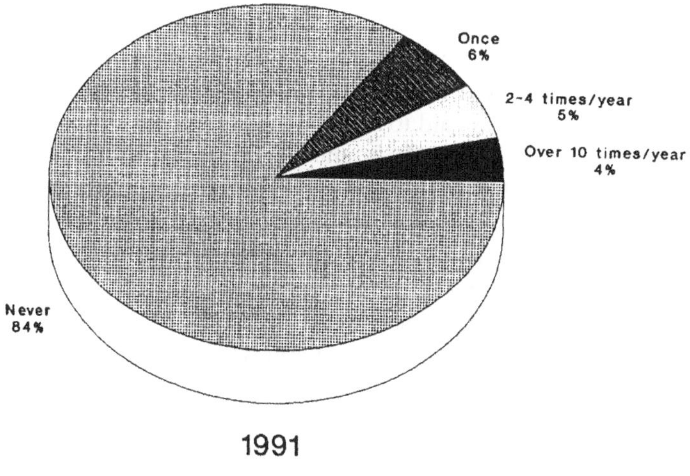

Fig. 1. Usage of video conferences (distribution of 96 R&D units in 12 Swedish MNCs). Source: own questionnaire data.

A somewhat curious, additional feature was that some totally new R&D departments, and which often represent a new direction in corporate R&D, were located *away* from existing firm subsidiaries.[21] The primary reason that was given by the firms in question was that well-established subsidiaries (or the headquarters) should not be allowed to dominate the new unit's activities.[22] Instead of the standard argument of the "locational pull" of corporate centers, here "locational push" is at work.

It should be emphasized, though, that the location of R&D activities in the interviewed firms was felt to be more of a process of fine-tuning and adaptation to necessity than active exploitation of particular locational opportunities. Another way of expressing it is that the firms' R&D strategy in terms of geographic location was perceived to be less than conscious, explicit and active.[23] Instead, it was generally argued that the main pressure for any locational change emanated from the desire to get more mileage out of an existing set of R&D establishments. Any expansion had most often been incremental, and any increased degree of internationalization was mainly a function of the firms being internationally active (including being active in acquiring other firms).

In all cases, the view was voiced that integration of international R&D had become easier, but was still by *no* means an easy task. The "critical mass" in terms of laboratory size had become less "critical", and the degrees of freedom were seen to have increased when it came to organizing corporate R&D. There was a fairly unanimous trend towards more local flexibility and greater room for peripheral entrepreneurship (if only the requisite resources could be secured). The explanations given tended to center on improved communications and (physical) transport having had a significant impact in the last decade or so, not least since the organizational structure and internal division of labor were finally being adapted in a more fundamental way to these new opportunities.

All the interviewed firms had data communications networks in place and it was reckoned that the R&D operations quickly would grind to a halt without these networks. The single exception was NSK, since most R&D still is carried out at a central location in Japan. Otherwise, differences in the basic dependence on the corporate communications network were slight. Possessing own communications capabilities in a very technical sense was said to be less crucial than has been found in other, broader studies (see OECD-BRIE, 1989; and Roche, 1992). Many, if not most, such services can be purchased. Even IBM, itself a supplier, purchases such services (e.g. virtual private network services from AT&T and basic transmission services for its European network from British Telecom's Syncordia). However, one firm had the control of its lines (including refusal to use public networks) elevated to a strategic issue because of security concerns.

Although usage of the corporate communications network was generally perceived as something positive that should be stimulated in order to foster integration, serious concerns about computer and communications costs were voiced in all R&D units that were visited. Much communication was indeed encouraged, but not all. Particularly worrisome is searching costs (communications and, more importantly, Central Processing Unit (CPU) costs). An opinion often encountered was that it was still "too easy" to use expensive CPUs, in spite of now having more work stations, as a function of improved communications possibilities and of the difficulty in individual pricing of CPU usage. In short, *inertia* in usage was seen as a problematic issue. There are some indications of moves towards individual charge-out, but so far the aim is only to make people more aware of the costs they incur. The experience is solidly that more cost information improves cost-consciousness even though no "penalty" is attached to the data made available. Of course, the problem here is not to stifle the productive communication (including maintaining personal contacts) through something like internal marginal pricing of communications services.

On a general level, we did find strong support among R&D personnel for the contention (cf. De Meyer, 1991 and 1992) that electronic communications were a complement to, and only a temporary replacement for, direct physical contact. Somewhat curiously, that complementary role had also been strengthened as travelling has become easier, since that has allowed for a wider network of personal contacts at all. Interestingly, initial electronic contacts were also reported by a number of interviewees to have a "paving the way"-effect in that expectations built up prior to actually meeting. One could thus talk of jacking up the initial level from which the decay, or half-life, sets in for personal contacts. In this context it should also be noted that even the large firms, the number of R&D personnel (and sites) is not so large that overview is lost. It was remarkable how often the number of people could be specified down to the individual when talking about R&D personnel, whereas other functions could only be given approximate numbers. In such readily identifiable environments, electronic communications tends to work well. There is some support for a hypothesis that this kind of communication is more widespread within R&D than in other functional areas.

The bottom line on cost, however, still is that although electronic communication does not come all that cheap, financial restrictions are more severe on travelling. Although travelling and electronic communication come out more as complements than substitutes, some astute interviewees pointed to a nontrivial source of saving: scheduled "necessary" trips can be made shorter and more efficient since electronic communication can improve the level of preparation for a meeting and can be used to "tie up loose ends", something that earlier would have required staying on site.

But there is more to electronic communications than supporting a fine web of personal contacts. It extends to support for specialization and modularization of R&D tasks. Independent development efforts could more easily receive required back-up from elsewhere in the organization paving the way for improvements in the quality of output. For instance, the centralized scanning of technical information would be impossible in Ericsson without their international communications capabilities (see also Bartlett and Ghoshal, 1990). In another of the interviewed firms, field tests could be monitored globally by the specialized unit saddled with the task of developing a particular process.

Interdependent systems development presents greater challenges of coordination if global or regional specialization is sought. A straightforward way of dealing with interdependencies in development is to modularize and specify the required interfaces.

Examples of this approach was found but with varying degrees of modularization. In hardware product development the task is particularly daunting with firms opting for as much independent specialization as possible. Still, several instances of hardware interfaces being specified (and modified as projects evolved) over the communications network were encountered. However, software[24] development is much more amenable to this kind of "wired modularization". Not only are interfaces typically defined in a format that is eminently suited for transmission via the data communications network, sections of developed software can be shipped around and tried for "fit" in the system under development. In one case, an overseas R&D unit was considerably bigger than its product mandate could motivate since the unit could capitalize on the local skills base by doing modularized software development work for other parts of the firm. In another company, there is an electronic market for software development man-hours; skills which were made available over the corporate network. A third example is a firm finding appropriate use for less sophisticated R&D personnel in a laboratory that existed solely as a result of host-country requirements by assigning software testing and debugging jobs to it; jobs delivered and collected over the data communications network.

An intriguing observation is that several firms reported a virtuous circle when trying to foster cooperation between dispersed R&D units. Instead of always getting the desired cooperation, internal competition tended to surface. The interesting twist here is that this competition was less destructive than it was constructive in that different units subsequently fought harder to keep within the agreed deadlines and specifications. The more that was found out about other units' activities, the more each unit tried to outdo their counterparts. Particularly in joint projects, the additional cost of coordination between units was in part offset by the gains from the internal "competitive climate" that was created.

Still, when comparing with other functions in the firm, R&D is something of a laggard when it comes to utilizing the potential for fragmentation, specialization and subsequent recombination of activities (cf. Hagström, 1990a). In practice, most R&D-generated traffic over the corporate data communications networks in the studied firms was regional rather than "global". With an analogy, a "hub-and-spoke-system" on a regional basis is the closest to coordination of global development projects apart from some isolated instances. In this image, the "hub" has the system responsibility and the "spokes" represent internally subcontracted development, typically of software.

Usage of CAD/CAM and common databases

Operationalizing the usage of common information systems in R&D, the relevant applications can be subsumed under the broad categories of common databases, including project databases, and of computer-aided-design/-manufacturing (CAD/CAM). Concerning the latter, the appropriate category is computer-aided design used in product development, but widespread practice favors using CAD/CAM as a single concept.

The survey data challenges the contention that the need for common CAD/CAM systems in R&D is a "must" for today's MNCs (see Fig. 2). Whereas some two-thirds of the respondents expressed strong agreement with a statement to that effect, a full 15% saw no such need (and with the remaining one-fifth having a "luke-warm" reaction). In the group being very sympathetic to this kind of standardization there is a strong bias toward firms with a high software content in their development efforts. Degree of internationalization or diversification did, on the other hand, not appear to influence the respondents' views.

There was no such ambiguity in the firms singled out for the case studies: without exception common CAD/CAM applications were judged to be so important that even being asked the question at all often caused surprise. However, it is noteworthy that Japanese firms, particularly in the manufacturing sector, have experienced considerable difficulties in implementing technical and manufacturing systems corporate-wide.[25] The same assessment was echoed in the non-Japanese firms in our case studies, and this lag was seen to constitute a competitive advantage *vis-à-vis* Japanese competitors in the respective industries.

Returning to the survey data, there is a fairly close correspondence between attitudes and actual usage (see Fig. 3). Still, there are evidently fewer respondents being convinced of the need for common applications than actually having been "blessed" with their arrival on the scene (65 vs. 73% in Fig. 2 and 3, respectively).

Figure 3 also distributes users of common CAD/CAM systems by the way in which they access them. All commonality indicates a measure of standardization, but how applications are accessed have additional implications. Network access improves the degree of interactive usage possible across great distances and tends to speed up the pace of upgrading computerized tools for product development. Still, a full one-third of respondents having common CAD/CAM systems reported local access only (see further below).

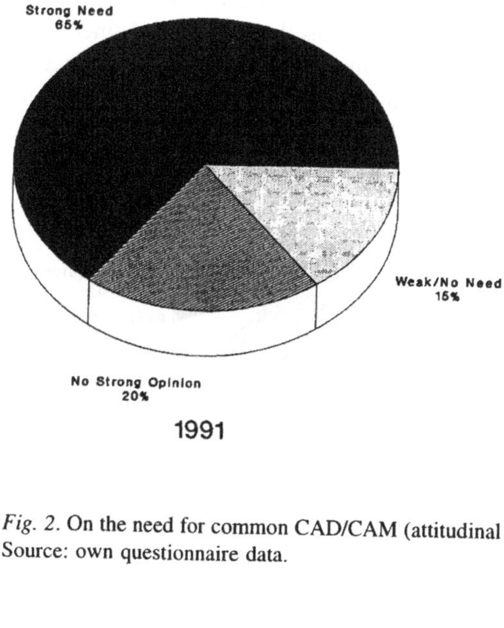

Fig. 2. On the need for common CAD/CAM (attitudinal data from 94 R&D units in 12 Swedish MNCs). Source: own questionnaire data.

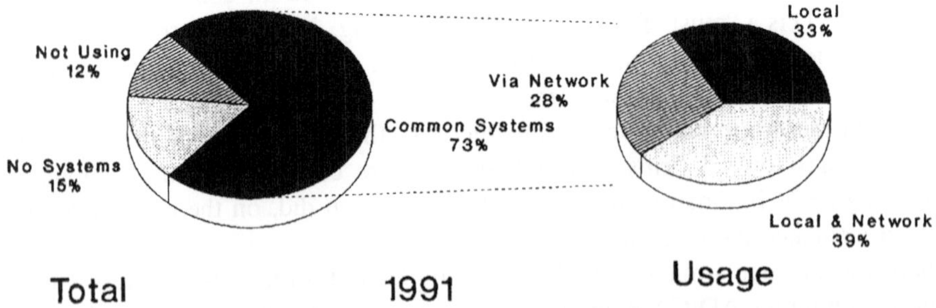

Fig. 3. Usage of common CAD/CAM/CAE-systems (distribution of 94 R&D units in 12 Swedish MNCs). Source: own questionnaire data.

Distributing usage of common systems by degree of "commonality" reveals a surprisingly high such "commonality" (see Fig. 4). A full four-fifths of the R&D units report a standardization that goes across business areas (the generic term "divisions" being used in Fig. 4). The indication here is one of CAD/CAM systems in use having a "general purpose" quality to them. That conclusion is, however, modified by the fact that Swedish MNCs are relatively less diversified in an international comparison.

The observed patterns in the case study firms shed some additional light on the survey findings. First, the positive attitude to using common CAD/CAM systems may well be underestimated. Respondents in the different R&D units are biased towards reflecting management views. Where interviews were carried out, the employees actually using the tools had a much more positive attitude to "commonality" than management. A second observation is the survey data obscuring the problem of the "smallest common denominator". Some integrative effects are lost when the CAD/CAM systems are dependent on input data from other applications, which may not quite be up to standard. Also, it is not uncommon to find that units in Anglo-Saxon countries have more sophisticated CAD/CAM-tools than other sites since those markets are the world-leading ones for such software. A third set of points concerns the role of an individual R&D unit. Where specialization was high and interdependence low, the need for common design tools was less readily accepted. However, that

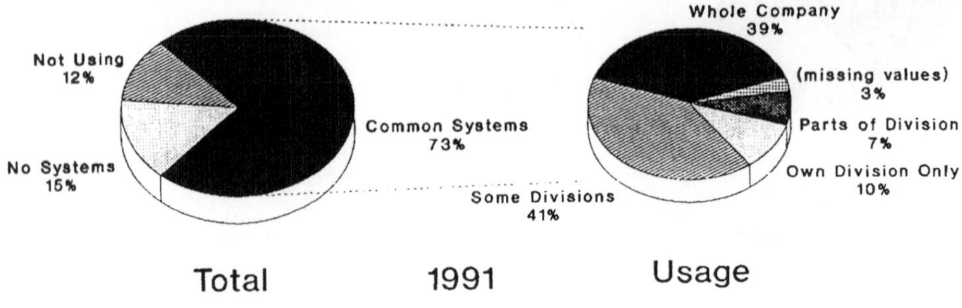

Fig. 4. Usage of common CAD/CAM/CAE-systems (distribution of 94 R&D units in 12 Swedish MNCs). Source: own questionnaire data.

is not an unequivocal observation in that the R&D mandate of a unit tends to be multidimensional and to vary over time. One should keep in mind that cross-sectional surveys only provide a snapshot of what is possibly changing practices. A partial, but significant illustration here, is the dramatic shift in favor of using work stations at the expense of mainframes. Even purely "local" access in Fig. 3 may well reflect that. In the case studies, the quick growth in usage of work stations tended to increase the share of "local" access without as much loss of standardization as previously would have been the case.

Data on common product/technical databases and databases on on-going projects show rather similar results to those of the more technical applications (see Figs. 5 and 6).

Common databases show up to be only slightly more popular than CAD/CAM systems in the surveyed Swedish MNCs (76 vs. 73% in Fig. 5 and 3 respectively). However, there are greater differences when it comes to access and the degree to which they are shared across businesses in the companies. A meagre 13% of the common database users report having a local access only (see Fig. 5). That should also be expected since the main idea with having a degree of commonality is to be able to share accumulated data, most often on product specifications and design characteristics. The converse situation is evident from whether databases are shared among different divisions, mostly product divisions (see Fig. 6). Databases are more likely to be shared across space and less likely to be shared across organizational boundaries than CAD/CAM systems. This also reflects the fact that databases typically are more product-specific than CAD/CAM tools. Thus the different tendencies for coordination across space (cf. access) and across businesses (cf. commonality) in different dimensions of R&D activity are illustrated. The point we wish to make based on the survey data is that sequential, joint, or, for that matter, specialized product development poses more of a spatial, i.e., communications network, challenge, whereas the problem of firm procedures on "how to do" things has more of a business coordination flair to it. Differently put, databases are more of an economies of scale issue, whereas CAD/CAM applications tend to be more associated with advantages of scope.[26]

Again, the case studies add insights to the observed patterns. The "smallest common denominator" problem reappears with a somewhat disturbing frequency.

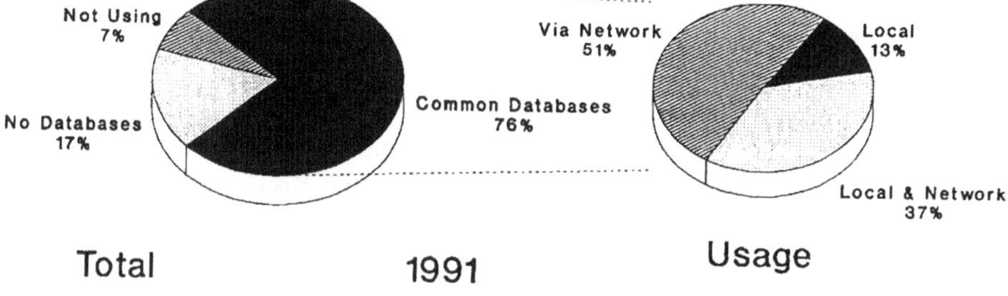

Fig. 5. Usage of common computerized databases (distribution of 94 R&D units in 12 Swedish MNCs). Source: own questionnaire data.

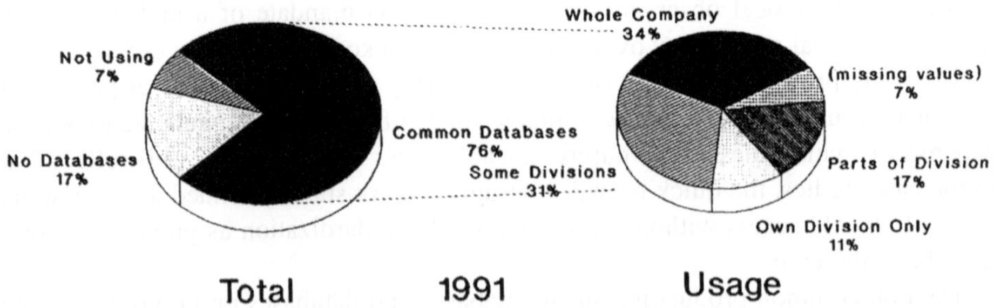

Fig. 6. Usage of common computerized databases (distribution of 94 R&D units in 12 Swedish MNCs). Source: own questionnaire data.

However, the problem is exacerbated by the limited incentives for units to make entries into common databases. They have a firm internal public goods nature to them in that from an individual's point of view entries from others are very helpful, but own data one has already. Increased use of powerful work stations has tended to undermine these already weak incentives. Procedures to induce interdependencies by managerial means[27] appears to be the best way to deal with the problem. That can also be deduced from the firms engaged in different kinds of modularized development, where the interdependencies "forced" participative use of common databases.

A slightly different, but related, issue is the one of security. Many, if not most, databases include commercially sensitive information such as engineering specifications and data on on-going projects concerning products under development. These are also the data where most could be gained from sharing. There are several problems associated with these security issues. The two most common ones were, first, the fact that critical information (e.g., objectives) were omitted from the database, and, second, the "catch 22" of having to know what is going on in projects in order to find out what is going on (since you cannot ask for information you are not aware of is available). No good solutions to these problems were encountered, but most firms were experimenting with greater differentiation and flexibility concerning access status for individuals in the R&D units. Perhaps a central "clearing-house" for such access authorizations is what is needed.

Implications for the management of international R&D

Rather than recalling issues and insights from the explorative discussion so far, we choose to highlight two areas where the empirical material indicates some challenge to widely held tenets of the management of international R&D in MNCs, viz.

1. The functionality of an international communications network.
2. The nature of the role(s) played by geographically dispersed R&D units in MNCs.

First, the international communications network offers a way to mitigate the difficulties encountered in managing R&D, which is argued (e.g., by Allen, 1977, and

De Meyer and Mizushima, 1989) to essentially be a question of communication; among the R&D staff and between the R&D staff and external sources of information.

Most of the research on R&D has been concerned with product R&D. This "hardware-orientation" tends to overemphasize the unstructured quality of R&D-related knowledge, be it embedded in a product or in the head of R&D personnel. On both accounts, physical transport (of goods or people) becomes a prime restriction when the degree of geographical dispersal that can be sustained is discussed. Still, we know that the software component in product offerings is increasing, not to mention the software component in service offerings. When software literally is software, it is already packaged for electronic transport. The idea of the "software factory" (Cusumano, 1991) illustrates how modularization of software components can remove the need for an integrated approach (see also Langlois and Robertson, 1992). *"Wired" modularization* not only of a "product" (e.g., a telephone exchange), but also of many associated R&D tasks (e.g., testing, record keeping and environmental scanning) constitutes an increasingly viable option for firms. In many cases, the modules can also immediately be tried for "fit" with the overall system through the communications network in which the very system may actually reside. These were clear trends observed in the interviewed firms, where the eminent *transportability* of computer-mediated activities was exploited. New, programmable technology relies on instructions and procedures that treat information as a separate entity (Sheil, 1983) and insofar as that kind of logic becomes more prevalent, transportability of R&D activities will improve further by virtue of new technology thus being codified "automatically" for electronic transfer.

The new possibilities for fragmenting (and subsequent recombination) of R&D activities translate into an increased freedom to exploit local advantages even of a temporary nature and to better utilize the capacity in a dispersed network of R&D facilities. An upshot of this argument is that a more diffused participation in a particular R&D effort will tend to offer some relief from the *"not-invented-here-syndrome"*, which has been found to all too often hinder that the full commercial advantage of completed R&D is realized in the MNC's different national markets (cf. Teece, 1976). Any improved local acceptance of new products or processes also reduces the time to market for new offerings.

The interviewed firms reported further benefits from using an international communications network such as "constructive competition" between dispersed R&D units, induced harmonization of product specifications and of procedures for R&D activities, and better possibilities for maintaining the network of personal, informal contacts. In short, the communications network makes a positive contribution to the integration of dispersed R&D. An interesting development in this context was that several interviewees indicated that more mature usage of the communications network was associated with a subtle shift in how information was acquired; rather than only being the passive recipients of information that was "sent" over the network, users would to a greater extent actively *search* for the relevant information. This may be seen as a way to deal with potential "information overload",[28] but is rather one indication of the changing role of far-flung R&D establishments in MNCs.

More recent, managerial theories of R&D in MNCs stress the need not to see R&D units as a homogenous group, but to *assign* them different roles depending on their specific capabilities and local external advantages (Bartlett and Ghoshal, 1989 and 1990; Ghoshal and Nohria, 1989; Gupta and Govindarajan, 1991; and Caeldries and Moenaert, 1993). Our discussion above has indicated increased opportunities for doing so with sophisticated usage of international information systems. However, two circumstances point to the need for modifying the picture. First, in recent theory the task of assigning of roles is given to corporate headquarters. For a complex MNC that is a gargantuan task if it is to be performed efficiently as it presupposes complete knowledge at the center; something that empirically is highly unlikely to be the case. Second, the R&D sites in the interviewed firms demonstrated a much more active role with "peripheral entrepreneurship", larger than "normal" R&D departments in subsidiaries, linkages to "own" external pools of knowledge, and increased internal competition for performing the firm's more modular R&D tasks. The flavor is one of also *assuming* roles in a differentiated network of R&D facilities. In addition, both assignment and assumption of roles is clearly becoming more fluid than commonly recognized. Temporal redistribution of roles in response to fast-changing local conditions is becoming more common, at least as evidenced by the experience of the interviewed firms.

In general, the strategic role of R&D appears to have deepened. In particular for manufacturing firms, some of the uncertainty as to the future was earlier absorbed by production and distribution. But with shorter lead times in production, less is manufactured on the basis of planning and more in close connection with customer orders. R&D may thus have its role as constituting an option on future states of nature relatively strengthened. A dispersed network of R&D facilities may then be less of a costly liability than a source of competitive advantage for MNCs in that in a less hegemonistic world, the MNC is exposed to more varied sources of information and is in a good position to exploit a wider range of locational advantages. A well developed corporate international communications infrastructure is an important ingredient if the most is to be made out of these opportunities.

Notes

1. See also Westney (1992) and the overviews in Caeldries and Moenaert (1993), Cheng and Bolon (1993), and Miller (1994). This increase has been observed for Swedish MNCs in particular by Håkanson (1990), and Håkanson and Nobel (1993a and 1993b). De Meyer and Mizushima (1989) note that European MNCs are more likely to conduct international R&D than US MNCs, which in turn have more international R&D than Japanese MNCs. Overall, the bulk of MNC R&D is, however, carried out in the home country (see Dunning, 1994 for a review, and also Patel and Pavitt, 1991).
2. For an authoritative overview, see Casson (1987).
3. Although commonly assumed, the empirical evidence for the existence of economies of scale in carrying out R&D is inconclusive (Granstrand et al., 1992; see also Pearce, 1989). This is mainly due to measurement problems.

4. Arguably, the first proponent of this basic idea is Vernon (1979). Important elaborations are Bartlett (1982 and 1986), and Kogut (1983).
5. The management challenge is argued by Bartlett and Ghoshal (1990) as constituting both enhancing efficiency and effectiveness of innovation processes and fostering conditions that bring out innovations through all different processes simultaneously. Similar arguments can also be found in Perrino and Tipping (1989), and in Pavitt (1991).
6. Pearce (1989) offers a wide-ranging literature review on the prevalence of dispersed R&D (see also Bartlett and Ghoshal, 1990; von Boehmer et al., 1992; and Cheng and Bolon, 1993). See also Håkanson and Zander (1986), Håkanson (1990), and Perrino and Tipping (1989) for further discussion and reporting of such findings.
7. Elaborating on Doz et al. (1986), Håkanson (1990), and De Meyer (1992).
8. For instance explicit rules, direct instructions, standards, pattern of authority, spending limits, reporting systems, budgets, plans, formal reviews and the like.
9. It is common to include meetings, boards, project teams, work groups, corporate culture, (non-financial) reward systems, recruitment, informal networks, and a whole host of socialization mechanisms. Martínez and Jarillo (1989) provides an excellent overview of informal means of coordination in particular and offers a review of research on their relative prevalence in MNCs over time.
10. See Teece (1976), and Bartlett and Ghoshal (1990).
11. Reported in De Meyer (1991 and 1992), but from slightly different angles. To our knowledge this is the only study to date that specifically explores the use of electronic communication in MNC R&D.
12. The preponderant view is that electronic communication is best suited for exchange within a known, shared context, whereas face-to-face communication is required for building that context (e.g., Zack, 1993).
13. These points are argued at greater length in Hagström (1991 and 1992).
14. This conclusion comes across indirectly from De Meyer (1991 and 1992), and Hagström (1991). The problem was highlighted in a state-of-the-art collective inquiry into firms' technology strategies, when the importance of "information and communications technology" was found to be beyond doubt, but when it was also noted that there was basically no research into the effects of corporations to operate information systems, as seen from a technology strategic perspective (Dodgson, 1989). Actually, neither internal workings of international firms, nor the role of information systems, are afforded mention even in recent, purportedly wide literature reviews (e.g., Cusumano and Elenkov, 1992, cf. also von Boehmer et al., 1992).
15. The questionnaire was sent to 18 of the largest Swedish firms with substantial R&D activities abroad, viz. *ABB*, *AGA*, *Alfa Laval*, *Atlas Copco*, Electrolux, *Ericsson*, *ESAB*, *Kabi Pharmacia*, Kanthal Höganäs, *Nobel Industries (Eka-Nobel)*, Perstorp, PLM, *SAAB-Scania*, Sandvik, *SKF*, Stora, *Tetra Pak*, *Trelleborg*, and *Volvo*.

    All R&D units of the above firms have been polled, giving a total sample of approximately 170. Here, the results from 105 such units (in the *italicized* firms) are reported. There does not appear to be any bias in the distribution of the units having reported to date. However, caution is recommended if extending the interpretation of the results as being valid for the whole sample (although one can also view the data on hand as constituting a 60% response rate).

    The questions relating to information systems usage constitute a part of a more extensive questionnaire, which addresses a much wider range of activities and assets in the R&D units.
16. Interviews focusing on the usage of computer-based information systems in R&D-related activities have been conducted in the following firms: *Air Products Plc., Ericsson, IBM (Networking Systems Business Unit and IBM Nordic), NSK, SKF, and Volvo.* Ericsson, IBM, and SKF have been singled out for more intensive study with the remaining firms serving as reference cases. The total number of interviews is in excess of 35. Illustrations in the text have been left anonymous unless the information is available from published sources.
17. See Hagström (1990a and 1992).

18. "Critical mass" has been popularized by Rogers (1983 and 1986) in relation to diffusion of technology. It is very often referred to as a main reason for concentrating firm R&D resources to a small number of locations, but there also seems to be diseconomonies of scale. For instance, Perrino and Tipping (1989), found support for the critical mass-idea, quantifying it at some 50–100 R&D personnel, also saw indications of an upper limit at some 250–300 people in an investigation of 16 "outstanding" international technology companies (see also Note 3, above).
19. Video conferencing is a technology that is on the cutting edge of present-day usage of international information systems and, as such, a reasonably robust indicator of firm-internal communications capabilities (cf. OECD-BRIE, 1989).
20. One rare exception is Bartlett and Ghoshal (1990), who stress the very high cost of globally linked innovation often being motivated by the requirement for transnational innovation (and who also specifically point to Ericsson as being a good illustration here).
21. Three major such cases in three different firms were found. In an additional case in a fourth firm the locational considerations were, however, somewhat more ambiguous but the same locational ingredient was present.
22. Another reason given was the higher productivity gained from R&D activities since less work needs to be done for local production. On the flip side there are problems, since new designs are more difficult to test in realistic environment and that it becomes more difficult to imbibe new R&D engineers with the requisite deep product knowledge. (For a general discussion on this kind of changing locational opportunity set, see Hagström, 1990b.)
23. With one firm being an exception.
24. Not to be confused with "firms in software industries" as the software component in product offerings of the "hardware firms" was considerable.
25. See Hosoda (1990), who is Executive Vice President of NSK, a company, it is argued, which has overcome these difficulties.
26. If the traditional view of scale economies as being linked to a particular product or function, then the communications network itself is a question of scope, since it can be used across basically all firm activities.
27. As found in one firm, where subsequent product testing was routinely dispersed although initial development may have been carried out at one site only.
28. "Information overload" is a *potential* problem in that improved information availability can make sorting and scanning difficult. However, and with an analogy, just because a common library has many books, it does not follow that one has to read all of them.

References

Allen, T.J., 1970. Communication Networks in R&D Labs. *R&D Management*, 1:14–21.

Allen, T.J., 1977. *Managing the Flow of Technology*, Cambridge, MA: The M.I.T. Press.

Bartlett, C.A., 1982. How multinational organizations evolve. *Journal of Business Strategy*, 3:20–32.

Bartlett, C.A., 1986. Building and managing the transnational: the new organizational challenge. In: Porter, M.E. (ed), *Competition in Global Industries*. Boston, MA: Harvard Business School Press, (pp. 367–401).

Bartlett, C.A., Ghoshal, S., 1989. *Managing Across Borders: The Transnational Solution*. Boston, MA: Harvard Business School Press.

Bartlett, C.A., Ghoshal, S., 1990. Managing innovation in the transnational corporation. In Bartlett, C.A., Doz, Y., Hedlund, G. (eds), *Managing the Global Firm*. London: Routledge, (pp. 215–255).

Behrman, J.N., Fischer, W.A., 1980. *Overseas R&D Activities of Transnational Companies*. Cambridge, MA: Oelgeschlager, Gunn and Hain.

Boehmer, A. von, Brockhoff, K., Pearson, A.W., 1992. The management of international research and development. In: Buckley, P.J., Brooke, M.Z. (eds), *International Business Studies: An Overview*, Oxford: Blackwell Publishers, (pp. 495–509).

Caeldries, F., Moenaert, R.K., 1993. International technology strategies: the role of innovation uncertainty. In: Rugman, A.M., Verbeke, A. (eds), *Research in Global Strategic Management*, Volume 4. Greenwich, CT: JAI Press Inc., (pp. 155—179).

Casson, M., 1987. *The Firm and the Market: Studies on Multinational Enterprise and the Scope of the Firm*, Oxford: Basil Blackwell.

Caves, R.E., 1982. *Multinational Enterprise and Economic Analysis*, Cambridge: Cambridge University Press.

Cheng, J.L.C., Bolon, D.S., 1993. The management of multinational R&D: a neglected topic in international business research. *Journal of International Business Studies*, 24:1—18.

Cusumano, M.A., 1991. *Japan's Software Factories: A Challenge to U.S. Management*. Cambridge, MA: MIT Press.

Cusumano, M.A., Elenkov, D., 1992. *Linking International Technology Transfer with Strategy and Management: A Literature Commentary*. Sloan School WP, Massachusetts Institute of Technology, Cambridge, MA, January 7 (mimeo).

De Meyer, A., 1991. Tech talk: how managers are simulating global R&D communication. *Sloan Management Review*, 32:49—58.

De Meyer, A., 1992. Management of international R&D operations. In: Granstrand, O., Håkanson, L., Sjölander, S. (eds), *Technology Management and International Business: Internationalization of R&D and Technology*, Chichester: John Wiley & Sons Ltd., (pp. 163—179).

De Meyer, A., Mizushima A., 1989. Global R&D management. *R&D Management*, 19:135—146.

Dodgson, M., 1989. Conclusions: realizing technological opportunities. In: Dodgson, M. (ed), *Technology Strategy and the Firm: Management and Public Policy*. Harlow: Longman Group UK Ltd., (pp. 219—230).

Doz, Y., Angelmar, R., Prahalad, C.K., 1986. Technological innovation and interdependence: a challenge for the large, complex firm. In: Horwitch, M. (ed), *Technology in the Modern Corporation: A Strategic Perspective*. New York, NY: Pergamon Press, (pp. 14—34).

Dunning, J.H., 1994. Multinational enterprises and the globalization of innovatory capacity. *Research Policy*, 23:67—88.

Ghoshal, S., Nohria, N., 1989. Internal differentiation within multinational corporations. *Strategic Management Journal*, 10:323—337.

Granstrand, O., Håkanson, L., Sjölander, S., 1992. Introduction and overview. In: Granstrand, O., Håkanson, L., Sjölander, S. (eds), *Technology Management and International Business: Internationalization of R&D and Technology*. Chichester: John Wiley & Sons Ltd., (pp. 1—18).

Gupta, A.K., Govindarajan, V., 1991. Knowledge flows and the structure of control within multinational corporations. *Academy of Management Review*, 16:768—792.

Hagström, P., 1990a. New information systems and the changing structure of MNCs. In: Bartlett, C.A., Doz, Y., Hedlund, G. (eds), *Managing the Global Firm*. London: Routledge, (pp. 164—185).

Hagström, P., 1990b. Unshackling corporate geography. *Geografiska Annaler*, Series B, Human Geography, 72(B):3—12.

Hagström, P., 1991. *The "Wired" MNC: The Role of Information Systems for Structural Change in Complex Organizations*. Stockholm: Institute of International Business.

Hagström, P., 1992. Inside the "wired" MNC. In: Antonelli, C. (ed), *The Economics of Information Networks*. Amsterdam: North Holland/Elsevier Science Publishers, (pp. 325—345).

Håkanson, L., 1990. International decentralization of R&D — the organizational challenges. In: Bartlett, C.A., Doz, Y., Hedlund, G. (eds), *Managing the Global Firm*. London: Routledge, (pp. 256—278).

Håkanson, L., Nobel, R., 1993a. Determinants of foreign R&D in Swedish multinationals. *Research Policy*, 22:397—411.

Håkanson, L., Nobel R., 1993b. Foreign research and development in Swedish Multinationals. *Research Policy*, 22:373—396.

Håkanson, L., Zander, U., 1986. *Managing International Research & Development*. Stockholm: Sveriges Mekanförbund.

Hewitt, G., 1980. Research and development performed abroad by US manufacturing multinationals. *Kyklos*, 33:308–327.

Hosoda, M., 1990. CIM at Nippon Seiko Co. *Long Range Planning*, 23:10–21.

Kogut, B., 1983. Foreign direct investment as a sequential process. In: Kindleberger, C.P., Audretsch, D. (eds), *The Multinational Corporation in the 1980s*. Cambridge, MA: The M.I.T. Press, (pp. 38–56).

Langlois, R.N., Robertson, P.L., 1992. Networks and innovation in a modular system: lessons from the microcomputer and stereo component industries. *Research Policy*, 21:297–313.

Martínez, J.I., Jarillo, J.C., 1989. The evolution of research on coordination mechanisms in multinational corporations. *Journal of International Business Studies*, 20:489–514.

Miller, R., 1994. Global R&D networks and large-scale innovations: the case of the automobile industry. *Research policy*, 23:27–46.

OECD-BRIE, 1989. Information networks and business strategies: an OECD-BRIE project on competitiveness and telecommunications policy. *BRIE Working Paper 38*, Berkeley Roundtable on the International Economy, University of California, Berkeley.

Patel, P., Pavitt, K., 1991. Large firms in the production of the world's technology: an important case of "nonglobalization", *Journal of International Business Studies*, 22:1–21.

Pavitt, K., 1991. Key characteristics of the large innovating firm. *British Journal of Management*, 2:41–50.

Pearce, R.D., 1989. *The Internationalization of Research and Development by Multinational Enterprises*. Basingstoke: Macmillan.

Perrino, A.C., Tipping, J.W. 1989. Global management of technology. *Research Technology Management*, 32:12–19.

Roche, E.M., 1992. *Managing Information Technology in Multinational Corporations*. New York, NY: Macmillan Publishing Company.

Rogers, E.M., 1983. *Diffusion of Innovations* (3rd edn). New York, NY: The Free Press.

Rogers, E.M., 1986. *Communication Technology: The New Media in Society*. New York, NY: The Free Press.

Ronstadt, R.C., 1977. *Research and Development Abroad by U. S. Multinationals*. New York, NY: Praeger Publishers.

Ronstadt, R.C., 1984. R&D abroad by US multinationals. In: Stobaugh, R., Wells Jr., L.T. (eds), *Technology Crossing Borders: The Choice, Transfer, and Management of International Technology Flows*. Boston, MA: Harvard Business School Press, (pp. 241–264).

Rubenstein, A.H., 1989. *Managing Technology in the Decentralized Firm*. New York, NY: John Wiley and Sons.

Sheil, B.A., 1983. Coping with complexity. *Office: Technology and People*, 1:295–320.

Teece, D.J., 1976. *The Multinational Corporation and the Resource Cost of International Technology Transfer*, Cambridge, MA: Ballinger Publishing Company.

Terpstra, V., 1977. International product policy: the role of foreign R&D. *Columbia Journal of World Business*, XII:24–32.

Vernon, R., 1979. The product cycle hypothesis in a new international environment. *Oxford Bulletin of Economics and Statistics*, 41:255–267.

Westney, E., 1992. Organizational change and the internationalization of R&D. In: Kochan, T.A., Useem, M. (eds), *Transforming Organizations*. New York, N. Y.: Oxford University Press, (pp. 245–260).

Zack, M.H., 1993. Interactivity and communication mode choice in ongoing management groups. *Information Systems Research*, 4:207–239.

Beyond Competition: The Future of Telecommunications.
Edited by D. Lamberton.

CHAPTER 21

Technology and systems competition in mobile communications

Sven Lindmark and Ove Granstrand
Chalmers University of Technology, Industrial Management and Economics, Göteborg, Sweden

Abbreviations: American Digital Cellular (ADC); American Telephone and Telegraph (AT&T); Advanced Mobile Phone Service (AMPS); Conférence Européenne des Postes et Télécommunications (CEPT); Centre National d'Etudes des Télécommunications (CNET); Cellular Telephone Industry Association (CTIA); Code Division Multiple Access (CDMA); Cordless Telephone System 1, 2, 3 (CT 1, 2, 3); Digital AMPS (D-AMPS); Digital Cellular System 1800 (DCS 1800); Digital European Cordless Telecommunications (DECT); European Telecommunications Standards Institute (ETSI); Federal Communications Commission (FCC); Frequency Division Multiple Access (FDMA); Future Land Mobile Telephone System (FLMTS); Frequency Modulation (FM); Global System for Mobile Communications (Groupe Spéciale Mobile) (GSM); Intelligent Network (IN); Integrated Services Digital Network (ISDN); Japanese Digital Cellular (JDC); Victor Company of Japan (JVC); Memorandum of Understanding (MoU); Nippon Automatic Mobile Telephone System (NAMTS); Nippon Electric Company (NEC); Nordic Mobile Telephone system (NMT); Nippon Telephone and Telegraph (NTT); Public Access Mobile Radio (PAMR); Private Branch eXchange (PBX); Personal Communications Network (PCN); Personal Communications Services (PCS); Personal Digital Cellular (PDC); Private Mobile Radio (PMR); Public Switched Telephone Network (PSTN); Post, Telegraph and Telephone administration in a country (PTT); Research and Development (R&D); Regional Bell Operating Company (RBOC); Radiocom 2000 (RC 2000); Stored Program Control (SPC); Total Access Communication System (TACS); Time Division Multiple Access (TDMA); Universal Mobile Telephone System (UMTS); Video Home System (VHS); Video Cassette Recorder (VCR).

1. Introduction

This paper addresses the phenomena of technology and systems competition, recognizing the general need for, as well as the difficulty of, multilevel competition analysis, not the least in telecommunications. The approach will be mainly qualitative, exploring these phenomena empirically, although an analytical and quantitative approach is tentatively used as well. To begin with, we will discuss, using some illustrative examples, different levels of competition, and next we will introduce a framework for multilevel competition analysis. Then, systems competition in cellular telephony will be analyzed empirically, based on historical accounts and data on different cellular systems. The international diffusion and competition of these different cellular systems will be described and analyzed, giving some tentative

---

Address for correspondence: Sven Lindmark and Ove Granstrand, Chalmers University of Technology, Industrial Management and Economics, S-412 96 Göteborg, Sweden. Tel. +46-31-772-3484/1209. Fax: +46-31-772-1237.

explanations for causes of success and failure of these systems. Furthermore, the technological competition, and possible substitution process, of wireless vs. wired telephony at the user end will be treated, as well as substitution of analog for digital cellular systems, using the simple logistic substitution model originally introduced by Fisher and Pry (1971). The model will also be used to provide a simple and speculative forecast of the future development of wireless systems. However, lack of data currently prevents an accurate econometric analysis. Since more data will be available as the competitive processes unfold, a number of hypotheses are formulated for further research.

2. Technology and multilevel competition analysis

2.1 General levels of competition

Competition is indeed a value-laden and many-faceted phenomenon, attracting a long-standing interest across a wide range of social and natural sciences (see e.g., Burke et al., 1988, for an overview). Various forms of interdependent competitive processes may be identified in an economy, which can be analyzed at several levels using a systems approach, quite in contrast to how competition has traditionally been analyzed in economic theory. This may be exemplified using audio/video/computer equipment. First, audio/video/computer equipment for consumer uses accounts for a certain share of the consumer's budget, and this type of equipment has to compete with other types of consumer spending.

Second, new product generations outcompete old ones: transistorized radios outcompeted vacuum-tube radios, color TV outcompeted black-and-white TV, etc. Typically, then, the technology base of the new product generation overlaps that of the old product generation, with some technologies becoming obsolete, some remaining unchanged and some being new, offering improved performance-to-cost ratios and/or entirely new functionalities. Usually a substitution implied by competition of this sort is not 100%, and the old product generation finds a certain niche, although perhaps small (thus contrasting with the principle of competitive exclusion in biology). This type of competition often also leads to intracompany competition or "cannibalism".

Third, two alternative technologies for essentially the same function may compete. Two examples are video cassette recorders (VCR) vs. video discs, that is, video tape vs. video disc technology, and electronic vs. chemical image processing in still photography. At subtechnology level, for example, different formats may compete (e.g., 8 mm vs. 16 mm tape). (For further examples see Arthur, 1988, among others.)

Fourth, there is the "ordinary" competition between different product or service brands from different companies, e.g., different types of VCRs (Sony, JVC, Philips etc.), or different company brands of essentially the same type of VCR (e.g., Nikon and Sony). From the company's point of view, this competition may be based on price and/or performance. Fifth, as a special case, there is sometimes competition between different product systems, around which sellers group themselves in a mixture of cooperation and competition. Examples are the "battle" between VHS (Video Home System) and the Betamax system in the consumer VCR area, and between IBM-compatible computers and non-IBM-compatible computers. This type of competition may be referred to as "systems competition" (see e.g., Katz and Shapiro, 1994, for further examples). The nature and means of such systems competition are not well understood, although apparently of increasing significance for future competitive conditions in many industries, not least in "network-based" industries like telecommunications (for an account of the "battle" between direct current and alternating current in the US, see David, 1987). An important means of gaining a competitive advantage in such systems competition is to achieve acceptance of one's product as the basis for a standard. An important means for this purpose may in turn be to have a large installed base, enjoying increasing returns by creating a strategic family of companies (e.g., through generous licensing or franchising agreements) with an internal mixture of cooperation and competition.

Sixth, a prominent type of competition in certain industries is competition among nations. Often national industries are promoted and/or protected by their governments, giving rise to a type of competition between different government/industry complexes. An "old" example is the competition between IBM and the Japanese mainframe computer industry, which was promoted by various Japanese government measures. To this exemplifying list of forms of competition may finally be added competition at group and individual level (as well as species level), a form of competition we are all familiar with, for better for worse.

We can now distinguish actor competition (between individuals, groups, companies, nations etc.) from the derived competition between objects like products, services, brand names, technologies, standards and systems. As these objects become adopted, diffusing throughout the economy, their diffusion processes interact in complementary, neutral or substituting ways as a result of more general interactive behavior among actors. These objects then share a "market" (suitably defined), just as companies do (cf. the notion of carrying capacity in biology).

Full substitution of one object for another we will regard as a special outcome of competition, which in turn will be regarded as a relation contrary to cooperation. It should now be clear what we mean by "technology competition", i.e., competition between technologies as distinguishable bodies of knowledge about techniques (as the word "technology" itself implies). The term "systems competition" refers to competition between products of a systems nature with many interrelated parts, or between services delivered by such a system. Typically the system has many interfaces between its parts, as well as with its operating environment, interfaces for which standardization is needed to avoid incompatibilities. These incompatibilities may also give rise to competing families of cooperating/competing companies, which reinforce the systems nature of the ensuing competition (see Bohlin and Granstrand, 1991, for a discussion of systems competition in this sense).

2.2 Multilevel competition in the telecommunications industry

In the telecommunications industry in general, competition could and should be analyzed at all the above-mentioned levels. Telecommunications accounts for an increasing share of the total spending in society, thus competing successfully for the consumer's budget. Various examples of new product generations outcompeting old ones can also be found, e.g., for cables and switches, which have undergone several technological transitions, with new product generations substituting for old ones.

Different technologies have competed in the area of transmission, e.g., intercontinental links, where radio transmission and undersea cables have competed for carrying traffic over (and under) the oceans. Introductions of new technologies and new product generations have tipped the competition in favor of either the one or the other transmission medium. Satellite technology gave radio transmission a competitive advantage, but cables based on fiber-optic technology have recently captured a large share of the intercontinental traffic.

Competition between companies and their product brands is an old phenomenon in the telecommunications equipment industry. Equipment suppliers have been active on the international market for several decades. In more recent years, liberalization and technology development have been driving the telecom operators to compete on international as well as domestic markets (see e.g., Johansson, 1994). Competition between branded telecom services is then a natural, although not yet fully developed, consequence. Finally, competition between nations and industry/government complexes can also be found, often linked with supporting regional standards and technologies.

2.3 Multilevel competition in mobile communications

"Mobile communications" or "wireless communications" usually refers to telecommunications, where the customer access, i.e., the first communication link, is wireless and uses radio transmission, thus making the user "mobile" (while making the call) or rather more mobile than if he were attached to a fixed network by wire.[1] "Mobile radio" usually refers to a private network for radio communications, while "mobile telephony" refers to radio voice communications using a network that is connected to the public switched telephone network (PSTN). "Cellular telephony" is a special case of mobile telephony (see section 3.2), where the area of coverage is divided into a cell-like structure with different cells using different frequencies. Traditionally, "cordless" technology has allowed the user mobility within a limited range from a base station connected to the PSTN. The scope of cordless telephony has now widened, enabling the user to access the PSTN via a public "telepoint" or a PBX.

In the mobile communications industry, both technology competition and systems competition seem intense compared to many other industries, making it necessary to analyze all the above-mentioned levels of competition. Some examples are provided here. First, the mobile communications industry accounts for an increasing share of the total telecommunications industry, thus competing successfully with fixed telecommunications for the total spending in society. Second, portable phones now account for almost all new mobile telephones sold, thus outcompeting car phones, and new generations of digital cellular systems are now challenging the analog ones (see section 4). The competition between those systems (NMT, TACS, GSM, etc., see below) is an example of systems competition in the sense that there is a market competition between systems rather than individual products (Katz and Shapiro, 1994). However, compared to the classical case of systems competition between VHS and Betamax in the VCR area, systems competition in mobile competition displays several differences. For example, the presence of a strong regulatory regime in telecommunications means that adoption of a system on the market of end users is essentially a two-stage adoption, since the system has to be adopted by the regulatory body, handing out concessions. The selection mechanism in the regulatory adoption stage is then different from a (pure) market selection mechanism, although it still could be considered as creating a form of competition.[2]

Furthermore, cellular technology competes successfully with other mobile communications technologies (e.g., paging, PMR and cordless technologies) in different applications. Data, for example, can be carried over specially designed mobile data networks but also over cellular networks. Several mobile data network systems have been competing internationally, supported by groupings of suppliers and operators, thus being, in our terminology, a case of systems competition.[3] Moreover, new technologies and developing standards, such as cordless technologies (e.g., DECT) offering increased capacity and functionality, as well as new mobile satellite systems (e.g., Iridium) offering worldwide coverage, threaten to substitute for cellular in some applications; but these technologies may also fuse, for example by using multimode terminals. New paging systems (e.g., ERMES) and trunked mobile radio/data systems (e.g., TETRA) are also potential substitutes for and complements to cellular telephony (and vice versa). All these (and other) existing and future technologies could be and are involved in a technological battle aiming for a future global personal communication system. Conceivably, these technologies are often within themselves under the influence of systems competition. Hence, there is an important but very complex interplay between technology competition and systems competition, which calls for some framework of analysis. Cellular technology is also competing with fixed network technology for customer access (see below). Finally, suppliers, distributors, operators, service providers and systems are competing for contracts, licenses and customers on the international market, in some cases supported by national governments.

Thus, the mobile communications service and equipment industry displays a complex structure of levels of competition. The general case of multilevel competition in telephony is graphically presented in Fig. 1. (Note that the level numbers in the figure do not correspond to the levels as numbered in section 2.1.) Technology competition occurs at levels III–V and systems competition occurs intra/interfirm/nation/technology-wise. At the level of intrafirm competition, important managerial challenges emerge in connection with "cannibalism". Not only does GSM cannibalize on NMT for Telia in Sweden or TACS for BT (i.e., Cellnet) in the UK, but it may

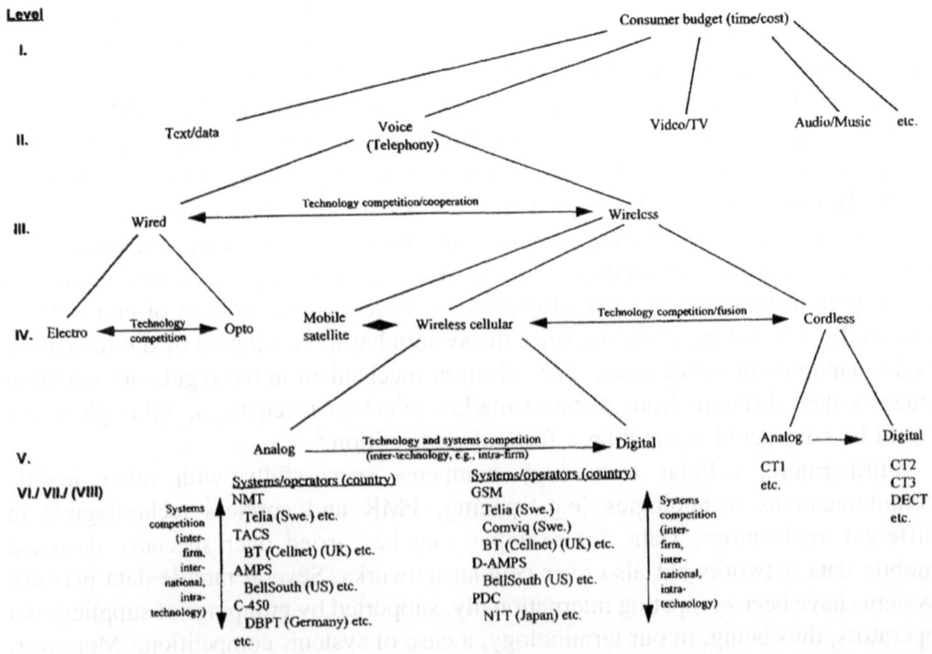

Fig. 1. Multilevel competition in the telecommunication access network. Note: Only technology relations of main relevance to voice telephony are shown here. Competitive processes are indicated by ↔, with → indicating substitution. Mobile satellite refers to high altitude systems providing wireless access at the user end, e.g., Inmarsat and proposed systems such as Iridium.

also cannibalize on wired telephony. Note also that clusters of actors may appear at other levels, e.g., manufacturers at levels IV–V.

This paper will not explicitly focus on all these relations but mainly on the systems competition occurring at levels VI–VIII (in section 3.2) and the technology competition and substitution occurring at levels III and V (in sections 4.2 and 4.3). In this context, an analytical framework will be outlined as a first attempt to come to grips with the complexities of multilevel competition in telecommunications.

2.4 Analytical framework

An analytical framework will be briefly outlined here, essentially using a systems approach. That is, a set of entities together with a set of typically dynamic relations will be identified (thereby making the so-called network approach a special case). The entities considered here could be actors (typically individuals, companies and nations) as well as their basis for economic subsistence (typically products and technologies) and the relations are, generally speaking, competition (adverse relation) and cooperation. Below we will especially illustrate the case of multilevel technology competition.

Figure 2 illustrates specifically three processes of technology competition taking

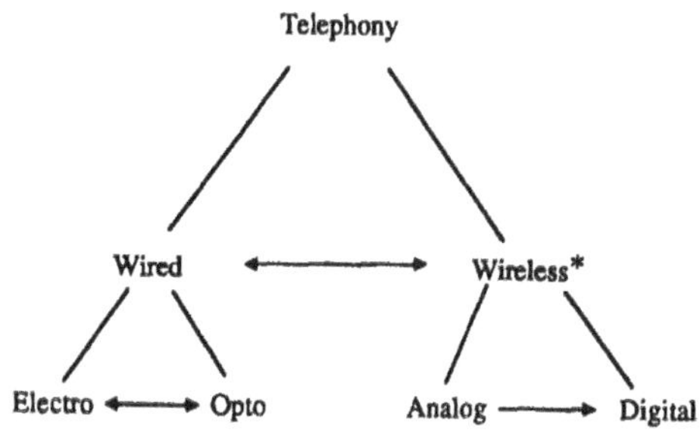

Fig. 2. Multilevel technology competition in telephony. *Only wireless cellular is considered for simplicity.

place at two levels: the competition between wired and wireless technology and, at a lower level, the competition between electro- and opto-cable wire technologies and the competition between analog and digital wireless technologies. Naturally more levels and technologies could be discerned (e.g., analog and digital signalling in wires).

Figure 3 gives another illustration of multilevel technology competition for the switching area. As buyers and sellers of products make their technology choices over time, they drive the process of technology competition. How can this be analyzed?

In the simple case of one-level competition between two technologies leading to substitution, this substitution has often empirically proved to be of the logistic or Fisher-Pry type (see further section 4.1). Also the case when several technologies substitute for each other in succession has been treated by logistic type modelling. The case of multilevel technology competition could be analyzed as indicated in Fig. 4.

We can look at a population b of adopters (buyers, sellers or more generally, number of lines or amount of traffic etc.), subdivided according to Figs. 2 and 4 (disregarding cordless for simplicity). Assuming logistic diffusion and substitution for the purpose of illustration gives interdependent dynamic equations such as (with constants k_{ij} and time t):

$$db_2/dt = k_1\ (b-b_2)b_2/b = k_1 b_1 b_2/b \tag{1}$$

$$db_{22}/dt = k_{21}(b_2-b_{22})b_{22}/b_2 = k_{21}b_{21}b_{22}/b_2 \tag{2}$$

If b is constant, $b_2(t)$ is the common logistic function, which gives the Fisher-Pry model. Similarly $b_{22}(t)$ is logistic if b_2 is constant. However, b_2 is not constant and the rate of substitution (transition) at level 1 affects the rate of substitution at level 2 (and in a more general case also vice versa). Still there are solutions in fairly simple closed form, even if b is allowed to change smoothly. In the case considered here (with b constant), the solution is a fractional polynomial of exponential functions of time.

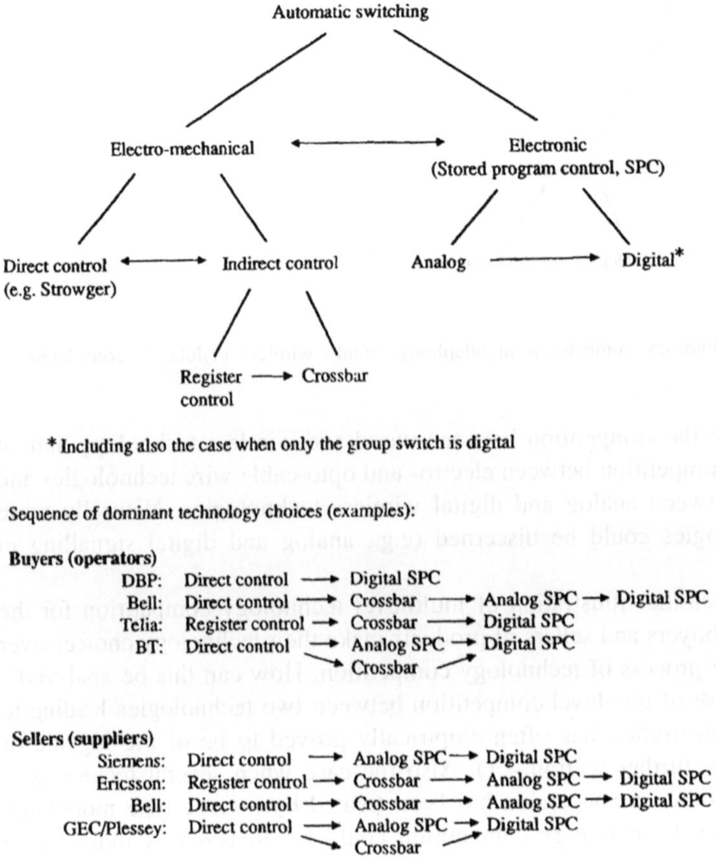

Fig. 3. Multilevel technology competition in public switching.

This framework can include general types of dynamics, e.g., of the general Lotka-Volterra type (see Granstrand 1994, Ch. 7), as well as it can be extended to even more general types of structures, e.g., more levels, more interrelations and more steps or stages with intermediate technologies in the transitions/substitutions.

Current lack of data has not allowed a more general case to be treated quantitatively in this paper. However, the framework here permits some qualitative reasoning. For example, a rapid transition from electro to opto cable (wire) corresponding to a migration of actors lowering b_{11} and increasing b_{12} would tilt the competition between wired and wireless in favor of wired, all else being equal, since opto cable offers more performance and service functionalities (e.g., multimedia). However, a rapid transition from analog to digital cellular technology in turn, corresponding to decreasing b_{21} and increasing b_{22}, will similarly result in tilting the competition between wired and wireless in favor of wireless, all else being equal. Moreover, if there is a transition from wire to wireless, it is likely to speed up any

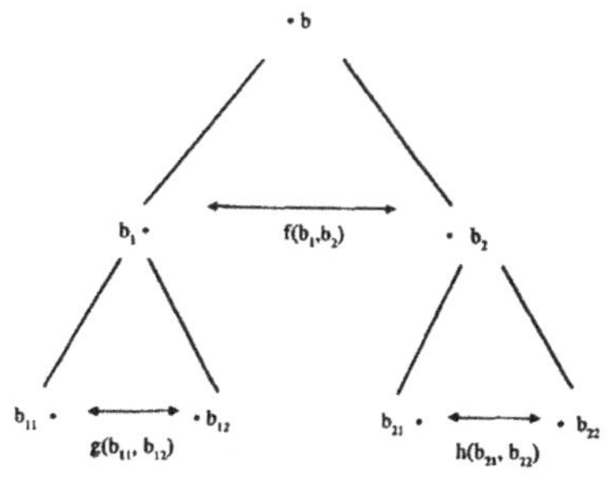

Legend:
b, b_1, b_2, etc. denote number of actors (e.g., subscribers).
f, g, h denote systems of dynamic equations (difference or differential) modelling the interaction (competition, cooperation) between the corresponding technologies (e.g., Lotka-Volterra equations).
Balancing equations:

$b = b_1 + b_2$
$b_1 = b_{11} + b_{12}$
$b_2 = b_{21} + b_{22}$

Fig. 4. Analytical framework for multilevel technology competition.

transition from analog to digital. Some empirical cases of this sort of interaction is given in section 4.

3. Technology and systems competition in cellular telephony

3.1 Historical overview[4]

This section traces the roots of cellular telephony and gives an overview of the development of different cellular system standards, setting the background for analyzing systems competition involving branded analog and digital systems as well as the technology competition between analog and digital cellular.

Precellular development
Cellular telephony is the result of a successful combination of wired telephony and radio technology, whose dates of historical origin roughly coincide. In 1873, a few years before A.G. Bell filed his famous patent application (in 1876), J.C. Maxwell published his theoretical discovery of radio waves, later confirmed by H. Hertz (in 1887). The Italian inventor G. Marconi was the pioneer in applying radio waves to practical use in the 1890s and 1900s. In 1906, R. Fessenden was able to transmit speech and music, which marked the beginning of voice radio communications. During the following decades, maritime mobile radio (1900s) and later land mobile radio (1930s) and maritime mobile telephony (1930s) developed, mainly driven by naval and police applications. Development of e.g., improved vacuum tubes and

frequency modulation (FM) made land mobile telephony feasible and, starting in the late 1940s, several countries introduced public land mobile telephone systems.

The cellular concept

Bell Labs is usually credited with having invented the design concept of cellular mobile telephony (in 1947), the main idea being to overcome radio spectrum congestion by combining space division with radio spectrum division. Within each spatial cell, there is a radio base station, and communication takes place on assigned frequencies, different from those of neighboring cells. Important features of a public cellular system are (a) handover (or hand-off), i.e., the transfer of the call between two base stations; (b) roaming, i.e., registering the location of the subscribers; (c) PSTN-access, that is, the cellular system is linked to the Public Switched Telephone Network, distinguishing it from stand-alone private radio networks which are sometimes ambiguously called mobile radio.

Technology to provide the proposed cellular system had not yet been developed[5] and mobile telephone systems with less functionality and capacity were being implemented in several countries (e.g., IMTS in the USA and MTD in Sweden). In the 1970s R&D on cellular systems gained momentum in parallel in a few countries (with the USA, the Nordic Countries and Japan as forerunners), which resulted in the introduction of eight cellular standards between the years of 1979 and 1985. The characteristics and functionality of these standards were basically the same.[6] Some facts about major cellular standards are given in Table 1.

AMPS (Advanced Mobile Phone Service)

By 1971, more than 20 years after the cellular concept was outlined, Bell Labs presented a detailed paper to FCC describing the cellular principles. In 1975 FCC reallocated a portion of the UHF television band to cellular service, and Bell-Labs started field-testing a small hexagonal cell grid, called the "cellular testbed". In 1977, FCC authorized two test licenses: (1) Illinois Bell Telephone (and Bell Labs) designed the Chicago Development Cellular System and (2) Motorola developed a similar test system in the Washington/Baltimore area, in partnership with the radio common carrier American Radio Telephone Service. After several years of disputes involving FCC, Bell and the radio common carriers, backed by Motorola, FCC opened 40 MHz of the 800 MHz band to cellular service, but a complicated licensing procedure further delayed the market introduction of AMPS. The first licenses were issued in 1982 and the first AMPS system was launched in 1983.

NMT (Nordic Mobile Telephone system)

In the mid-1960s, Televerket Radio carried out an investigation of future mobile telephony. The investigation proposed a country-wide system, and in 1969, at the Nordic[7] telecom conference in Kabelvåg, Lofoten, Norway, the idea of building a Nordic system was introduced and accepted. The Nordic PTTs developed a temporary system to satisfy the immediate needs of mobile telephony. In 1973, a report from the Nordic working group outlined a complete technical solution, stating that the system

Table 1. Some data and characteristics of selected cellular systems.

System	AMPS	TACS	NMT 450	NMT 900	C-450	GSM	D-AMPS	PDC
Main developers	Bell, Motorola	Ericsson, Motorola	Nord. PTTs, mainly Televerket	Nord. PTTs, mainly Televerket	Siemens	WE. PTTs, later manufacturers	a.o. Ericsson, Motorola, AT&T, Qualcomm	NTT
Home market	USA	UK	Nord.	Nord.	Germany	WE	USA	Japan
Technology	Analog	Analog	Analog	Analog	Analog	Digital	Digital	Digital
Access method	FM	FM	FM	FM	FM	TDMA	TDMA/CDMA	TDMA
Frequency band (MHz)	800	900	450	900	450	900/1800	800	800/1500
Introduction year	1983	1984	1981	1986	1985	1991	1991	1993
Foreign entry year	1984	1985	1981	1987	1986	1993	1992	–
Countries (end 1993)	51	25	36	11	3	26	7	1
Subscribers (thousands, end 1993)	20839	4851	1581	1789	851	1419	n.a.	n.a.
Infr. suppliers (end 1991)	8	4	5	5	2	10	n.a.	3
Term. suppliers (end 1991)	30	28	15	25	7	20	n.a.	6
Foreign subscribers (thousands, end 1993)	5150	2893	917	585	48	n.a.	n.a.	0
Major foreign markets (end 1993)	Canada, Australia, Mexico, Korea, Taiwan	Italy, Japan	Malaysia, France, Thailand	Switz., Neth., Thailand	Portugal, South Africa	Australia, Hong Kong, Singapore	Canada	–
Main regions for foreign entries	NA, LA, C, AP	WE, AP	WE, AP, EE, S	WE		AP, ME, EE, S	NA, LA, AP	

Note: Infr. = Infrastructure, Term. = Terminal, Nord. = Nordic countries, Switz. = Switzerland, Neth. = The Netherlands, NA = North America, LA = Latin America, C = Caribbean, WE = Western Europe, EE = Eastern Europe, S = former Soviet Union republics, ME = Middle East, AP = Asia Pacific. Note also that Cable & Wireless operates an AMPS system (Boatphone) in the Caribbean, here counted as one country. NMT infrastructure supplies include both NMT 450 and NMT 900.

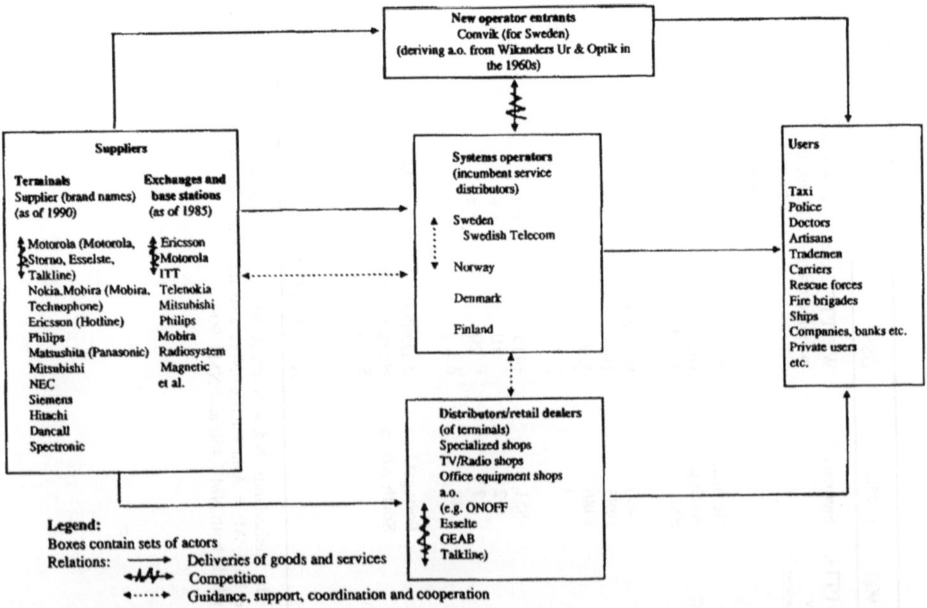

Fig. 5. The actor clusters in the development of the public land mobile telephone system NMT. Source: Granstrand (1993).

should utilize new SPC switching technology and roaming. During 1975–1978 systems development and test work were carried out, mainly at the radio laboratory at the Swedish PTT, Televerket (now Telia), and in 1978 a test system was successfully demonstrated. In 1980 a final technical specification was issued and on October 1, 1981, the system was officially introduced.

In the mid-1980s congestion occurred in the 450 MHz band, actualizing the need for a system with higher capacity. After some discussions on whether to wait for the GSM system or not, the Nordic PTTs introduced the NMT 900 system in late 1986. NMT 900, principally based on the NMT 450 technology, offered better functionality – portable phones and increased capacity – mainly as a result of the higher frequencies used. Since the introduction of NMT, the Nordic countries have shown the highest penetration rates in the world, constantly outstripping forecasts. Central to this development, and comparative success, was a cluster of cooperating teleservice providers (the Nordic PTTs), acting as a kind of "hub family". This "hub family" was subjected to competition from new entrants, allowed access to the market through regulated liberalization. The PTT family organized and cooperated with other families in turn for equipment supply and distribution. However, within the latter families competition prevailed. (Granstrand, 1993). Figure 5 depicts the actor competition and cooperation in connection with the development of NMT.

TACS (Total Access Communication System)

In the early 1980s the old mobile telephone system was to be replaced in the UK.

The only fully functioning cellular system at that time (besides Japanese NAMTS) was NMT 450. The British and French PTTs agreed on introducing an NMT system (at 900 MHz) which would include roaming between the two countries. However, this cooperation was never finalized, due to government intervention and deregulation of the British telecom market. Instead the government licensed two operators[8] to provide cellular service, based on a quickly developed, slightly modified and improved 900 MHz version of the AMPS system called TACS.

Other analog systems
When BT withdrew from the NMT cooperation with the French PTT, Matra developed RC 2000, a hybrid[9] system lacking automatic roaming, which was put into operation in Paris in 1985. In Germany, Siemens developed Netz C (C-450). Field tests began in 1978 and the system was commercialized in 1985. Another system, C-900, also developed by Siemens, was never commercialized.

In Japan NTT commenced R&D on mobile telephony systems around 1961. At first this R&D centred around a 400 MHz system. Later, it became clear that such a system would not meet demand, and in 1967 development effort was redirected towards an 800 MHz system with technology similar to AMPS. This so-called NAMTS was commercially introduced in 1979, being the first cellular system in the world, and some years later a high-capacity version went into operation to overcome congestion in Tokyo (Kuwabara, 1985).

GSM (Global System for Mobile communications)
In June 1982, CEPT (the European post and telecom administrations) appointed a working group with the task of specifying an automatic pan-European, public mobile communication system to operate in the 900 MHz band. This action was based on a proposal from the Scandinavian and Dutch telecom administrations, which pointed to the risk of losing the opportunity of European harmonization. The group was named Groupe Spéciale Mobile (GSM), in accordance with an earlier appointed group, GSI, working with ISDN. The main goal was to have completed the design of the basic outline of a standard for a fully automated public mobile communication system, including European roaming, in 1986. From 1982 to 1986 there was a discussion on basic design and technological alternatives, e.g., the choice between an analog and a digital system, the latter eventually prevailing. In 1984, initiated by the French and German telecom administrations, four Franco-German consortia were set up to develop digital cellular experimental systems, which turned out to be based on broadband TDMA. In parallel, Nordic PTTs and suppliers[10] had developed their own technical solutions based on narrowband TDMA. A prestigious technical debate was developing.

In late 1986, French CNET carried out tests, known as the "Paris trials", on GSM solutions proposed by the major manufacturers as well as Elab of Norway and Televerket of Sweden, under the supervision of the Permanent Nucleus.[11] The choice was mainly between a narrowband and a broadband solution. After a few months of heated discussions, the latter solution was finally chosen, and representatives of 15 countries signed the Memorandum of Understanding (MoU) on GSM, thereby sending signals to the industry of the commitment of the operators to implement the system. The standardization work was transferred from CEPT to ETSI (in 1989), and was divided into certain phases (phase 1, phase 2, phase 2+, etc.). Phase 1 was finished, under stress, in 1990–91 and the name of the system was changed to "Global System for Mobile communications". The GSM system was scheduled to commence operation on July 1, 1991, but most operators did not launch their systems until 1992.[12]

In 1990, at the request of the United Kingdom, where three consortia had been licensed to operate PCN networks in the 1800 MHz band, specifications of an 1800 MHz version of GSM, later named Digital Cellular System 1800 (DCS 1800), were included in the standardization work. The first DCS 1800 system went into operation in London in 1993.

D-AMPS (Digital AMPS)

In 1988 the US decided to introduce a digital standard. The major driving force was the expected congestion in the analog networks. AT&T proposed an FDMA system at an early stage, probably hoping to get it approved as a standard. As it turned out, the Cellular Telephony Industry Association's (CTIA) approach was to confront two technologies, FDMA and TDMA. The major industry actors backing FDMA were AT&T and later Motorola, while Ericsson and Northern Telecom were backing TDMA. In January 1989 a decision was taken with a small margin in favor of TDMA technology. Standardization then progressed fairly quickly and in 1990 CTIA published a first draft of the so-called IS-54 standard.

IS-54 includes dual-mode terminals, functioning in both analog and digital environments. The first systems were launched in 1991, but the system did not become commercially viable until 1993, and so far adoption has been slow.[13] Some suppliers have had problems delivering reliable systems, and several operators have been reluctant to adopt IS-54, waiting for the CDMA standard, developed by Qualcomm, to be available. In July 1993 CDMA was also approved as an official standard (IS-95), but it is not expected to become operational before 1995.[14]

PDC (Personal Digital Cellular)

In 1989, following the USA's decision, the Japanese authorities decided to introduce a digital system based on TDMA. Standardization progressed rapidly, most of the efforts being made in 1989 and 1990. Officially MPT was responsible, but most of the work towards standardization was performed by NTT. Nevertheless, large foreign infrastructure suppliers, such as Ericsson and Motorola, were allowed to participate. No controversies occurred openly, in contrast to the GSM and D-AMPS standardization. PDC is standardized in both an 800 MHz and a 1500 MHz version. As the first Japanese operator, NTT launched JDC (as it was called at that time) in 1993. JDC has changed its name to PDC as part of an international marketing effort so far (1995) unsuccessful.

3.2 International diffusion patterns of cellular systems

Some of the analog as well as digital system standards have diffused internationally. Figures 6 and 7 show this diffusion, in terms of number of subscribers and number of countries where the systems have been implemented, for those cellular systems that have been implemented outside their home markets.

The NMT 450 system, being the first Western cellular system and the first internationally adopted system, is widely diffused. During the mid-1980s, the system made entries into Western Europe,[15] but also into the Middle East and southeast Asia. NMT 450 benefited from early mover advantages, being the only cellular alternative.[16] Moreover, it was actively marketed, at least in the initial phase, a fact that most certainly contributed to its diffusion. When the UK chose to implement TACS instead of NMT, it marked the beginning of the end for NMT's future as *the* leading cellular system in the world. Instead all four major economic powers of Europe (Germany, France, Italy and the United Kingdom) chose to develop and implement their own national standards. NMT 450 was, however, the only system in the 450 MHz frequency band until C-450 was introduced in 1985, a time when the NMT system's diffusion was gaining momentum. When C-450 proved not to be a success in Germany, NMT 450 increasingly became the only viable alternative in the 450 MHz band. NMT 450 is more and more being considered a low-cost alternative, offering quick wide-area coverage and not occupying frequency spectrum in the 900 MHz band (often reserved for GSM systems or for the military as in Eastern Europe). This explains some of the diffusion of the NMT 450 system in the early 1990s, being implemented in most countries in Eastern Europe and the former Soviet Union, resulting in an increased interest among suppliers and the introduction of hand-portable phones.

Interestingly enough, compared with NMT 450, the diffusion of NMT 900 has been very modest, the main reason probably being that competition from AMPS and

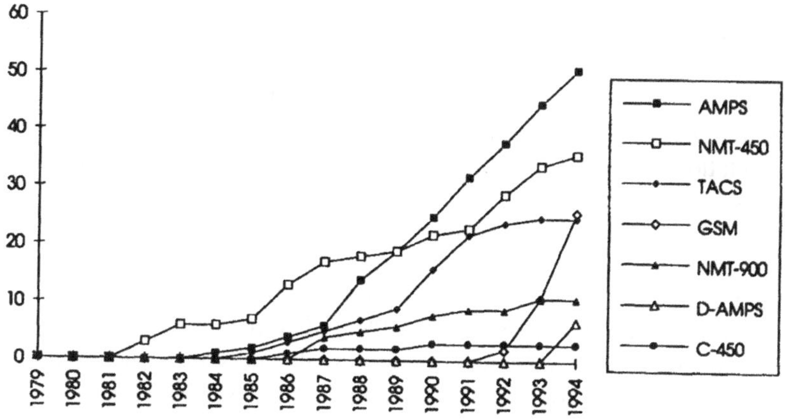

Fig. 6. Accumulated number of countries where selected cellular systems have been commercialized. Note: Beginning of years.

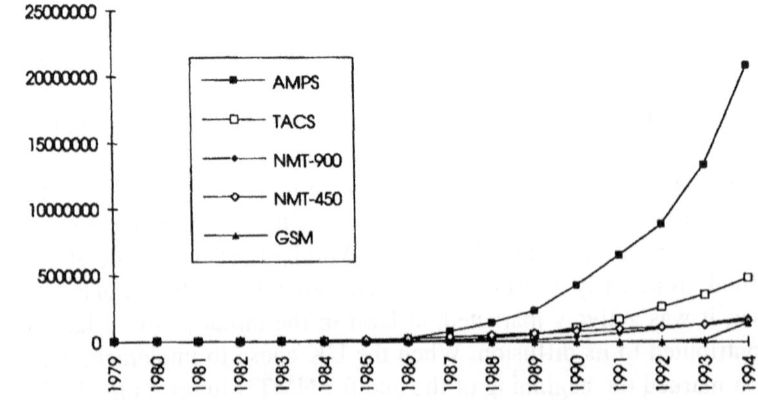

Fig. 7. Accumulated subscribers for selected cellular systems. Note: Beginning of years.

TACS was far more severe regarding NMT 900 in the late 1980s than was the competition from C-450 for NMT 450 in the mid-1980s. NMT 900 basically used the same technology as NMT 450 – technology which some industry experts state was out of date at the time, so that TACS and AMPS would therefore have shown a better price/performance ratio. However, NMT technology included one important feature, the possibility of international roaming, at least to neighboring countries.[17]

The AMPS system dominates the international market, especially in terms of subscribers. The fast subscriber development is explained not only by the size of the large home market (the USA), but also by some large markets in the Asia-Pacific region and Canada. AMPS has a monopoly position among analog systems over the whole American continent including the Caribbean islands, and is widely diffused in Asia-Pacific countries. At about the same time, as decision-makers in potential adopting countries were faced with the reality of incompatible national standards in Europe, AMPS diffusion was gaining momentum as the accumulated customer base began to outgrow the other systems. Since learning by producing/dynamic economies of scale leads to functional improvement and consequently to a more rapid price reduction, expectations that this would happen in the case of AMPS may have boosted the sales of this system.

TACS's international diffusion had a relatively slow start, but gained momentum around the turn of the decade. The TACS system can be found all over the world except for America. TACS seems to have been promoted by the British Government but also by Motorola.[18] Furthermore, the AMPS and TACS systems are similar enough to allow manufacturers some economies of scale and scope when manufacturing both AMPS and TACS compared to manufacturing AMPS/TACS and NMT. This in turn would lead to comparatively more supplier-induced improvements of performance-to-cost ratios of AMPS and TACS, all else being equal.[19]

Another feature distinguishing NMT from e.g., AMPS and TACS is the open interface between switches and radio base stations, a fact which reduces the operator's switching costs. This implies competition among suppliers, and would probably bias both the authorities' and operators' decisions in favor of such a system. However, at the same time one might suspect that the powerful infrastructure suppliers would prefer a system where they can "lock in" the customer, perhaps at the risk of a mutual lock-in.

A similar systems competition has emerged between the different digital standards. GSM was developed with the intention of being a European standard, but is rapidly spreading throughout Eastern Europe, the former Soviet Union republics, Africa, the Middle East, Asia Pacific and even Latin America. The GSM standard has so far (as of December 1994) become operational in some 26 countries.[20] One could compare this with the, so far, modest diffusion of the USA's D-AMPS (operational in seven countries) and Japanese PDC (not diffused at all). Further to this development, DCS 1800 is one of the standards proposed to be used for PCS in the United States.

PDC and GSM have been actively promoted outside their home territories. JDC was changed to PDC when the Japanese authorities, as well as NTT and several suppliers, began to promote the system (so far unsuccessfully) throughout Southeast Asia. The proponents of the Japanese system have been compelled to market their system, since GSM and D-AMPS have secured a much larger base in their home markets (and internationally).

Timing has been crucial, since GSM has been perceived as the only viable and proven digital alternative, thus benefiting from early mover advantages. Moreover, the roaming functionality will enable adopters to use subscriptions and terminals in adjacent and other countries.[21] These are some explanations for the very rapid diffusion of GSM so far – a success which makes some GSM proponents believe that GSM is well on its way to becoming a de facto global standard.

There are, however, some factors that might promote the other digital standards or at least slow down GSM diffusion. D-AMPS uses dual-mode terminals and shares infrastructure with analog AMPS, thereby locking in operators already using AMPS. Everything else being equal, this would imply that there will be at least as many countries using D-AMPS as AMPS. Furthermore, PDC exists in a 1500 MHz version which may lead to the opening up of some markets, providing the opportunity of using an extra bit of spectrum. Finally, there is a possibility that CDMA proves to be considerably more effective than TDMA. If so, in its CDMA variant D-AMPS will have a competitive advantage over all TDMA standards. We should not forget what the situation looked like in 1986–1987 and claim that the battle is won for GSM.

3.3 Empirical summary and discussion

Although the initial cellular idea originated in the United States, incompatible analog cellular systems developed more or less in parallel in the USA, Japan and the Nordic countries. Some of these systems have diffused internationally, the most widely diffused being NMT 450, AMPS and TACS. Regulatory decisions, which have been influenced by a large set of technical, economic and political factors such as frequency availability, size of and success on home market of the system, timing of commercialization, lack of aggressive international marketing and performance-to-price ratios, increasing returns in service provision, and economies of scale and scope in manufacturing, have been of particular importance behind this international diffusion. The complexity of the influential factors is apparently due to the systems nature, regional differences and the political context.

Digital systems have also diffused internationally, although any patterns have not yet fully developed. Not surprisingly, digital systems are diffusing much faster than did the analog systems, with a significant convergence towards GSM. An early vision and a dedicated development effort led to early mover advantages for the European digital cellular system GSM, so far (as of March 1995) being the most widely diffused digital cellular system. However, lower switching costs may lead to a rapid diffusion of D-AMPS.

Competition between cellular systems has been somewhat restrained, because of the importance of the first-stage adopters – the regulators and/or operators (see section 2.3). Nevertheless these system standards have diffused internationally and in some cases been competing for adoption in a country. Thus, so far, there has been a "restrained systems competition" making it preferable for manufacturers (e.g., Ericsson) and operators (e.g., AirTouch) with international ambitions to support several systems standards – i.e., pursuing a multisystem strategy.

It may also be noted that the USA analog cellular market, characterized by a single system standard and market competition in service provision, shows higher penetration rates than the European analog market (on an aggregated level), which has been essentially characterized by multiple incompatible standards and national monopolies. In the case of digital systems the situation is quite different, with multiple standards in the USA and low penetration compared to Europe, having a single standard coupled with introduction of market competition. From this it may be hypothesized that the number of system standards has a negative impact on diffusion rate, while market competition (within certain limits) affects diffusion rate positively.

Systems competition may acquire even increasing future importance in the cellular industry. Manufacturers have already been "competing" in early phases of the systems specification and standardization process to get their technical solutions accepted as standards (e.g., broadband vs. narrowband TDMA for GSM, and FDMA vs. TDMA and later TDMA vs. CDMA for D-AMPS). Moreover, the US authorities have not chosen one single standard for the digital systems, either for D-AMPS or for the PCS-licenses. Consequently there are already signs of manufacturers and operators clustering around different technical solutions. In this case the systems competition is becoming more closely coupled with technology competition. Besides, introduction of new digital wireless systems, e.g., PCS and DCS 1800, will give long-distance operators the possibility to bypass the RBOCs in the local loop (as in the USA) and will allow competing PSTN operators wireless customer access at low cost (as with Mercury in the UK and possibly Tele2–Comviq in Sweden), which in turn may speed up a transition from analog to digital wireless as well as from wired to wireless access.

4. A tentative forecast of the technology competition

4.1 Introduction to the model

Numerous forecasting models and models for describing technological diffusion, substitution and competition have been proposed over time (see e.g., Mahajan and Petersen, 1985). However, assuming logistic diffusion and substitution (see section 2.4), we will use the Fisher-Pry (1971) model, in testing for substitution, mainly because of the simplicity and generality of the model and because it has proved to describe substitution patterns fairly well (see e.g., Porter et al., 1991).

Fisher and Pry (1971) give a simple model for technological substitution which can be described by the following expression (where f= the market share of the new technology, $t_{0.5}$= the time at which f(t)=0.5, i.e., the substitution midpoint, and with a constant k and time t):[22]

$$f(t)=1/\{1+\exp[-k(t-t_{0.5})]\} \tag{3}$$

When trying to forecast the technology competition using a model for substitution, we must first determine if a substitution is actually taking place. Testing for this, we have used the following simple and speculative approach: (1) Estimate the parameters for k, $t_{0.5}$ and L,[23] on the basis of data up to the introduction of the new system. (2) Based on these figures, forecast the growth of the old system (A_F). (3) Compare these figures with those of the actual outcome. The difference between the forecast and the actual outcome of the old system (A) will tentatively, with a certain probability,[24] be due to a substituting effect if this difference is positive, ($A_F - A > 0$) and accordingly be due to a complementary effect if the difference is negative (i.e., $A_F - A < 0$). If the total growth (i.e., including both systems) is larger than the forecasted growth (i.e., $A_F - (A+D) < 0$) of the old system, this may explained by both complementary effects and the fact that the new system attracts new customer groups, e.g., because of lower price and/or more functionality (which, in turn is driving the substitution).[25]

4.2 Results

4.2.1 Digital vs. analog cellular

The first question we will try to answer is whether digital cellular is substituting for analog cellular. In this paper we have tested for substitution by using subscriber statistics.[26]

Table 2 shows indications of substitution in Germany and Denmark in 1993 and Denmark, Sweden and Europe on an aggregated level in 1994. However, using other estimates (not shown here) of the upper bound (L) results in indications of substitution also in Portugal and France in both 1993 and 1994 and for Germany in 1994. The results are, however, distorted by the following circumstances: (1) A temporary decrease in growth rates during 1991 and 1992, probably due to the economic recession at that time. (2) The analog systems show an extremely high

Table 2. Substitution indicators for selected markets in Europe.

Market	Analog system	Analog growth (1993)	Digital growth (1993)	Forecast A_f (1993)	A_f–A (1993)	A_f–(A+D) (1993)	Analog Growth (1994)	Digital growth (D) (1994)	Forecast A_f (1994)	A_f –A (1994)	AF–(A+D) (1994)
Europe	Various	1509	1161	3233	1724	563	2563	2827	4660	***2097***	–731
Sweden	NMT	118	49	256	138	89	184	374	325	***140***	–234
Germany	C-450	21	786	662	***641***	–145	–76	776	1144	1220	444
Portugal	C-450	5	54	55	550	497	–3	93	2507	1952	2010
Finland	NMT	103	20	147	44	24	79	92	193	114	22
Denmark	NMT	41	60	69	***28***	–32	19	174	85	***66***	–108
Italy	TACS	404	6	1239	835	829	964	69	2535	1571	1502
France	RC2000 +NMTF	30	91	222	192	101	–46	368	337	383	15
UK	TACS	494	42	721	226	184	1102	422	987	–116	–537

Note: Forecasting according to section 4.1. Since GSM systems in the countries concerned started up in 1992, the forecast is simply the mean of forecasts; one based on data up until the end of 1991 and the other based on the data up until the end of 1992. All figures are yearly subscriber growth rates in thousands A_f = Forecast of growth in analog system(s). A= Actual growth of digital system, ***Bold italics*** indicate substitution.

growth rate just after the digital systems' introduction. This extreme growth might be explained as the combined effect of an upward economic trend and extensive marketing of the old analog system. The latter is probably a result of analog system monopolist operators trying to "squeeze out" as much as possible from the analog systems, avoiding "cannibalism" when subjected to competition from competing GSM operators, before switching to GSM themselves. Thus a sort of "market sailing effect" could be observed in the technology transition, analogous to the often observed "sailing effect" in technical parameters, that is, improvements of an old technology when subjected to competition from a new one. Yet another tentative explanation is that increased marketing from new GSM operators spills over to the analog networks, at least initially. Therefore a substitution might take place on other markets as well, which could have been detected with a more refined forecasting model, taking initial complementarities, substitution delays and recessionary fluctuations into account.

Assuming that a substitution is after all taking place, according to the logistic model, the estimates of the substitution midpoint ($t_{0.5}$), as well as the actual outcome, and the substitution takeover time ($t_{0.9}$–$t_{0.1}$), are shown in Table 3.

Table 3 shows that the substitution rate is extremely high and that digital systems will reach 50% of the total European cellular market in September 1995. Intuitively, this will not come true, perhaps not even if the substitution process starts to gain momentum on the fast-growing British and Italian markets. Moreover the simple Fisher-Pry model seems to overestimate the rate of substitution (see e.g., the projected and actual values for $t_{0.5}$ in Denmark and France), a fact that calls for testing of other variables as well as other models,[27] perhaps better describing the underlying dynamics of this specific substitution process.

To summarize, there are indications that the substitution of analog for digital cellular systems is well under way in several countries,[28] as well as on an aggregate European level, and there are indications of a fairly typical logistic substitution with rapid rate. It is hypothesized here that a substitution will eventually take place but has not yet developed clearly due to (a) initial complementary effects and (b) a market sailing effect and (c) recessionary fluctuations. The market sailing effect can moreover be hypothesized to be stronger for countries and companies with a high penetration rate for their analog systems (e.g., Sweden).

4.2.2 Wireless vs. wired telephony

The second question we will try to answer is whether wireless technologies are substituting for wired technologies. A further note on measurement units: when we study competition between the fixed network and the wireless network, we compare mobile telephony subscribers (disregarding cordless telephones) vs. main lines, which should well reflect competition in terms of subscribers. Intuitively, a comparison made between traffic volumes would show another substitutional pattern as well as earlier signs of substitution. Unfortunately, traffic volume statistics were not fully available when this paper was prepared.

The logistic model could not, so far, be successfully applied to data on the development of the fixed network. Therefore, the proposed approach to look for

Table 3. Estimated substitution rates for selected markets in Europe.

Market	Analog system	Actual	Based on data until the end of 1993				Based on the data until the end of 1994			
		$t_{0.5}$	$t_{0.5}$	$t_{0.9}-t_{0.1}$	r^2	Digital system market share (%)	$t_{0.5}$	$t_{0.9}-t_{0.1}$	r^2	Digital system market share (%)
Europe	Various		1995.05	2.74	1.00	15.4	1995.78	4.28	0.97	29.4
Denmark	NMT		1994.24	0.75	0.97	20.2	1994.88	2.60	0.89	46.7
Finland	NMT		1995.52	2.22	0.79	4.2	1996.06	3.01	0.98	11.3
France	RC 2000 + NMT	1994.9	1994.24	1.59	0.97	16.2	1994.89	2.38	0.98	52.0
Germany	C-450	1993.9	1993.18	2.76	1.00	54.9	1994.00	3.87	0.96	70.8
Italy	RTMS + TACS		1996.18	1.79	0.95	0.5	1996.21	1.57	0.98	3.4
Portugal	C-450	1993.4	1993.46	2.54	0.95	67.3	1993.49	3.69	0.92	84.9
Sweden	NMT		1994.86	1.32	0.99	6.5	1995.40	2.07	0.99	30.6
UK	TACS		1994.34	0.39	1.00	0.7	1995.99	2.37	0.97	13.2

Note: Estimates are based on quarterly subscriber growth data until 1993 and 1994 respectively. r^2 = the proportion of the variation explained by the model. GSM (including DCS 1800)is the digital system in all the above-mentioned countries.

substitution tendencies was not applicable here. However, when studying the growth pattern of main lines[29] in the fixed network and subscribers in the cellular networks, some interesting observations could be made.[30]

Until 1991, the growth rate of cellular networks was steadily increasing, while the growth of the fixed network for the same period was steady, perhaps showing a slight downward trend. In 1991 and 1992 cellular growth slowed down, while fixed network growth decreased sharply. In 1993 cellular growth increased again, to a level higher than in 1991, while main line growth was actually negative. In 1994 cellular growth virtually exploded, mainly driven by growth in the digital systems, while main line growth somewhat recovered.

These results may partly be explained by recessional fluctuations but also by migration to the wireless network. If we assume that there has been a substitution process going on since the start, using the Fisher-Pry-model as a forecasting tool, substitution will be half-way completed ($t_{0.5}$) in the year 2000, and the takeover time ($t_{0.9}-t_{0.1}$) will be 15 years. We may compare this with the controversial but visionary scenario outlined by Swedish Telecom Radio (now Telia Mobitel) in Swedish media where it was stated that "nine out of ten Swedes" will have a "personal pocket telephone" in the year 2000 (see e.g., Swedish Telecom Radio, Annual Report, 1991). Simply forecasting the mobile telephony growth by using a logistic model, and estimating the upper bound to the total population, 90% of the Swedish population will have a mobile telephone in the year of 2006.

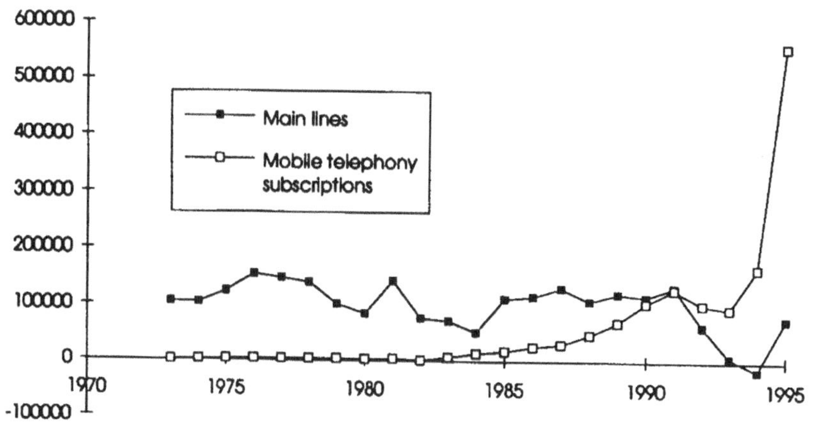

Fig. 8. Annual growth in wired and wireless networks in Sweden. Note: Beginning of years. Mobile subscriptions include the mobile telephone systems MTA, MTB, MTD, NMT 450, NMT 900 and GSM.

However, numerous factors must be taken into account when judging this type of forecast. First, cordless phones are not considered in the forecast and the inclusion of these may lead to a more rapid transition from wired to wireless. Moreover, transition to wireless is probably aided by the transition from analog to digital cellular systems, which was dealt with above. This transition is probably also aided by the introduction of market competition on the digital wireless market. However, wired telephony might "fight back" with the aid of a substitution for opto wire (although at present copper wire is quite successful in holding its own with the help of new data compression techniques, etc.), which would allow for new and superior applications. Also the complementarities in the hybridization of wired and wireless must be accounted for, that is, the usual link-up of wireless to the PSTN. Here, improvements in traditional cordless access, giving less limited mobility, would disfavor traditional cellular telephony, everything else being equal. However, cordless and cellular are fusing, blurring the competition at this level. Satellite technology, on the other hand, may quite conceivably be combined with cellular telephony in new hybrid networks, possibly also weakening the complementarity between cellular and the traditional wired PSTN, e.g., in case satellite transmission is used for long-distance traffic e.g., for remote rural areas or for intercontinental traffic. Finally, as always, technologies have natural limitations. The radio spectrum has often been considered to have natural limitations and there is a possibility that these limits may be reached (from time to time), e.g., in interactive video applications. Still, the notion of the spectrum as a fixed natural resource has been grossly misleading. In practice, numerous ways to economize on the spectrum have been developed, cellular design being just one, currently progressing into e.g., micro- and nano-cellular designs to deal with urban congestion when wireless takes off on a larger scale.

5. Summary and Conclusions

The telecommunications sector is subjected to many forms of competition and cooperation, operating between various entities (individuals, companies, nations, products and technologies) at different levels, often with strong interdependencies. This paper recognizes a general need in economics to elaborate on various forms of competition and their interdependence, and focuses on two special forms of competition in the telecommunications sector, taking place at three interdependent levels: technology competition between wired and wireless telephony in the access network at the user end; technology competition between analog and digital wireless technology; and systems competition between mobile cellular systems. A framework for analyzing multilevel technology competition is outlined, together with a model, which can generate traditional logistic substitution as a special case. However, data availability has so far only allowed application of the model to the substitution of analog for digital cellular systems.

Incompatible cellular systems developed more or less in parallel in the USA, Japan and some European countries, and some of the systems became commercially successful on their respective home markets. Systems competition on the international market was then a natural consequence. However, systems competition in mobile communications displays several differences compared to one of the classical cases of systems competition, i.e., between VHS and Betamax in VCR. The most important difference is that a mobile communication system first has to be adopted by a regulatory body or an operator and then by an end user, thus constituting a two-stage adoption process. The diffusion process of cellular systems has been much influenced by regulatory decisions, in turn influenced by a large set of technical, economic and political factors including: frequency availability, timing, national fragmentation in Europe with concomitant incompatible standards, lack of aggressive international marketing, performance-to-price ratios, size of home market, increasing returns in service provision, and economies of scale and scope in manufacturing. The systems nature and regional differences and political context account for this complexity of influential factors. The most commercially successful systems so far have been AMPS, TACS, NMT and GSM.

Regulated liberalization alters the prerequisites for this systems competition, and is likely to accelerate a transition from analog to digital cellular. In section 4 it is shown that such a transition is well under way in several European countries, as well as it is under way in Europe at aggregate level. There are also indications of a fairly typical logistic substitution with a rapid rate, a substitution which has not yet clearly developed, however. This is due to (a) initial complementary effects (b) a "market sailing effect" and (c) recessionary fluctuations.

Current lack of data has not allowed for a more general case to be treated quantitatively in this paper, but the framework outlined in section 2.4 has permitted some qualitative reasoning. The paper has hypothesized that there will be a general transition from wired to wireless telephony (at the user end). It has also been argued that the pace of transition from analog to digital wireless is positively correlated to the pace of transition from wired to wireless telephony.

Acknowledgements

Financial support from Telia and the assistance of various people both in- and outside Telia Mobitel is gratefully acknowledged. This ongoing research project is part of the research program "Economics and Management of Technology" and carried out under the auspices of the Institute for Management of Innovation and Technology (IMIT) at Chalmers University of Technology. The Jan Wallander and Tom Hedelius Foundation for Social Science Research is also gratefully acknowledged for financial support for the theoretical work. An earlier version of this paper was presented at the ITS biannual conference in Sydney, 3–6 July 1994. The comments from various conference participants are acknowledged. We finally thank Åke Lundqvist, Ola Wallinder and especially Erik Bohlin for valuable comments on earlier versions of this paper.

Notes

1. The fixed network may, in turn, include radio transmission between fixed points, e.g., microwave and satellite links.
2. In the VHS vs. Betamax case, MITI of Japan intervened at an early stage, attempting to have JVC and Sony agree on a common standard. However, the otherwise so mighty MITI failed to persuade the two consumer electronic firms to comply; see Granstrand (1984).
3. MOBITEX, jointly developed by Ericsson and Telia and internationally supported by RAM Mobile and BellSouth, has been successful in this competition, now being the most widespread mobile data communications standard on the international market.
4. Data and statistics, used throughout this section, have been collected and compiled from several sources, e.g., Financial Times Mobile Communications, Mobile Communications International, Cellular Business, EMC World Cellular Reports and Ericsson Reference Lists. The same data are used throughout this chapter; see e.g., Figs. 5 and 6. Whenever possible sources have been cross-checked. The NMT development is mainly extracted from Granstrand (1993). For a more detailed description of the development of NMT, see also Hultén & Mölleryd (1993). For a description of GSM, see for example Mouly and Pautet (1992). For an entertaining story on the emergence of the cellular industry, from the perspective of Ericsson, see Meurling and Jeans (1994).
5. Examples from the technological development that made cellular telephony possible are: (a) Trunking, i.e., combining several channels into a single set, allowing the mobile unit to be connected to any unused channel. (b) Low-cost frequency synthesizers allowed trunking to be incorporated into multiple user requirements. (c) Microprocessors made sophisticated call-switching technology possible.
6. The standards are: NAMTS, NMT, AMPS, TACS, C-450, RC-2000, RTMS and Comvik, although some experts state that RC-2000, RTMS and Comvik are not fully functional cellular systems.
7. Nordic countries are Denmark, Finland, Iceland, Norway and Sweden.
8. Cellnet (60% owned by BT and 40% by Securior) and Vodaphone (majority-owned by Racal Electronics at that time). Motorola and Ericsson were chosen as infrastructure suppliers.
9. RC 2000 has two levels of service – a car telephone service and a trunked PMR service.
10. Since 1977, Ericsson and Televerket had cooperated in developing digital cellular technology.

11. The Permanent Nucleus, created in 1986, consisted of a team of full-time members, coordinating the work of the GSM working parties.
12. One of the main problems was to develop and to have GSM phones tested within a reasonable time.
13. 25 000 dual-mode terminals are estimated to have been sold in during 1993. More recent figures indicate a TDMA take-off during 1994.
14. To complicate the picture further, Hughes Network Systems has developed E-TDMA which is an enhanced variant of TDMA.
15. e.g., Benelux countries, Spain and Austria.
16. Besides NAMTS.
17. This could imply development of "NMT clusters" where neighboring countries adopted NMT systems, which to a certain extent has happened, e.g., in the Nordic countries, Benelux countries, Spain and Andorra, Eastern Europe, and the Baltic states.
18. e.g., in Japan where a modified version of TACS, called JTACS or NTACS, was introduced.
19. Given this resemblance one might expect a higher correlation of AMPS/TACS manufacturers than for instance of NMT/TACS or NMT/AMPS manufacturers, a hypothesis not yet tested.
20. 45 countries have so far signed the GSM MoU, with another 21 applying.
21. A fact which, also here, suggests "GSM clusters".
22. Fisher and Pry give the equation in another form which can be converted to (3) by some straightforward algebraic manipulation. Another useful characteristic of the model is the substitution takeover time defined as: $t_{0.9} - t_{0.1} = 4.4/k$.
23. L is the upper bound given by $f(t)=L/\{1+\exp[-k(t-t_0)]\}$. As noted in e.g., Porter et al. (1991), using data to estimate the upper bound, is risky since initial data are relatively insensitive to the upper bound. The model also seems to underestimate the upper bound (L) and overestimate the rate of substitution. Therefore, the results should be interpreted with caution. This approach has also been tested using an exogenously estimated L (equaling the population on the specific market), a test which yielded less conservative forecasts. However, when penetration has not progressed far and observations are few, this approach tends to overestimate growth rates dramatically (see e.g., Portugal and Italy in Table 2). Finally, L might not be a constant (see section 2.4).
24. Fisher-Pry regression were, so far only tested for r^2 (the proportion of the variation explained by the model), yielding results typically between 0.9 and 1.0.
25. In those cases where the total growth is lower than the forecasted growth (i.e., AF – (A+D) > 0) of the old system, this approach provides no explanations.
26. When comparing analog systems with GSM, we have used subscriber statistics, bearing in mind that, in GSM, customers are able to buy several subscriptions. Traffic volume statistics would have been another possible measurement unit. Yet another possible measure is the terminal (telephone equipment) unit, which would probably not yield the same result since it would measure the substitution on the terminal market and not on the subscriber market. A natural extension is then to measure infrastructure equipment sold. Thus, in an extended analysis we could measure substitution, equally indicative, on three different markets (infrastructure equipment, terminals and services). However, in this paper we will only use subscriber statistics.
27. e.g., Gompertz or exponential decay; see Porter et al. (1991) for more examples.
28. A preliminary analysis suggests that the national variations in the rate of substitution of analog for digital cellular is affected by a set of factors including: market structure for analog cellular, market structure for digital cellular, quality of analog network, saturation of analog network, switching costs, technical and commercial competence of operators etc. For example one may compare the rapid substitution in Germany (analog system monopoly, modest success of C-Netz, introduction of competition) and Italy (successful analog network with extreme growth rates, no introduction of competition so far, etc.).
29. A main line is a telephone line connecting the subscriber's terminal equipment to the public switched network and which has a dedicated port in the telephone exchange equipment (ITU, 1992).
30. Recent statistics were only available from Sweden. Assuming there is or will be a migration from fixed to mobile telephony, this would first show in markets where the penetration rate is high.

Sweden has the highest penetration (15% at the end of 1994) of cellular subscribers in the world and would accordingly show signs of substitution at an early stage.

References

Arthur, W.B. 1988. Competing technologies: An overview. In: Dosi, G., Freeman, C., Nelson, R., Silverberg, G. and Soete, L. (eds), *Technical Change and Economic Theory* (590–607). London: Pinter Publishers.

Bohlin, E., Granstrand O., 1991. Strategic options for monopolies in transition, the case of Swedish Telecom. *Telecommunications Policy*, 15:453–476 (Oct).

Burke, T., Genn-Bash, A., Haines B., 1988. *Competition in theory and practice*. London: Croom Helm.

David, P.A., 1987. The hero and the herd in technological history: Reflections on Thomas Edison and the Battle of Systems. *CEPR Publication No. 100*, Stanford: Center for Economic Policy Research Program, Stanford University.

Fisher, J.C and R.H. Pry 1971. A simple substitution model of technological change. *Technological Forecasting and Social Change*, 3:75–88.

Granstrand, O., 1984. The evolution of the video tape recorder industry and the main frame computer industry in Japan. *CIM Working Paper*, No: 1984 WP:07, Department of Industrial Management and Economics, Chalmers University of Technology, Goteborg, Sweden.

Granstrand, O., 1993. The evolution of Nordic mobile telephony, *CIM Working Paper*, No: WP 1993:06, Department of Industrial Management and Economics, Chalmers University of Technology, Goteborg, Sweden.

Granstrand O., 1994. Technological, Technical and Economic Dynamics – Towards a Systems Analysis Framework. In: Granstrand, O. (ed). *Economics of Technology*, Elsevier Science.

Hultén S., Mölleryd B., 1993. Cellular Telephony in Retrospect – Foundations of Success. *Tele*, (1):2–7.

International Telecommunication Union, 1992. *Yearbook of Common Carrier Statistics Telecommunication Statistics*, 19th edn. Geneva: UIT.

Johansson, O., 1994. Internationalization and diversification of technology-based services: The case of telecommunications. In: Bohlin, E. and Granstrand, O. (eds), *The race to European eminence – Who are becoming the teleservice multinationals?* North-Holland: Elsevier Science Publishers B.V.

Katz, M., Shapiro C., 1994. Systems Competition and Network Effects. *Journal of Economic Perspectives*, 8(2):93–115 (Spring).

Kuwabara, M., 1985. Present and Future of Mobile Communications. *Japan Telecommunications Review*. October: 208–218.

Mahajan, V., Petersen R.A., 1985. *Models for Innovation Diffusion*. London: Sage Publications.

Meurling, J., Jeans R., 1994. *The Mobile Phone Book*. Communications Week International, London.

Mouly M., Pautet M-B., 1992. *The GSM System for Mobile Communications*. published by the authors, 49, rue Louise Bruneau, F-91120, France.

Porter, A.L., Roper, A.T., Mason, T.W., Rossini, F.A., Banks J., 1991. *Forecasting and Management of Technology*. John Wiley & Sons.

Swedish Telecom Radio, Annual Report, 1991.

Beyond Competition: The Future of Telecommunications.
Edited by D. Lamberton.

CHAPTER 22

Employment implications of information and telecommunication technologies in the Danish banking sector

Knud Erik Skouby and Anders Henten
Center for Tele-Information, Technical University of Denmark, Lyngby, Denmark

1. Introduction

The aim of the paper is to study the effects of the implementation of new information and communication technologies (ICT) with relation to the development of employment (both quantitatively and qualitatively) and organizational structure in the Danish banking sector.

Clearly, the most difficult task in relation to that aim is to separate the technology conditioned effects from the effects of the financial crisis and the extensive wave of mergers in the banking sector in the beginning of the 1990s. The relationship between these three elements (technology, financial crisis and mergers) is complicated and forms part of a broader context which is illustrated in Fig. 1.

Figure 1 has 3 levels:

1. Technology, internationalization and liberalization.
2. Financial crisis and mergers.
3. Organization, qualifications and employment.

Level 1 consists of the most fundamental driving forces in the development of the banking sector at present. In this paper, we will not describe the relationship between the three elements (technology, internationalization and liberalization) but will limit ourselves to note that the liberalization of the financial sector with regard to the compulsory separation between different parts of the financial sector (banking, insurance, real estate) and the rules for borrowing, to a great extent, has contributed to the acute character of the financial crisis in Denmark and that the internationalization has been an important motive behind the mergers between the large banks in Denmark.

At *level 2*, it is illustrated that the financial crisis may cause mergers and acquisitions and that mergers, on the other hand, may be a contributing factor with regard to the extent of the financial crisis. By the inclusion of the technology element, it is illustrated that the development of information technology is affected

Address for correspondence: K. Skouby, Center for Tele-Information, Technical University of Denmark, Building 371, DK-2800 Lyngby, Denmark. Tel.: +45-4587-1577. Fax: +45-4596-3171.

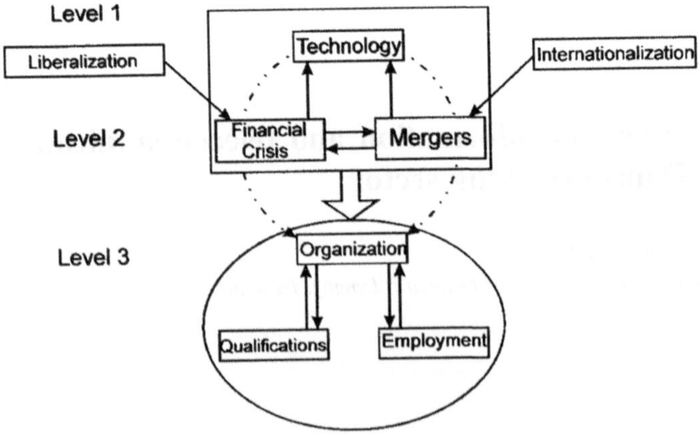

Fig. 1. Technology, financial crisis and mergers in the banking sector.

both by the financial crisis and the mergers. Concerning the financial crisis, the technology may constitute both an offensive and a defensive tool in a strategy for crisis management, and, furthermore, mergers will affect the speed and character of technology implementation.

It could be argued that technology is a factor in constituting both the financial crisis and the strategy for mergers. Technology is certainly an important element in the crisis involving the international financial markets. This international financial crisis, however, has no direct relation to the national financial crises which have hit the Nordic countries severely in the early 1990s. Furthermore, it is true that the technology constitutes an element in conditioning the climate for mergers. However, there is no necessity involved.

In relation to our issue, it is more important that both the financial crisis and the mergers can be said to trigger off the effects of information technology on the organizational structure of banks, on employment and requirements for qualifications. Through a number of years, an extensive rationalization and increase in productivity, conditioned by the use of technology, has been going on; but until lately, this has not resulted in any decreasing employment which is a consequence of constantly growing markets for financial services. However, when these markets are in crisis and mergers are taking place, the effects of rationalizations are realized. (This is illustrated by the dotted line in Fig. 1).

Level 3 deals with the organizational, employment and qualification related consequences. As illustrated in Fig. 1, these three elements are closely related and constitute an organic entity. This interrelated entity, on the other hand, affects technology development, the financial crises and the strategy for mergers; but such effects will not be dealt with in this paper. Here, we will solely be concerned with organizational structures, employment and qualification requirements as results of a development, not as factors affecting this development.

2. Delimitation of the area

The paper deals with banks, savings banks and cooperative banks. The financial sector, moreover, comprises insurance companies, real estate institutions, pension funds, etc. In the course of the 1980s, a certain convergence between different branches of the financial sector has taken place. However, in this paper we will concentrate on banking whose core areas consist of intermediation of capital (borrowing and lending) and payments. Banking comprises approximately a quarter of the total number of employees in the financial sector.

3. Crisis in the banking sector

In 1992, Danish banks had a deficit of some Dkr 11 billion before tax, and in the period 1990–1992, losses constituted approximately 2% of total lending and guarantees.[1] The reasons for this crisis in the Danish banking sector were:

1. A general economic recession which caused extraordinarily large provisions for bad debts.
2. The transition from high inflation to low inflation which has put many customers with outstanding debts into difficulties.
3. Liberalization of the financial sector which has given financial institutions possibilities for entering new business areas which, again, is part of the background for a number of losing investments.

This crisis in the banking sector has brought about a realization of the potential rationalization effects of the implementation of new ICT in the 1970s and 1980s leading to decreasing employment even though the red figures in the accounts are not directly caused by salary expenses for employees. The above mentioned Dkr 11 billion approximately equalled total annual salary expenses in the banks. Expenses for a place of work covers, of course, much more than just the pure salaries; but a strategy for savings, primarily based on redundancies, would, all the same, lead to a totally unrealistic decrease in the number of employees and would completely undermine the business of banks. The number of employees may, with the given extent of activities, be reduced as a consequence of the implementation of new technology.

4. Mergers

In 1990, the two largest banks in Denmark were created through mergers: Den Danske Bank (Den Danske Bank, Handelsbanken and Provinsbanken) and Unibank (SDS, Privatbanken and Andelsbanken). In 1992, these two banks had a market share in Denmark of 56%.[2] Formerly, it was primarily smaller banks and savings banks that

Table 1. The number of banks and branches (1985–1992).

	1987	1988	1989	1990	1991	1992	1993	1994
Banks	203	157	155	151	119	113	112	113
Branches	3476	3318	3221	3079	2656	2467	2340	2245

Note: Including banks lending out more than Dkr 100 million.
Source: Finanstilsynet.

were merged or taken over, but with the mergers between the larger banks, the number of branches will continue to decrease radically in the coming years. The number of banks and branches is shown in Table 1.

5. Technology development in the banking sector

Banks were among the first to utilize new information and communication technology and are, today, among the largest and most advanced users. This is caused by the fact that banking services are information intensive and suitable for electronic media. Banking services are, furthermore, relatively simple and uniform which also eases the use of information technology. Additionally, it should be mentioned that the banks, with the evolvement of country-wide networks of branches, had an apparent need for a management tool.

The technology development in the banking sector constitutes a clear illustration of the theory of the reversed product cycle in the service area.[3] According to this theory, technological innovations at first take place in relation to the production process and thereafter in relation to the services produced. In the banking sector, information technology was implemented first in the so-called back-offices (e.g., batch processing of accounts). Later on, the technology reached the customers in the front-offices.

The first phase of the informatization began in the 1960s where the processing of mass data were put on mainframe computers.

The second phase began in the 1970s when on-line terminals were installed at the counters in the branches.

The third phase began in the 1980s with self-service technology (ATM, EFTPoS and home banking).[4]

Self-service banking technology was implemented relatively late in Denmark. The first (internal) Automatic Teller Machine (ATM) in Denmark was mounted in 1980 whereas the first machine in Europe was set up in 1967 in the UK. The number of ATMs is, furthermore, relatively small in Denmark today because of an agreement between the banks only to set up a limited number of machines. Electronic payments via Dancard-terminals in shops (EFTPoS), on the other hand, is much more developed in Denmark than in most other countries which is a consequence of a common system – based on the Dancard – developed by the banks which made the technology more user-friendly.[5]

Among the two above mentioned fundamental functions of banks (payment and credit intermediation), the information technology has, until now, had the largest impact on payment intermediation. Credit intermediation is a task that requires a certain amount of evaluation and is, therefore, less suitable for automatization. Decision support systems (expert systems) are being used here at a growing extent, and uncomplicated credit intermediation will, in the future, become full-automatic. So far, though, credit intermediation requires a great deal of human labor and, furthermore, the banks will presumably also in the future try to keep up the personal contact with customers in this area with the purpose of retaining the customers and being able to offer advisory services and a broader spectrum of other services.

The development of self-service technology in the retailing market is reflected in Table 2 concerning the number of Dancards and the number of shops where it is possible to pay with Dancards.

The rise in the number of Dancards and Dancard-shops is levelling out as the market is becoming saturated. The number of transactions, however, is still increasing steeply which is reflected in Table 3.

In Table 3, it is shown that the total number of Dancard transactions is increasing at a fast pace. Growth is primarily concentrated in the EFTPoS system. The number of ATM transactions administered by PBS grew during the 1980s, but stagnated around 1989–1991 and fell thereafter. The number of PBS administered ATM transactions never became very high, partly because the number of PBS administered ATM terminals by agreement between the banks never exceeded 201 terminals. In 1993, the common PBS administered ATM terminals were phased out and in 1994

Table 2. Number of Dancards and Dancard-shops (1985–1994).

Year	Number of Dancards	Number of Dancard-shops
1985	363,000	—
1986	685,000	18,100
1987	957,000	23,500
1988	1,320,000	29,500
1989	1,610,000	38,500
1990	1,870,000	45,700
1991	2,104,000	53,600
1992	2,299,000	61,000
1993	2,412,000	66,500
1994	2,552,000	70,000

Note: Approximately one third of Dancard-shops are using the EFTPoS system. The remaining two thirds use manual transactions.
Source: PBS.

Table 3. Number of transactions in the Dancard system, 1983–1994.

Year	EFTPoS	Manual transaction	ATM adm. by PBS	Total
1983	0	78,500	0	78,500
1984	5,500	575,500	255,000	836,000
1985	106,000	1,060,000	887,000	2,053,000
1986	884,000	1,809,000	3,934,000	6,627,000
1987	6,632,000	3,015,000	8,484,000	18,131,000
1988	21,834,000	6,262,000	12,319,000	40,515,000
1989	42,872,000	12,107,000	15,476,000	70,465,000
1990	65,783,000	15,803,000	15,553,000	97,119,000
1991	92,034,000	18,842,000	15,024,000	125,900,000
1992	119,264,000	22,755,000	13,826,000	155,845,000
1993	147,314,000	24,868,000	5,809,000	177,991,000
1994	181,456,000	26,284,000	0	207,740,000

Source: PBS.

all ATM terminals were bank specific. However, another reason for the relatively low use of ATM terminals in Denmark is that the EFTPoS system not only is used for payments but also as a means for getting cash in connection with shopping.

The growth of electronic payment intermediation is also expressed in the rising market share of PBS in relation to the total number of payments, see Table 4.

In 1986, approximately 217 million checks were issued in Denmark. In 1994, the number of checks issued is lower than 100 million.[6] However, the circulation of coins and notes has increased – during the 1980s at approximately Dkr 1 billion per year, but in 1994 with approximately 3 billion.[7] Electronic payments have not out-competed cash. There seems to be a consistent preference for cash which in 1994 has been combined with an upturn in the economy.

ATMs and the EFTPoS system alone is far from sufficient as an explanation for the decrease in employment in the banks. A realistic calculation from the beginning of the 1990s of the potential savings in personnel in relation to these two systems shows that the total savings are around 450 employees.[8]

Other technologies have also had an effect on the employment development. We will mention here the automatization of back-office functions that continuously take place; office automation occurring in the banks as in other businesses; decision support systems and finally Electronic Data Interchange (EDI) with corporate customers – the two last mentioned technological areas being only in their "infancy".

Table 4. Market share 1988–1991 (%).

	Cash	Checks	PBS	Other
1988	69.1	11.4	7.4	12.1
1990	64.8	9.5	11.9	13.8
1991	64.1	8.1	13.8	14.0

Source: PBS.

The banks desire the highest possible share of self-service regarding simple transfer of payments and information. In Den Danske Bank, it is estimated that approximately 20% of the labor time of employees in branches is occupied by such simple transactions.[9] At the same time, banks wish to develop their advisory functions. It is not the opinion of banking specialists — as it was in the 1980s — that it is possible or even desirable to standardize the majority of banking services. Today, the code word is adaptation to customers. This is an expression of the fact both that the illusion of a complete standardization has faded, and that banks aim at keeping close contact with their customers that otherwise might diffuse into the discount banks.

In the self-service area, a larger number of ATMs (both internal and external) are mounted. The common ATMs (common to all banks) have been dismantled in 1993 and the banks have agreed to mount (up to) 667 new ATMs (distributed among the different banks). Furthermore, a greater number of customers will use the telebanking systems, primarily by means of telephones but also, to a certain extent, by means of PCs. In the Spring of 1993, for instance, Den Danske Bank had 170,000 customers on their telephone system that was used approximately half a million times each month.[10] This will also contribute to a limitation of simple operations at the banking counters.

However, the effects of technology cannot be analyzed in isolation. The technology must be seen in a greater context with a number of organizational factors. The technology, e.g., gives a greater flexibility and a decreasing specialization which may also be a contributing factor in lowering of the employment in the banking sector.

6. Development of employment

The number of employed in banks has been decreasing for the last 5–6 years, but it is only during the recent years that the decrease has been pronounced, as seen in Table 5.

Apparently, there is a relation between the use of information technology and the number employed: with increasing use of information technology, the number of employed is reduced. But this has not been a permanent trend. Information technology was introduced long before the present downward wave in employment. As shown in Table 5, the first half of the 1980s even showed increasing numbers in employment in banks. This trend is connected to the larger range of services offered by banks. The union of employees in the financial sector has argued that "the debate on the future number employed in the financial sector, can be based only on the strategic plans of the banks, as they are decisive for the character and extension of the future work assignments".[11] There are many signs, however, that the banks now are under pressure from the financial crisis to such an extent, that the information technology no longer is used only to broaden the range of services offered, but also to cut down the personnel. It can be argued that the financial crisis has released potentials inherent in the information technology.

Table 5. Number of employees in banks (transformed to full-time employed) by year-end 1980–1994.

Year	Number of employees
1980	42,422
1985	46,343
1986	49,744
1987	51,766
1988	50,793
1989	50,010
1990	49,466
1991	50,788
1992	47,560
1993	45,465
1994	44,865

Note: Including Girobank, founded in 1991 with 2,400 employees.
Source: Finanstilsynet (Financial Authority): Pengeinstitutter mv., various years.

In a sense it is true, of course, that the choice of banks between upgrading the level of services and cuts in personnel is a strategic choice. But much points to, that the limits for the choice now and in a foreseeable future are rather narrow.

The tendency to cuts in employment are likely to continue in the coming years. An important element behind this is the growing importance of discount-banks. Discount-banks are offering limited services against a narrower margin of interest rates. The competition on the margin of interest rates is a new phenomenon promoted by the new information technology. This opens up for opportunities for "remote service" of customers without a comprehensive network of branches. The resulting lower level of costs allows the discount-banks to offer higher interest on deposits and to demand lower interest on loans. This development is likely in future to put pressure on banks for a generally lower level of costs.

7. Qualifications

Table 6 shows, as an illustration of the development in the level of qualifications within banks, the relative distribution of staff on categories in the savings banks.[12] Figures for the commercial banks are not substantially different from those shown in the table. The table shows a declining ratio for the category "clerk" and increasing ratios for the categories "Head of section" and "Higher". Together with this, there is a decrease in the ratio of part-time employees. Taken together, this indicates an increasing qualifications profile for persons employed in the banking sector.

With regard to the gender distribution, as more than 90% of the part-time employees were women during the 1980s, it follows that the majority of workplaces, that have disappeared in this category, were for women. Also "Clerk", another decreasing category, has had more than 90% female employees. This development is, however, modified by fast growing ratios of women in the categories "Head of

Table 6. Relative distribution of salary earners on job categories in the savings banks 1982–1992.

Job category	1982	1983	1984	1985	1986	1987	1988	1989	1990	1991	1992
Clerk	19.3	17.8	16.8	16.9	17.0	17.8	16.0	14.4	12.9	11.8	10.9
Assist. head of section	26.4	17.7	28.0	26.3	24.8	22.5	24.8	28.5	29.9	28.8	29.5
Head of section	18.8	19.1	19.7	20.1	20.7	22.3	23.2	24.2	25.5	26.7	28.1
"Higher"	9.4	9.5	9.5	9.8	10.3	11.0	11.5	11.8	12.6	13.4	13.8
Service staff	2.3	2.3	2.3	2.1	2.1	2.0	2.0	1.7	1.9	1.8	1.7
Total full-time employed	76.1	76.4	76.4	75.2	75.0	75.5	77.5	80.6	82.8	82.4	84.0
Trainees	5.3	4.3	4.6	6.4	7.2	8.0	6.8	4.2	2.7	3.2	3.0
Part-time employed	18.6	19.3	19.0	18.4	17.8	16.5	15.6	15.2	14.5	14.4	13.1
Grand total	100.0	100.0	100.0	100.0	100.0	100.0	100.0	100.0	100.0	100.0	100.0

Source: Employer's Association of the financial sector.

section" and "Higher", resulting in the ratio of women in the total number of full-time employment (transformed numbers) being stable at just above 50%.[13]

The decrease in part-time employees and in the category with the lowest level of qualifications can be taken as an indication of a change in work tasks in banks. Increasingly, the routine tasks are disappearing as self-service and automation are introduced. In parallel the banks are emphasizing counselling the customers, with the implication that the staff is required to be able to perform a broader range of functions.

8. Organizational structure

Banks only became coherent firms when the branch networks were tied together by information technology. The organizational implications of this technology have, thus, been exhaustive, both in relation to external and internal structures.

The first step was a change of the internal production processes – the back office functions – with the introduction of electronic computing of standardized large-volume data on mainframes to substitute manual handling. Due to two special circumstances, the small size of even the bigger banks and very well established cooperation in industry-associations, an organizational implication of the new technology in Denmark was construction of a few EDP-centers, owned and used by several banks, but also cooperating in a network using common standards. One of the spectacular results was the establishment of the debitcard system, the Dancard, where every bank mutually recognized each others debitcard and used the same standards. This cooperation was an important factor in the above-described growth in the use of electronic payment.

The tendency for centralization today is modified by the development of the information technology opening up for distributed solutions built on mini's and PCs connected in telecommunications networks. But the tendency is also modified by a simultaneous concentration of the Danish banks making most of them big enough to use the new technology in its present phase with the implication that the mutually recognized debitcard system is disintegrating.

With the spread of the information technology from back-office functions to the front-office, the work tasks at the counter in banks have been changed. Before automation, a customer typically had to go both to a transactions desk and to the cashier. Now these functions are integrated and handled by one person.

In a later stage, when the technology moves over the counter and develops into self-service technology, the tendencies mentioned above will show a decrease in the number of transactions over the counter, opening up the possibility of cutting the number of branches. The emerging discount banks are obvious examples of these possibilities.

The development from big, centralized computing centers to self-service technology is seen as a paradigmatic example demonstrating that the technological development in the service sectors has a tendency to start as a rationalization of the production processes, rather than being driven by innovation of products. The general development in the banks is, however, also an example showing that the chain of innovations typically reaches a level, where the product cycle "returns to normal". The banks today focus increasingly on marketing the many *products* made possible by the information and telecommunications technology. The basic function of the products is often the same (e.g., payment transfer or credit allowance); but such a function can be offered using many different kinds of products, depending on the actual category of customer.

Another consequence of the changing technology is reflected in the position of the different departments in the banks. During the period of centralized computing centers, the EDP departments were rather dominant, whereas today, sales and marketing are functions with high priority. This tendency is reflected also in the movement from standard to customized products and in the general working situations in the branches where routine functions tend to be replaced by counselling. Division of work in the branches also is becoming less pronounced. The individual employee has to be able to perform in an increasing number of functions. This again demands a higher level of qualifications, but also availability to the individual employee of the necessary information in electronic form. The internal automated office and information systems of the banks have to be oriented to perform this task.

9. Conclusion

We have above discussed some of the effects of the new information and telecommunications technologies on employment, demand for qualifications, and organizational structures in banks. We will stress, however, that employment, qualifications and organization are different aspects of the realized development in which one cannot isolate the implication on employment of technology in itself as effective changes in this component demands changes in qualification profiles and organizational structures.

Other important factors than change in technology to influence employment, level of qualifications and organizational structures have been the crisis in the financial sector and mergers. Also with regard to these factors it is not possible to isolate the effects of the single elements – we are clearly dealing with interrelations between the elements. But also, it seems rather clear from the development that the crisis and the mergers have been accelerating agents in the process of realizing the potentials for rationalization in the technological development of the recent decades. Only, did the 1980's not show a decrease in employment because of the expansion of the demand for financial services. Seen in this perspective the ongoing decrease in employment is an expression of strategic choices of the managements of the banks. Instead of cutting employment a strategy of enlarging the range of services could have been chosen. There are many signs, however, that the margins for such choices are narrow in the present market situation. This has presently implied a general tendency towards use of information and telecommunications technologies as part of defensive rationalization strategies rather than more aggressive expansion strategies.

Concluding the analysis, three main sets of trends from the use of ITT technologies in banks can be seen.

1. Presently they have a limiting effect on the number of jobs in banks. The decrease is primarily felt among the least qualified groups, especially women working part-time.
2. The structure of organization in branches is becoming less hierarchical, because the individual staff-members have to perform more functions.
3. Banks are aiming at more self-service in simple transactions and information services both in relation to private and business customers, but are emphasizing counselling functions, implying a higher level and range of qualifications among the staff.

An effective realization of these tendencies demands development of on-line information systems both to private and business customers. In relation to business customers the focus will be on broadening and refining the use of EDI applications, whereas the counselling functions increasingly will use decision support systems. Finally, the general office functions – as in other sectors – will be increasingly automated.

Notes

1. Danmarks Nationalbank: "Beretning og regnskab 1992" (Report and Account p. 46).
2. Danmarks Nationalbank: Op. cit. (p. 47).
3. See Barras, R., 1986. Towards a Theory of Innovation in Services. *Research Policy*, No. 15.
4. It should be emphasized that the development of information technology has many other functions in the banks, including internal administrative functions. The development from back-office to front-office and towards self-service, though, describes a paradigmatic development in the technology use in banks. In this paper, we will focus on the so-called Dancard (the common Danish payment card) because it clearly illustrates such a development.
5. It is PBS (Pengeinstitutternes BetalingsSystemer) that operates the Dancard-system.
6. Formerly, PBS collected statistics on checks. They do not any more as they do not consider checks to be a serious competitor any longer.
7. Den Danske Bank, 1995. *Aktionærnyt (Shareholder News)*, March:20–21. In 1983, Dkr 16 billion in coins and notes were in circulation. In 1994, the figure was approximately Dkr 33 billion.
8. Our own calculations.
9. See *ComputerWorld (Danish edition)*, 1993. No. 15, 23 April:32.
10. Ibid. (p. 32).
11. Nyhedsbrev fra Finansforbundet (News Bulletin from the Union of Employees in the Financial Sector), Jan. 1993, no. 1. An article discussing the number of employed in the financial sector in Denmark relative to the population size, compared to the situation in a number of other countries.
12. The table only includes salary earners and excludes the skilled and unskilled workers employed.
13. Information from the Employers Association in the Financial Sector.

Index of authors

Allen, D. 13

Bowles, D. 299
Briffaut, J.P. 187

Cave, M. 227
Cracknell, D. 97

Eliasson, G. 335

Granstrand, O. 377

Hagström, P. 359
Henten, A. 405
Holcer, N. 279
Hultén, S. 47

Joram, D. 147

Karunaratne, N.D. 315
Kurisaki, Y. 57

Lamberton, D.M. 3
Lindmark, S. 377

Menou, M.J. 71
Mölleryd, B.G. 47

Nightingale, J. 249
Nyberg, A. 47

Ouwersloot, H. 161

Saether, J.P. 33
Sharma, Y. 127
Skouby, K.E. 405
Snow, M.S. 207
Spitz, F. 187
Svendsen, B. 33

Taylor, J. 111

Williams, H. 111

Keyword index

access 16, 111, 367
and network economics 134
lines per employee 307
accountancy paradigm 117
accounting
management 114
rate 227
acquisitions 363
activity-based
business model 187, 189, 197
model 189
additional and useful insight 181
ADSL 127, 139, 141, 143, 144
ADSL-enhanced networks 139
Advanced Mobile Phone Service 386
see also AMPS
advertising 103
all-fibre networks 136, 138
Alvey project 125
amplification techniques 128
AMPS 387, 386, 400
anomalous state of knowledge 77
antitrust 207-213, 218, 219, 223
activism 209
asymmetric digital subscriber line 127
see also ADSL
asymmetry 67
in regulation 62, 65
AT&T 207-215, 217-224, 364
divestiture 207
ATM terminals 409
AUSTEL 256
Australia 14
Australian information economy 9

bandwidth usage 33
banking sector 408
Bayesian learning 251
Bell System 208
best practice 271
betweenness 51
BISDN 130, 135
bounded rationality 337
British Telecom 364
broadband
network 7, 147
services 33, 36
BT's pricing programme 104
bus network 131
business
communication 34
demand 312

cable television companies 119
call pricing 97
call reversion 230
calls market 97
capital 4
capital stock 307
carrier resale 237
causality 317
CCITT 125, 231
CEE countries 8, 279
cellular 377, 395
networks 48
systems 9
telephony 377, 380, 385
COCOM 279
cognitive 17-19, 23
cointegration 317
collective information resources 79
Comité Consultatif International de
Télégraphique et Téléphonique 114
see also CCITT
Commission of European Communities 327
common databases 369
communication activity 163
characteristics 163
communication-intensive economic structures
188
communications 18, 19, 26, 29
community 19, 20, 22, 24-27
competence 7, 38
competition 6, 13, 15, 16, 25, 28, 36, 42, 58,

68, 345, 378
among nations 379
and consensus 20
and cooperation 21
competitive process 253
complexity 174
compression 42
technology 139
computable risks 343
computer-aided design and manufacturing (CAD/CAM) 362
computer-services market 200
computerized information systems 361
computing systems 66
concentration 74
concepts of competition 3
conceptual model 168
concomitance of two technologies: 155
conditional growth model 323
confidentiality 174
consensus 15, 19, 28
Consent Decree 211
consumer research 98
continuous 169
control 362
convergence 3
cooperation 9, 54, 112
cooperative networking architecture management (CNA-M) 115
coordination 360
corporate
communications network 359
headquarters 372
management 359
users 59
cost function subadditivity 213
costs 116, 119, 159, 173, 203, 228, 251
creative destruction 3, 261
creativity 24
critical mass 7, 27, 31, 156, 362, 374
cross-subsidies 150
cross-subsidization 154, 315
Czech Republic 283, 290, 294

D-AMPS (Digital AMPS) 390
Dancard 409
Danish banking sector 10, 405
data communication 6, 35, 37, 41
services 33
degree of point 51
demand 43, 161, 234, 301
for information 301, 312
deployment of optical fiber 153
deregulation 217, 328
deregulatory 216, 220, 221
design 113
determination 173
developing countries 73, 77, 86, 88-91
development of employment 405
development plans 291
diffusion 377, 383, 391, 400
of information technology 72, 73
digital technologies 112
digitizers 42
disadvantaged groups 91
discrete choice 162, 164, 182
diseconomies of scope 149
divestiture 207, 208, 211, 213-224
division
of labour 38
of work 415
dominance 16
dynamic 22, 23, 28
efficiency 258
dynamics of information exchange 7

economic
and social consequences 79
development 8, 80
growth 38
economics
establishment 16
of communications 29
economies
of scale 118, 150, 360
of scale and scope 117
of scope 23, 152
effects of characteristics 180
efficiency threshold 148
EFTPoS system 410
elasticities of demand 228
elasticity of demand 311
electronic
payment 410
super highways 75
elimination models 166
Elimination-by-Aspect (EBA) model 162
e-mail 36
employment 10, 411
endogenous growth 39

entrepreneur 49, 262
entrepreneurial activities 50
entrepreneurship 364
equal access requirement 215
Ericsson 365
estimate parameterization 181
evolution 20, 21, 23, 26
evolutionary economic theory 261
evolutionary economics 6, 261
example of media choice situation 170
experimentally organized economy 343
externalities 23, 155, 269

face-to-face communication 361
facilities management 191
facsimile 36
fax 36
FCC's First Computer Inquiry 210
fiber optic
  systems 40
  technology 151
fibre-in-the-loop (FITL) 127
fibre-to-the-curb (FTTC) 137
fibre-to-the-home (FTTH) 138
financial
  crisis 406
  markets 352
Finland 14
flexibility 62
flexible regulation 62
flow of information 68
forecasting 395
  methods 33
  models 34
foreign trade 306
formal information 77
frame relay 65
frequencies of media use 171
functionality 112

game-theory perspective 118
GDP per capita 305
gender distribution 412
geographic location 364
geographical dispersion 38
Global System for Mobile communications 389
  see also GSM
globalization 219
GPRS 48
graph centrality 50
GSM 387, 389, 393, 400

HDTV 28
high-tech sectors 322
homogenous 173
household size 306
hubbing 239
human dimensions 83
Hungary 280, 287, 292
hybrid networks 136, 137

IBM 364
ICT 405
idiosyncrasies 80
impact of information 77, 80, 85, 88, 89, 90
income elasticity 44, 308
independent of circumstances 166
index of political freedom 307
industrialised countries 40
industry
  policy 254
  structure 27, 28
inertia 365
informal information 77
information 76
  asymmetry 5
  costs 339
  economics 4
  economy 315
  exchange 161
  impact assessment 78
  management 190, 197
  measures 80
  overload 371, 374
  problems 253
  processing 29, 350
  richness 85
  society 71, 91
  systems 190, 359
  systems and services 88
  technology 190
  technology and systems 197
  technology managers 187
  theory 80, 81, 301
  use environments (IUEs) 84
information-as-contents 81
information-intensive firms 219
infrastructure 5, 57, 112, 316, 327
  competition 57

requirements 301
innovation 250, 374
institutional memories 79
integration 74, 364
intellectual property rights 117
intelligent company 72
inter-LATA interexchange market 214
interactive service provision 127
interconnection 116
interdependence 368
interdisciplinary study 5
internal organization 5
international
Development Research Center 7, 89
diffusion patterns 391
R&D activities 9, 359
resale 8
research network 90
simple resale 227
Telecommunications Union 304
internationalization 360, 361
internet 28, 221
ISDN 7, 28, 36, 130, 147, 158
IT intensive 71

Japan 13, 22, 58, 59, 364
job loss 217
job reductions 123

knowledge 71, 74, 77, 79, 80, 83, 87, 91, 360
base 77, 79, 81, 86, 87
leakage 198
of pricing structure 100
knowledge-based information economy 337

large business users 129
large firms 52
LDCs 8
learning 336, 361
theory 342
leased circuits (LCs) 57, 60, 235
liberalization 7, 57, 321, 400
literacy 74
local exchanges 128
locations 173
logit model defined as a special case 168

management
accounting for control of IT-IS resources 196
structures 121
manufacturing industry 287
market
democracy 75
failure 22, 117
penetration 36
structure 265
markets and hierarchies 340
media 163
choice 161, 172
preferences 164
mediacy 74
mergers 406
methodological individualism 15
MFJ 214-218, 220-222, 224
Microsoft 28
Minitel 150, 158
minorities 91
mobile
communications 9, 33, 380
data 47
data networks 47
telephony 47
Modification of Final Judgment (MFJ) 211
modularization 365
multilevel competition 377, 378, 381
multimedia 74
multinational
companies 242
corporations (MNCs) 359
multinomial discrete-choice models 164

National Communications Infrastructure 18
National Information Infrastructure 221
natural monopoly 118, 152, 207, 212, 213, 221, 222
nested
hierarchy 78
MNL (NMNL) model 166
network 33
analysis 50
capacity 33
costs 98
management technologies 7
monopoly 149
of networks 31
network-management technologies 111, 124

networking 49
new services 36
new technologies 42
New Zealand 14
NII 75
NISDN 130, 135
NMT 387, 386, 391, 400
no access 173
Nordic Mobile Telephone system 386
Norway 37
NSK 364

OECD 304
member countries 61
Office for Telecommunications (Oftel) 119
Oftel 97
see above
oligopoly 4, 14, 249
on-line information systems 416
Open Network Provision (ONP) 67
operationalized EBA model 181
ORANI 255
organizational
structures 112
capabilities 361
capital 5
change 339
structure 405, 414
outperform EBA model 181
outsourcing 8, 187, 191, 193, 199
of IT-IS resources 200
overheads 188

paradigmatic structure of knowledge base 81
parameterization 173
parsimonious 179
partial 161
patents 208, 263
path dependency 3
PDC (Personal Digital Cellular) 390, 393
peak load pricing 98
perceptions and feelings 173
performance 21
management 114
personal
characteristics 173
communicators 48
digital assistants 48
point centrality 50
Poland 282, 289, 293

policy formation 5
political freedom 307
pooled time-series 101
population density 307
postal services 36
price-cap
regulation 8, 215
system 257, 273
price 42, 43, 63, 67
elasticity 44, 101, 310
of telephone service 304
pricing
of interconnection 111
perequation 156
structure 100, 106
principal-agent theory 249
private networks 7, 57
probabilities 179
problem-solving process 77
product variation 361
productivity 360
change 249
Project Sovereign 121, 122
protection 316
public demand 58
putty-clay 274

qualification requirements 406
qualitative changes 71

R&D 47, 119, 271
RACE 125
Ramsey prices 249, 275
Random Utility Maximization models (RUM) 164
rate rebalancing 64
rational behavior 85
RBOCs 217
rebalancing 98, 264
receiver competence 335
redundancy reduction 42
redundant 176
refile 239
regulated competition 215
regulation 3, 8, 140, 151, 249
regulatory policy 111, 250
relevance and appropriateness of information systems and services 88
replacement of the network technology 151
resale 64, 68

research and development 359
see also R&D
reseller's costs 232, 238
residual factor 39
rich and flexible model 169
ring network 133
risk 303
robust 173
rush 51

scale 16
scanners 42
Schumpeter 50
searching costs 364
secondary school 312
security management 114
self-service technology 414
service providers 57
settlement rates 229, 243
share of the labor force in telecommunications 307
simple resale 58
Slovak Republic 295
Slovakia 285, 290
SMDS 65
SME 234
social
construction of reality 84
purpose 250
socioeconomic transformation 80
software
development 366
factory 371
special pricing offers 100
special-offer price elasticities 105
specialization 38, 365
specialized networks 149
speculative investment 55
stakeholders 78
standards 48, 385
star network 131
states of information 76
stationarity 317
status symbols 73
stock
of knowledge 86, 351
of paradigms 82
store 76
strategic
alliances 7, 47, 50, 55
behaviour of firms 111
structure of the economy 305
subadditivity of cost function 147
subscriber networks 128
subsidiaries 364
substitution 9, 378, 379, 383, 395, 400
superhighway 28
sustainability 148
of information systems 89
switched video services 7
switching costs 247
system dynamics 34
systems
approach 378, 382
competition 9, 377, 379, 394, 400
development 365

TACS 387, 388, 400
tales 84
TANDOORI 103
technical progress 39
technological
change 259
lock-in 269
progress 251, 254
regime 269
revolutions 71
technology 377
competition 379, 395, 400
push 219
transfer 361
telecommunication 33
industry 47
investments 38
telecommunications 207, 209, 211, 212, 215, 216–224, 256
deregulation 216
infrastructure 38
manufacturing industry 8
operators 57
performance 274
policy 6, 13, 26, 28, 59, 68, 324
technology 315, 321
telephone
demand 101, 299
networks 131
penetration 8, 34, 217, 299, 304, 311
telephony 97
prices 7, 97
TETRA 48

TEUREM 240
time series econometric modelling 101
time-of-day call pricing structure 103
Total Access Communication System 388
    see also TACS
total factor productivity 258, 265
trade policy 30
traffic distribution 75
transactions costs 353
transfer mechanism 265
transit arrangements 239
tree structure 166
two-layer process 177
typology of IT applications 72

UK 14
uncertainties 49, 302
uniform quality 150
urgency 172, 175
USA 14
use of information 88, 91
user demand 33
USSR 25

value added
    networks 69
    services 69
value chain 200, 201
value of information 302
VANs 65
VAR innovation accounting 320
versatility of information 72
viability conditions of the broadband network 159
video
    cameras 42
    conferences 363
    conferencing 374
video-on-demand (VOD) 127
videotex 7, 147, 158
virtual companies 38
VOD 127, 140

Walker Commission 209
welfare theory 252
wisdom 71, 85-87
work stations 369
workers' skills 305
World Bank 280, 297, 304

X-inefficiency 251

year of introduction 36
Yugoslavia 25

Printed and bound by CPI Group (UK) Ltd, Croydon, CR0 4YY
16/09/2026
14963399-0005